WHAT IS REAL?

WHAT IS REAL?

THE UNFINISHED QUEST FOR THE MEANING OF QUANTUM PHYSICS

ADAM BECKER

JOHN MURRAY

First published in Great Britain in 2018 by John Murray (Publishers)
An Hachette UK company

1

A CIP catalogue record for this title is
available from the British Library

Hardback ISBN 978-1-473-66135-6
Trade paperback ISBN 978-1-473-67858-3
Ebook ISBN 978-1-473-66134-9

Printed and bound by Clays Ltd, St Ives plc

John Murray policy is to use papers that are natural, renewable
and recyclable products and made from wood grown in sustainable forests.
The logging and manufacturing processes are expected to conform to the
environmental regulations of the country of origin.

John Murray (Publishers)
Carmelite House
50 Victoria Embankment
London EC4Y 0DZ

www.johnmurray.co.uk

For Elisabeth, who always knew.

The author acknowledges with gratitude
the support of the Alfred P. Sloan Foundation
in the research and writing of this book.

The soundest fact may fail or prevail in the style of its telling.

—Ursula K. Le Guin

Contents

Introduction

The objects in our everyday lives have an annoying inability to appear in two places at once. Leave your keys in your jacket, and they won't also be on the hook by the front door. This isn't surprising—these objects have no uncharted abilities or virtues. They're profoundly ordinary. Yet these mundane things are composed of a galaxy of the unfamiliar. Your house keys are a temporary alliance of a trillion trillion atoms, each forged in a dying star eons ago, each falling to Earth in its earliest days. They have bathed in the light of a violent young sun. They have witnessed the entire history of life on our planet. Atoms are epic.

Like most epic heroes, atoms have some problems that ordinary humans don't. We are creatures of habit, monotonously persisting in just one location at a time. But atoms are prone to whimsy. A single atom, wandering down a path in a laboratory, encounters a fork where it can go left or right. Rather than choosing one way forward, as you or I would have to do, the atom suffers a crisis of indecision over where to be and where not to be. Ultimately, our nanometer Hamlet chooses both. The atom doesn't split, it doesn't take one path and then the other—it travels down both paths, simultaneously, thumbing its nose at the laws of logic. The rules that apply to you and me and Danish princes don't apply to atoms. They live in a different world, governed by a different physics: the submicroscopic world of the quantum.

Quantum physics—the physics of atoms and other ultratiny objects, like molecules and subatomic particles—is the most successful theory in all of science. It predicts a stunning variety of phenomena to an extraordinary degree of accuracy, and its impact goes well beyond the world of the very small and into our everyday lives. The discovery of quantum physics in the early twentieth century led directly to the silicon transistors buried in your phone and the LEDs in its screen, the nuclear hearts of the most distant space probes and the lasers in the supermarket checkout scanner. Quantum physics explains why the Sun shines and how your eyes can see. It explains the entire discipline of chemistry, periodic table and all. It even explains how things stay solid, like the chair you're sitting in or your own bones and skin. All of this comes down to very tiny objects behaving in very odd ways.

But there's something troubling here. Quantum physics doesn't seem to apply to humans, or to anything at human scale. Our world is a world of people and keys and other ordinary things that can travel down only one path at a time. Yet all the mundane things in the world around us are made of atoms—including you, me, and Danish princes. And those atoms certainly are governed by quantum physics. So how can the physics of atoms differ so wildly from the physics of our world made of atoms? Why is quantum physics only the physics of the ultratiny?

The problem isn't that quantum physics is weird. The world is a wild and wooly place, with plenty of room for weirdness. But we definitely don't see all the strange effects of quantum physics in our daily lives. Why not? Maybe quantum physics really is only the physics of tiny things, and it doesn't apply to large objects—perhaps there's a boundary somewhere, a border beyond which quantum physics doesn't work. In that case, where is the boundary, and how does it work? And if there is no such boundary—if quantum physics really applies to us just as much as it applies to atoms and subatomic particles—then why does quantum physics so flagrantly contradict our experience of the world? Why aren't our keys ever in two places at once?

Eighty years ago, one of the founders of quantum physics, Erwin Schrödinger, was deeply troubled by these problems. To explain his concerns to his colleagues, he devised a now-famous thought experiment: Schrödinger's cat (Figure I.1). Schrödinger imagined putting a cat in a box along with a sealed glass vial of cyanide, with a small hammer hanging over the vial. The hammer, in turn, would be connected to a

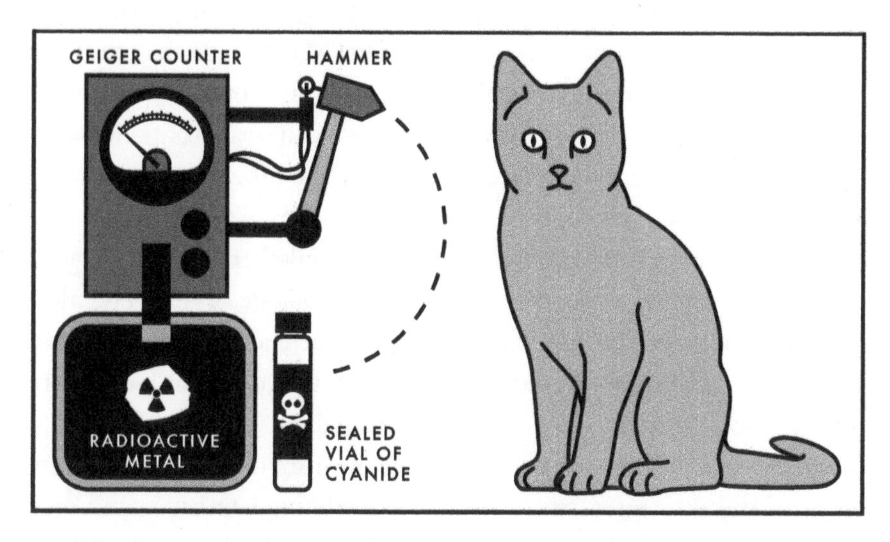

Figure I.1. Schrödinger's cat. When the metal gives off radiation, the Geiger counter will register it and drop the hammer, releasing the cyanide and killing the cat.

Geiger counter, which detects radioactivity, and that counter would be pointed at a tiny lump of slightly radioactive metal. This Rube Goldberg contraption would be set off the moment the metal emitted any radiation; once that happens, the Geiger counter would register the radiation, which would release the hammer, smashing the vial and killing the cat. (Schrödinger had no intention of actually conducting this experiment, to the SPCA's relief.) Schrödinger proposed leaving the cat in the box for a certain period of time, then opening the box to find the cat's fate.

The radiation emitted by the lump of metal is composed of subatomic particles, breaking away from the atoms in the metal and flying off at high speeds. Like all sufficiently tiny things, those particles obey the laws of quantum physics. But, instead of reading Shakespeare, the subatomic particles in the metal have been listening to the Clash—at any particular moment, they don't know whether they should stay or they should go. So they do both: during the time the box is closed, the indecisive lump of radioactive metal will and won't emit radiation.

Thanks to these punk-rock particles, the Geiger counter will and won't register radiation, which means the hammer will and won't smash the vial of cyanide—so the cat will be both dead and alive. And this, Schrödinger pointed out, is a serious problem. Maybe an atom can travel down two paths at once, but a cat certainly can't be both dead and alive. When we open the box, the cat will be either dead or alive, and it stands to reason that the cat must have been one or the other the moment before we opened the box.

Yet many of Schrödinger's contemporaries piled on, denying exactly that point. Some claimed that the cat was in a state of dead-and-alive until the moment the box was opened, when the cat was somehow forced into "aliveness" or "deadness" through the action of looking inside the box. Others believed that talking about what was going on inside the box before it was opened was meaningless, because the interior of the unopened box was unobservable by definition, and only observable, measurable things have meaning. To them, worrying about unobservable things was pointless, like asking whether a tree that falls in the forest makes a sound when nobody's around to hear it.

Schrödinger's concerns about his cat weren't allayed by these arguments. He thought that his colleagues had missed the point: quantum physics lacked an important component, a story about how it lined up with the things in the world. How does a phenomenal number of atoms, governed by quantum physics, give rise to the world we see around us? What is real, at the most fundamental level, and how does it work? Yet Schrödinger's opponents carried the day, and his concerns about what was actually happening in the quantum world were dismissed. The rest of physics simply moved on.

Schrödinger was in a minority, but he wasn't alone. Albert Einstein also wanted to understand what was really happening in the quantum world. He debated Niels Bohr, the great Danish physicist, over the nature of quantum physics and reality. The Einstein-Bohr debates have entered into the lore of physics itself, and the usual conclusion is that Bohr won, that Einstein's and Schrödinger's concerns were shown to be baseless, that there is no problem with reality in quantum physics because there is no need to think about reality in the first place.

Yet quantum physics is certainly telling us something about what is real, out in the world. Otherwise, why would it work at all? It would be very difficult to account for its wild success if it had no connection to anything real in the world. Even if the theory is simply a model, surely it's modeling something and doing a reasonably good job of it. There must be some *thing* that ensures the predictions of quantum physics come to pass, with phenomenally high precision.

But figuring out what quantum physics is saying about the world has been hard. This is, in part, due to the sheer weirdness of the theory. Whatever is in the world of the quantum, it is nothing familiar at all. The seemingly contradictory nature of quantum objects—atoms that are here and there at the same time, radiation that has both been emitted and remains latent in its source—isn't the only alien aspect of the theory. There are also instantaneous long-distance connections between objects: subtle, useless for direct communication, but surprisingly useful for computation and encryption. And there does not appear to be any limit to the size of object that is subject to quantum physics. Ingenious devices built by experimental physicists coax larger and larger objects to display strange quantum phenomena almost monthly—deepening the gravity of the problem that no such quantum phenomena are seen in our everyday lives.

These phenomena aren't the only challenge to deciphering the message of quantum physics. They're not even the largest challenge. Despite the fact that every physicist agrees that quantum physics works, a bitter debate has raged over its meaning for the past ninety years, since the theory was first developed. And one position in that debate—held by the majority of physicists and purportedly by Bohr—has continually denied the very terms of the debate itself. These physicists claim that it is somehow inappropriate or unscientific to ask what is going on in the quantum realm, despite the phenomenal success of the theory. To them, the theory needs no interpretation, because the things that the theory describes aren't truly real. Indeed, the strangeness of quantum phenomena has led some prominent physicists to state flatly that there is no alternative, that quantum physics proves that small objects simply do not exist in the same objectively real way as the objects in our everyday lives do. Therefore, they claim, it is impossible to talk about reality in quantum physics. There is not, nor could there be, any story of the world that goes along with the theory.

The popularity of this attitude to quantum physics is surprising. Physics is about the world around us. It aims to understand the fundamental constituents of the universe and how they behave. Many physicists are driven to enter the field out of a desire to understand the most basic properties of nature, to see how the puzzle fits together. Yet, when it comes to quantum physics, the majority of physicists are perfectly willing to abandon this quest and instead merely "shut up and calculate," in the words of physicist David Mermin.

More surprising still is that this majority view has, time and again, been shown not to work. Despite the popular view among physicists, Einstein clearly got the better of Bohr in their debates and convincingly showed there were deep problems that needed answering at the heart of quantum physics. Simply dismissing questions about reality as "unscientific," as some of Schrödinger's opponents did, is an untenable position based on outdated philosophy. And some dissenters from the majority have developed alternative approaches to quantum physics that clearly explain what is going on in the world without sacrificing any of the theory's accuracy.

The existence of these viable alternatives puts the lie to the idea that we are forced to give up on reality in quantum physics. Yet most physicists still subscribe to some form of this idea. It's still what's taught in classrooms, and it's still the picture that's usually painted for the public. Even when the alternatives are mentioned, they are mentioned as just that—alternatives to the default, despite the fact that the default is entirely unworkable. Thus, nearly a century after quantum theory was first developed—after it has thoroughly altered the world and the lives of every single human in it, both for better and worse—we still don't know what it's telling us about the nature of reality. This thoroughly strange story is the subject of this book.

This is an astonishing state of affairs, and hardly anyone outside of physics knows about it. But why should anyone else care? After all, quantum physics certainly works. For that matter, why should *physicists* care? Their mathematics makes accurate predictions; isn't that enough?

But science is about more than mathematics and predictions—it's about building a picture of the way nature works. And that picture, that story about the world, informs both the day-to-day practice of science and the future development of scientific theories, not to mention the wider world of human activity outside of science. For any given set of equations, there's an infinite number of stories we could tell about what those equations mean. Picking a good story, and then searching for holes in that story, is how science progresses. The stories told by the best scientific theories determine the experiments that scientists choose to perform and influence the way that the outcomes of those experiments are interpreted. As Einstein pointed out, "The theory decides what we can observe."

The history of science bears this out over and over again. Galileo didn't invent the telescope—but he was the first to think of pointing a good one at Jupiter, because he believed that Jupiter was a planet, like Earth, that went around the Sun. After that, telescopes were used regularly to look at everything from comets to nebulae to star clusters. But nobody bothered to use a telescope to find out whether the Sun's gravity bent starlight during a solar eclipse—not until Einstein's theory of general relativity predicted just such an effect, over three centuries after Galileo's discovery. The practice of science itself depends on the total content of our best scientific theories—not just the math but the story of the world that goes along with the math. That story is a crucial part of the science, and of going beyond the existing science to find the next theory.

That story also matters beyond the confines of science. The stories that science tells about the world filter out into the wider culture, changing the way that we look at the world around us and our place in it. The discovery that the Earth was not at the center of the universe, Darwin's theory of evolution, the Big Bang and an expanding universe nearly 14 billion years old, containing hundreds of billions of galaxies, each containing hundreds of billions of stars—these ideas have radically altered humanity's conception of itself.

Quantum physics works, but ignoring what it tells us about reality means papering over a hole in our understanding of the world—and ignoring a larger story about science as a human process. Specifically, it ignores a story about failure: a failure to think across disciplines, a failure to insulate scientific pursuits from the corrupting influence of big money and military contracts, and a failure to live up to the ideals of the scientific method. And this failure matters to every thinking inhabitant of our world, a world whose every corner has been reshaped by science. This is a story of science as a human endeavor—not just a story about how nature works but also about how people work.

Prologue

The Impossible Done

John Bell first encountered the mathematics of quantum physics as a university student in Belfast, and he was not happy with what he found. To Bell, quantum physics was a vague mess. "I hesitated to think it was wrong," said Bell, "but I *knew* it was *rotten*."

The godfather of quantum physics, Niels Bohr, talked about a division between the world of big objects, where classical Newtonian physics ruled, and small objects, where quantum physics reigned. But Bohr was maddeningly unclear about the location of the boundary between the worlds. And Werner Heisenberg, the first person to discover the full mathematical form of quantum physics, was no better. Bohr and Heisenberg's approach to quantum physics—known as the "Copenhagen interpretation," named after the home of Bohr's famous institute—was pervaded by the same vagueness that Bell had found in his quantum physics courses.

Shortly before Bell graduated from university in 1949, he stumbled upon a book by Max Born, another architect of quantum physics. Born's book, *Natural Philosophy of Cause and Chance*, made quite an impression on Bell—especially the discussion of a proof by the great mathematician and physicist John von Neumann. According to Born, von Neumann had proven that the Copenhagen interpretation was the only possible way of understanding quantum physics. So either the Copenhagen interpretation was correct or quantum physics was wrong. And, given the wild success of quantum physics, it seemed that Copenhagen and its vagueness were here to stay.

Bell couldn't read von Neumann's original proof himself—it had been published only in German, which Bell didn't speak. But after reading Born's description of the proof, Bell "got on with more practical things" than his concerns about the Copenhagen interpretation. He went to work on Britain's nuclear energy program and put his doubts about quantum physics aside. But, in 1952, Bell "saw the impossible done." A new paper shattered his short-lived complacency about the problems of the Copenhagen interpretation.

Somehow, despite von Neumann's proof, a physicist named David Bohm had found another way to understand quantum physics. How? Where had the mighty von Neumann gone wrong, and why hadn't anyone seen it before Bohm? Bell couldn't answer these questions without reading von Neumann's proof. And by the time von Neumann's book was published in English three years later, life had intervened: Bell had gotten married and gone off to Birmingham to get his PhD in quantum physics. But Bohm's paper "was never completely out of my mind," Bell said. "I always knew that it was waiting for me." Over a decade later, Bell finally returned to it—and made the most profound discovery about the nature of reality since Einstein.

Part I

A Tranquilizing Philosophy

The people of Tlön are taught that the act of counting modifies the amount counted, turning indefinites into definites. The fact that several persons counting the same quantity come to the same result is for the psychologists of Tlön an example of the association of ideas or of memorization.

—Jorge Luis Borges, "Tlön, Uqbar, Orbus Tertius"

This epistemology-soaked orgy ought to come to an end.

—Albert Einstein, letter to Erwin Schrödinger, 1935

1

The Measure of All Things

Two great theories shook the world and shattered the earth in the first quarter of the twentieth century, scattering the remains of the physics that had come before and forever altering our understanding of reality. One of these theories, relativity, was developed in true science-fiction fashion, by a lone genius working in splendid isolation, who had left the academy only to return triumphant with profound truth in his hand—this was, of course, Albert Einstein.

The other theory, quantum physics, had a more difficult birth. It was a collaborative effort involving dozens of physicists working over the course of nearly thirty years. Einstein was among them, but he was not their leader; the closest thing this disorganized and unruly band of revolutionaries had was Niels Bohr, the great Danish physicist. Bohr's Institute for Theoretical Physics in Copenhagen was the mecca of quantum physics in its infancy, with nearly every big name in the field for fifty years studying there at one point or another. The physicists who worked there made profound discoveries across nearly every field of science: they developed the first genuine theory of quantum physics, found the underlying logic of the periodic table of the elements, and used the power of radioactivity to reveal the basic workings of living cells. And it was Bohr, along with a group of his most talented students and colleagues—Werner Heisenberg, Wolfgang Pauli, Max Born, Pascual Jordan, and others—who developed and championed the "Copenhagen interpretation," which rapidly became the standard interpretation of the mathematics of quantum physics. What does quantum physics tell us about the world? According to the Copenhagen interpretation, this question has a very simple answer: quantum physics tells us nothing whatsoever about the world.

Rather than telling us a story about the quantum world that atoms and subatomic particles inhabit, the Copenhagen interpretation states that quantum physics is merely a tool for calculating the probabilities of various outcomes of experiments. According to Bohr, there isn't a story about the quantum world because "there is no quantum world. There is only an abstract quantum physical description." That description doesn't allow us to do more than predict probabilities for quantum events, because quantum objects don't exist in the same way as the everyday world around us. As Heisenberg put it, "The idea of an objective real world whose smallest parts exist objectively in the same sense as stones or trees exist, independently of whether or not we observe them, is impossible." But the results of our experiments are very real, because we create them in the process of measuring them. Jordan said when measuring the position of a subatomic particle such as an electron, "the electron is forced to a decision. We compel it *to assume a definite position;* previously, it was, in general, neither here nor there. . . . We ourselves produce the results of measurement."

Statements like these sounded ludicrous to Albert Einstein. "The theory reminds me a little of the system of delusions of an exceedingly intelligent paranoiac," he said in a letter to a friend. Despite his crucial role in the development of quantum physics, Einstein couldn't stand the Copenhagen interpretation. He called it a "tranquilizing philosophy—or religion" that provides a "soft pillow to the true believer . . . [but it] has so damned little effect on me." Einstein demanded an interpretation of quantum physics that told a coherent story about the world, one that allowed answers to questions even when no measurement was taking place. He was exasperated with the Copenhagen interpretation's refusal to answer such questions, calling it an "epistemology-soaked orgy."

Yet Einstein's pleas for a more complete theory went unheard, in part because of John von Neumann's proof that no such theory was possible. Von Neumann was arguably the greatest mathematical genius alive. He had taught himself calculus by the age of eight, published his first paper on advanced mathematics at nineteen, and earned a PhD when he was twenty-two. He played a crucial role in building the atomic bomb, and he was one of the founding fathers of computer science. He was also fluent in seven languages. His colleagues at Princeton said, only half-joking, that von Neumann could prove anything—and anything he proved was correct.

Von Neumann published his proof as part of his textbook on quantum physics in 1932. There's no evidence that Einstein was even aware of this proof, but many other physicists were—and for them, merely the idea of a proof from the mighty von Neumann was enough to settle the debate. The philosopher Paul Feyerabend experienced this firsthand after attending a public talk given by Bohr: "At the end of the lecture [Bohr] left, and the discussion proceeded without him. Some speakers attacked his qualitative arguments—there seemed to be lots of loopholes. The Bohrians did not clarify the arguments; they mentioned the alleged proof by von Neumann and that settled the matter . . . like magic, the mere name of 'von Neumann' and the mere word 'proof' silenced the objectors."

At least one person did notice a problem with von Neumann's proof shortly after it was published. Grete Hermann, a German mathematician and philosopher, published a paper in 1935 criticizing von Neumann's proof. Hermann pointed out that von Neumann failed to justify a crucial step, and thus the whole proof was flawed. But nobody listened to her, partly because she was an outsider to the physics community—and partly because she was a woman.

Despite the flaw in von Neumann's proof, the Copenhagen interpretation remained totally dominant. Einstein was painted as an old man out of touch with the rest of the world, and questioning the Copenhagen interpretation became tantamount to questioning the massive success of quantum physics itself. And so quantum physics continued for the next twenty years, piling success upon success, without any further questions about the hole at its heart.

Why does quantum physics need an interpretation? Why doesn't it simply tell us what the world is like? Why was there any dispute between Einstein and Bohr at all? Einstein and Bohr certainly agreed that quantum physics worked. If they both believed the theory, how could they disagree about what the theory said?

Quantum physics needs an interpretation because it's not immediately clear what the theory is saying about the world. The mathematics of quantum physics is unfamiliar and abstruse, and the connection between that mathematics and the world we live in is hard to see. This is in stark contrast with the theory quantum physics replaced, the physics of Isaac Newton. Newton's physics describes a familiar and simple world with three dimensions, filled with solid objects that move in straight lines until something knocks them off their paths. The math of Newtonian physics specifies the location of an object using a set of three numbers, one for each dimension, known as a vector. If I'm on a ladder, two meters off the ground, and that ladder is three meters in front of you, then I could describe my position as (zero, three, two). The zero says that I'm not off to one side or the other, the three says I'm three meters in front of you, and the two says I'm two meters above you. It's fairly straightforward—nobody runs around deeply worried about how to interpret Newtonian physics.

But quantum physics is significantly stranger than Newtonian physics, and its math is stranger too. If you want to know where an electron is, you need more than three numbers—you need an infinity of them. Quantum physics uses infinite collections of numbers called *wave functions* to describe the world. These numbers are assigned to different locations: a number for every point in space. If you had an app on your phone that measured a single electron's wave function, the screen would just display a single number, the number assigned to the spot where your phone is. Where you're sitting right now, the Wave-Function-O-Meter™ might display the number 5. Half a block down the street, it'd display 0.02. That's what a wave function is, at its simplest: a set of numbers, fixed at different places.

Everything has a wave function in quantum physics: this book, the chair you're sitting in, even you. So do the atoms in the air around you, and the electrons and other particles inside those atoms. An object's wave function determines its behavior, and the behavior of an object's wave function is determined in turn by the Schrödinger equation, the central equation of quantum physics, discovered in 1925 by the Austrian physicist Erwin Schrödinger. The Schrödinger equation ensures that wave functions always change smoothly—the number that a wave function assigns to a particular location never hops instantly from 5 to 500. Instead, the numbers flow perfectly predictably: 5.1, 5.2, 5.3, and so on. A wave function's numbers can go up and down again, like a wave—hence the name—but they'll always undulate smoothly like waves too, never jerking around too crazily.

Wave functions aren't too complicated, but it's a little weird that quantum physics needs them. Newton could give you the location of any object using just three numbers. Apparently, quantum physics needs an infinity of numbers, scattered across the universe, just to describe the location of a single electron. But maybe electrons are weird—maybe they don't behave the way that rocks or chairs or people do. Maybe they're smeared out, and the wave function describes how much of the electron is in a particular place.

But, as it turns out, that can't be right. Nobody's ever seen half of an electron, or anything less than a whole electron in one well-defined place. The wave function doesn't tell you how much of the electron is in one place—it tells you the *probability* that the electron is in that place. The predictions of quantum physics are generally in terms of probabilities, not certainties. And that's strange, because the Schrödinger equation is totally deterministic—probability doesn't enter into it at all. You can use the Schrödinger equation to predict with perfect accuracy how any wave function will behave, forever.

Except that's not quite true either. Once you do find that electron, a funny thing happens to its wave function. Rather than following the Schrödinger equation like a good wave function, it collapses—it instantly becomes zero everywhere except in the place where you found the electron. Somehow, the laws of physics seem to behave differently when you make a measurement: the Schrödinger equation holds all the time, except when you make a measurement, at which point the

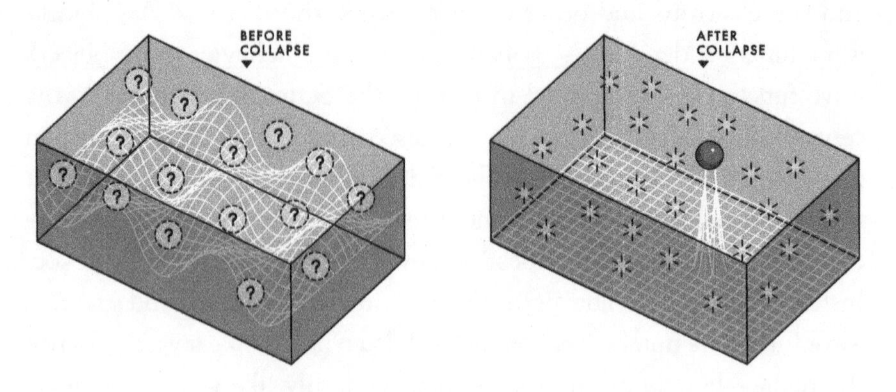

Figure 1.1. The measurement problem.

Left: The wave function of a ball in a box undulates smoothly, like ripples on the surface of a pond, governed by the Schrödinger equation. The ball could be anywhere in the box.

Right: The ball's location is measured and found in a particular spot. The wave function immediately and violently collapses, radically disobeying the Schrödinger equation. Why does the Schrödinger equation—a law of nature—apply only when measurements are not occurring? And what counts as a "measurement" anyhow?

Schrödinger equation is temporarily suspended and the wave function collapses everywhere except a random point. This is so weird that it gets a special name: the *measurement problem* (Figure 1.1).

Why does the Schrödinger equation only apply when measurements aren't happening? That doesn't seem to be how laws of nature work—we think of laws of nature as applying all the time, no matter what we're doing. If a leaf detaches from a maple tree, it will fall whether or not anyone is there to see it happen. Gravity doesn't care whether anyone is around to watch.

But maybe quantum physics really is different. Maybe measurements do change the laws that govern the quantum world. That's certainly strange, but it doesn't seem impossible. But even if that's true, it still doesn't solve the measurement problem, because now we have a new challenge: what is a "measurement," anyhow? Does a measurement require a measurer? Does the quantum world depend on whether it has an audience? Can anyone at all collapse a wave function? Do you need to be awake and conscious for it, or can a comatose person do it? What about a newborn baby? Is it limited to humans, or can chimps do it too? "When a mouse observes, does that change the [quantum] state of the universe?" Einstein once asked. Bell asked, "Was the world wavefunction waiting to jump for thousands of millions of years until a single-celled living creature appeared? Or did it have to wait a little longer for some more highly qualified measurer—with a Ph.D.?" If measurement has nothing to do with living observers, then what does it involve? Does it just mean that a small object, governed by quantum physics, has interacted with a big one, which is somehow exempt from quantum physics? In that case, doesn't that mean that measurements are happening basically all of the time, and the Schrödinger equation should almost never apply? But then why does the Schrödinger equation work at all? And where's the divide between the quantum world of the small and the Newtonian world of the large?

Finding this Pandora's box of weird questions lying at the heart of fundamental physics is disturbing, to say the least. Yet despite all this weirdness, quantum physics is wildly successful at describing the world—much more so than simple old Newtonian physics (which was already pretty good). Without quantum physics, we wouldn't have any understanding of why diamonds are so hard, what atoms are made of, or how to build electronics. So wave functions, with their numbers scattered across the universe, must somehow be related to the everyday stuff we see around us in the world, otherwise quantum physics wouldn't be any good at making predictions. But this makes the measurement problem even more urgent—it means there's something about the nature of reality that we don't understand.

So how should we interpret this strange and wonderful theory? What story is quantum physics telling us about the world?

Rather than answering that question—which seems like it would be difficult—we could deny that it's a legitimate question at all. We can claim that making predictions about the outcomes of measurements is all that matters in quantum physics. Now we don't have to worry about what's happening when we're not making measurements, and all these difficult questions melt away. What is the wave function? How is it connected to the objects in the world around us? Easy, comforting solutions are at hand: the wave function is merely a mathematical device, a bookkeeping tool to allow us to make predictions about measurements. And it has no connection to the world around us at all—it's merely a useful piece of mathematics. It doesn't matter that wave functions behave differently when we're not looking, because between measurements, nothing matters. Even talking about the existence of things between measurements is unscientific. This, strangely enough, is the orthodox view of quantum physics—the "soft pillow" of the Copenhagen interpretation.

These suspiciously easy answers raise another question, one without an obvious solution. Physics is the science of the material world. And quantum theory purports to be the physics governing the most fundamental constituents of that world. Yet the Copenhagen interpretation says that it's meaningless to ask about what's actually going on in quantum physics. So what is real? Copenhagen's reply is silence—and a look of stern disapproval for having the temerity to ask the question in the first place.

This is, at best, a profoundly unsatisfying answer. But this is also the standard answer. The physicists who pursued the question anyhow—physicists like Einstein, and later on, Bell and Bohm—did so in open defiance of Copenhagen. So the quest for reality is also the story of that rebellion, a rebellion as old as quantum physics itself.

2

Something Rotten in the Eigenstate of Denmark

The call had finally come for Werner Heisenberg. The fresh-faced physicist, all of twenty-four years old, had been invited to give a talk at the University of Berlin, the center of physics in Germany and arguably the world. He would be explaining his astonishing new ideas in front of Einstein himself.

"Since this was my first chance to meet so many famous men, I took great care to give a clear account of the concepts and mathematical foundations of what was then a most unconventional theory," Heisenberg recalled decades later. "I apparently managed to arouse Einstein's interest, for he invited me to walk home with him so that we might discuss the new ideas at greater length."

As they walked to his apartment on that spring day in 1926, Einstein innocuously asked Heisenberg about his education and background, careful not to turn the subject to Heisenberg's new theory. He waited until they were safely indoors before he sprung the trap.

Heisenberg's "most unconventional theory" was an enormous breakthrough. It promised to solve the outstanding scientific challenge of his day: the nature of the quantum world. Physicists had known for nearly three decades that something was wrong, that a change was desperately needed to understand what was happening in the world of the very small—the world of atoms. But they were working blind. Atoms are simply too small to see through any normal microscope, no matter the magnification. The wavelength of visible light is thousands of times larger than the size of an individual atom. But atoms do give off different colors of light when heated, and different kinds of atoms each have their own distinct spectrum of colors, like a fingerprint. While physicists of the late nineteenth and early twentieth centuries learned to recognize these fingerprints, they didn't understand what kind of internal atomic structure could be producing these spectra. There were hints of mathematical regularity among the spectra, and every so often someone managed to devise a way of understanding part of one of them—most notably, Niels Bohr.

In 1913, inspired by the experimental work of New Zealand–born physicist Ernest Rutherford, Bohr proposed a "planetary" model of the structure of an atom, with a tiny yet massive nucleus surrounded by orbiting electrons. In Bohr's model, the electrons were restricted to a particular set of allowed orbits. Electrons could never be between Bohr's allowed orbits, but they could "jump" from one orbit to another. Each orbit corresponded to a different energy, and, as the electrons jumped, they would emit or absorb light equal to the change in their energy, producing the spectrum seen in the lab. These discontinuous jumps of certain energies were known as *quanta*, from the Latin for "how much"—hence the new science of the atomic world came to be known as "quantum physics."

Bohr's model worked astonishingly well for the simplest kind of atom, hydrogen—so well that Bohr won the Nobel Prize for his idea in 1922. Bohr's model seems simple in retrospect, but that's a result of how profoundly it altered and shaped the idea of an atom. When you hear the word "atom," the cartoon image of electrons orbiting a nucleus that pops into your head is almost entirely due to Bohr. His model was a truly brilliant and original insight into the workings of nature. But it was also incomplete, as Bohr well knew. His model utterly failed to predict the correct spectrum of colors for any other kind of atom, even helium, the next simplest after hydrogen. And even for hydrogen, Bohr's model could only explain so much. It could explain the colors in hydrogen's spectrum but not the relative brightness of those colors. It incorrectly predicted single colors where pairs or triplets of closely spaced colors appeared instead. And finally, atomic spectra were susceptible to external influences, some of which couldn't be fully accounted for by Bohr's model. Put an atom in a magnetic field, and its spectrum changed. Put it in an electric field, and its spectrum changed in a different way. Colors shifted, blurred, and split, dimmed and brightened, with no larger pattern in sight—until Heisenberg.

In June 1925, Heisenberg came down with a hideous case of hay fever. Sneezing and nearly blind, with tears streaming down his impressively swollen face, the desperate young physicist took two weeks' vacation to the island of Heligoland, a small barren island in the North Sea, utterly devoid of trees and flowers. After several days on the island, he recovered and resumed his research. Ignoring everything Bohr's model said about the orbits of electrons in atoms, Heisenberg focused on what he could actually see: the spectrum of light emitted from the jumps between energy levels themselves. Working alone at three in the morning, in a shack on a rock battered by a frigid sea, his hands shaking, excitedly fumbling over "countless arithmetical errors," Heisenberg made a breakthrough. "I had the feeling that, through the surface of atomic phenomena, I was looking at a strangely beautiful interior, and felt almost giddy at the thought that I now had to probe this wealth of mathematical structures nature had so generously spread out before me." Heisenberg had developed a strange new mathematics on the fly, one in which simple statements like "three times two equals two times three" were not always true. Using this unwieldy math, Heisenberg had found a way to predict the spectrum of a quantum oscillator—a tiny pendulum—which, in turn, allowed him to predict how atomic spectra respond to magnetic fields.

When Heisenberg returned to his job at the University of Göttingen, he cautiously sent a draft of his new theory to his friend, the brilliant physicist Wolfgang Pauli—"generally my severest critic," as Heisenberg recollected years later—who greeted the new theory with effusive praise. "[Heisenberg's ideas offer] a new hope, and a renewed enjoyment of life. . . . Although it is not the solution to the riddle, I believe that it is now once again possible to move forward," said Pauli.

Max Born, Heisenberg's supervisor, agreed. Born and his student Pascual Jordan helped Heisenberg to elucidate the structure and implications of his new theory, which Born dubbed "matrix mechanics" after the unfamiliar mathematical objects at its heart. Heisenberg's matrix mechanics was technically forbidding and impossible to visualize—but it offered the prospect of a theory not just for atomic spectra but for the entire quantum world.

Einstein had started his own revolution in physics twenty years earlier, when he was Heisenberg's age—and in a kind of isolation as well, though it wasn't brought on by hay fever. In 1905, while working as a patent clerk in Switzerland, Einstein published his theory of special relativity, resolving a long-standing debate over the nature of light. Before Einstein, light was thought to be a wave in some kind of as-yet-undetected medium with the (spectacularly nineteenth-century) name of *luminiferous aether.* But in 1887, physicists Albert Michelson and Edward Morley had attempted to detect the motion of the Earth through the aether, and failed. Increasingly complex ad hoc ideas were thrown around to account for the results of the experiment. One physicist suggested that the results could be a sign that the aether compressed objects as they moved through it. Another physicist pointed out that wouldn't be enough—the aether would also have to slow down all physical processes in objects moving through it! Yet allowing the aether to have these strange properties, all while maintaining its insubstantial nature, was increasingly difficult to believe or understand.

Einstein resolved the confusion in a brilliant stroke, the kind that is obvious only in hindsight. The aether, he proposed, was difficult to imagine because it didn't exist at all. Light was simply a wave of electromagnetic fields, with no medium necessary, always traveling at a constant speed. From that simple assumption, Einstein spun out an entire theory of motion, the theory of special relativity. Special relativity was able to account for the negative result of the Michelson-Morley experiment, and it derived from first principles all of the strange effects—length contraction, time dilation—that others had only been able to assume.

Special relativity also made novel predictions. One consequence of the theory was that the speed of light was an absolute speed limit: no object or signal could go faster than the speed of light in a vacuum. The mathematics of special relativity dictated that any object approaching the speed of light would require an infinite amount of energy to get there. And an object that did somehow manage to travel faster than light could, in theory, travel into its own past and prevent itself from leaving in the first place—a paradox. Light speed is still plenty fast—about 300,000 kilometers per second—but Einstein had discovered that speed was the fastest any object could travel, signal, or influence any other object.

In a follow-up paper the same year, Einstein extended his theory of relativity to modify Newton's laws of motion, discovering in the process his famous equation that shows mass is a form of energy: $E = mc^2$. And these were just two of the papers Einstein published during his "miraculous year" of 1905. He also published two more seminal papers, on the behavior of atoms and the interaction of light and matter—the work for which he later won a Nobel Prize.

In his work on relativity, Einstein was guided, in part, by the work of the Austrian physicist and philosopher Ernst Mach. Mach believed that science should be based on descriptive laws that don't make any claims about the true nature of the world—he dismissed such claims as unnecessary for the practice of science. To Mach, one of the worst offenders was the great god of physics, Isaac Newton himself. Newton's masterwork, the *Principia,* opened with the assumption that space and time were absolute entities unto themselves, with real existence out in the world. This "conceptual monstrosity of absolute space" was, in Mach's view, "purely a thought-thing which cannot be pointed to in experience." Mach thought that a proper science of mechanics would dispense with these kinds of ontological claims—claims about what things actually exist in the real world—and instead simply lay down descriptive, mathematical laws that accurately predict the observed motion of all objects. Good theories, according to Mach, were about connecting observations, not about positing things that couldn't be observed at all.

The laws of thermodynamics, developed in the early 1800s, were the very model of a modern physical theory, according to Mach. As laid down by Carnot, Joule, and others, thermodynamics simply quantified the observable behavior of heat in steam engines and elsewhere in the world, allowing prediction without positing any extraneous unobservable ideas about the nature of heat itself. Thermodynamics didn't rely on abstruse, unverifiable ideas about what was actually in the world—it simply described the world.

Einstein had read Mach's *History of Mechanics* as a student and was deeply impressed with his criticism of the Newtonian ideas of absolute space and time. "This book exercised a profound influence on me," he wrote decades later. Taking Mach's ideas about eliminating extraneous unobservable entities to heart, Einstein had tackled the problem of the aether, finding it to be an unnecessary hypothesis in special relativity. And, better still, special relativity also consigned to oblivion the absolute space and time that Mach had so despised.

Einstein had, in short, used Mach's ideas to brilliant effect. Machians took inspiration from his work for years, thinking that relativity's success vindicated their approach to the world. Mach's views, they figured, were obviously shared by Einstein, since they played such an important role in his most famous and profound work. But when Mach's followers actually spoke with Einstein himself, they were surprised to find that he was not a dogmatic Machian after all—far from it. Although his theory of relativity dismissed the idea of absolute space and time, it replaced those notions with a different absolute: *spacetime*, a combination of space and time, which is the same for all observers. And the name "relativity" itself, which suggests a rejection of absolutes, was introduced by the physicist Max Planck, not Einstein—Einstein disliked the name "relativity" precisely because it connoted a kind of relativism. He preferred the name "invariant theory," which conjures up a very different set of associations. (The "invariants" in relativity are quantities like spacetime that all observers agree upon—and there are many of these in the theory.) And, later in life, Einstein himself said repeatedly that he did not think that Mach's ideas were to be taken too seriously. "Mach's epistemology . . . appears to me to be essentially untenable," Einstein wrote. "It cannot give birth to anything living. It can only exterminate harmful vermin." While Mach believed that physics was merely about organizing perceptions of the world, to Einstein, physics was about the world itself. "Science," he said, "has the sole purpose of determining what *is*."

But perhaps most convincing and revealing about Einstein's true stance toward Mach in 1905 are his two other celebrated papers from that year. In one, Einstein explained Brownian motion, the random motion of microscopic dust motes in a fluid. The botanist Robert Brown had noticed this phenomenon nearly eighty years earlier (and Jan Ingenhousz, the discoverer of photosynthesis, had seen it forty years before that) but nobody had been able to satisfactorily explain it. Einstein did so masterfully—and he did it by rejecting Mach's approach to physics. Instead, Einstein adopted the approach of Mach's nemesis, Ludwig Boltzmann, who claimed that the world was made of a phenomenal number of tiny atoms. Mach had loudly and repeatedly proclaimed that he did not believe in atoms, as they were too small to be observable in principle. But Boltzmann had managed to show that the statistical behavior of massive numbers of atoms led directly to the laws of thermodynamics that Mach had been so eager to simply assume. (There was also evidence for atoms from chemistry, which had by then accepted the existence of atoms for over half a century.) Mach was unconvinced by Boltzmann's arguments. But Einstein found them compelling and elegant, and happily subscribed to the existence of atoms in order to solve the problem at hand. Using Boltzmann's statistical methods, Einstein showed that Brownian motion was caused by the dust motes bouncing off of the atoms in the fluid. In one stroke, Einstein not only explained a century-old puzzle, but conclusively demonstrated that Boltzmann's statistical, atom-based approach to physics was both sound and useful.

As bad as Einstein's Brownian motion paper was from a Machian perspective, his other paper was even worse. In it, Einstein again proposed a solution for an old puzzle, the photoelectric effect, in which shining light on a metal plate could cause a current to jump through the air to a nearby wire. The puzzling thing about the photoelectric effect was that the color of the light involved seemed to matter: if the light was too far toward the red end of the spectrum, then no matter how bright the light was, no current was seen. Einstein accounted for this strange behavior by proposing light was composed of a totally new particle, the photon. This was an audacious hypothesis that not only flew in the face of Machian philosophy but also seemingly contradicted a century of experimental evidence that light was a wave, not a particle. Einstein certainly knew that light was an electromagnetic wave—the idea was crucial inspiration for his theory of relativity—but was nonetheless proposing that light was somehow also a particle, or had some kind of particle-like nature. In defense of this strange idea, Einstein could only point to the photoelectric effect itself, along with a strange quirk of the "black-body radiation law" discovered by German physicist Max Planck five years earlier. For nearly two decades, almost nobody other than Einstein believed in photons. Even Planck himself didn't think his work suggested that light was made of particles (though, years later, Planck's work was hailed as the start of the quantum revolution). Only when Arthur Compton actually caught photons in the act of bouncing off of electrons, in 1923, did the physics community finally come around to Einstein's way of thinking—and even then there were a few holdouts.

But Einstein was accustomed to isolation. He had changed the world in 1905 working alone in a Swiss patent office and continued that habit for the rest of his life. Einstein once said he went through life as a "one-horse cart"; he rarely collaborated with other physicists and almost never took on students of his own. He was eternally suspicious of the status quo, both scientifically and elsewhere; he characterized common sense as the collection of prejudices accumulated by the age of eighteen. So when Heisenberg's astonishing new theory arrived on the scene in 1925, Einstein was unsurprisingly skeptical. "Heisenberg has laid a big quantum egg," he wrote to his friend Paul Ehrenfest shortly after Heisenberg's ideas were first published. "In Göttingen they believe in it. I don't." Presented with the opportunity to interrogate Heisenberg at close quarters, Einstein pounced.

Safely ensconced in his apartment, Einstein finally asked Heisenberg what he really wanted to know. "You assume the existence of electrons inside the atom, and you are probably quite right to do so. But you refuse to consider their orbits. . . . I should very much like to hear more about your reasons for making such strange assumptions."

"We cannot observe electron orbits inside the atom," replied Heisenberg. He pointed out that only the spectrum of light from an atom is really observable and concluded with a rather Machian statement. "Since a good theory must be based on directly observable magnitudes, I thought it more fitting to restrict myself to these."

In Heisenberg's later retelling of this encounter, Einstein was shocked at this. "But you don't seriously believe that none but observable magnitudes must go into a physical theory?"

"Isn't that precisely what you have done with relativity?" replied Heisenberg.

"Possibly I did use this kind of reasoning, but it is nonsense all the same," said Einstein. "On principle, it is quite wrong to try founding a theory on observable magnitudes alone. In reality the very opposite happens. It is the theory which decides what we can observe." Einstein then went on to explain that the information about the world around us that we receive from scientific instruments—or even from our own senses—would be totally incomprehensible without some kind of theory about the way the world works. When you use a thermometer to test the temperature of a chicken you've cooked in the oven, you're assuming the thermometer accurately indicates the temperature inside of the chicken—and that the light that reflected off of the thermometer and entered your eyes accurately indicates the reading of the thermometer. In other words, you have a theory about how the world works, and you're using that (very well-justified!) theory to inform your use of the thermometer. Similarly, Einstein pointed out to Heisenberg that, when looking at the spectrum of an atom, "you quite obviously assume that the whole mechanism of light transmission from the vibrating atom to the spectroscope or to the eye works just as one has always supposed it does."

Heisenberg was "completely taken aback by Einstein's attitude," as he recalled later. Falling back on the seemingly solid ground of Mach's philosophy, Heisenberg replied, "The idea that a good theory is no more than a condensation of observations . . . surely goes back to Mach, and it has, in fact, been said that your relativity theory makes decisive use of Machian concepts. But what you have just told me seems to indicate the very opposite. What am I to make of all this, or rather what do you yourself think about it?"

"Mach rather neglects the fact that the world really exists, that our sense impressions are based on something objective," Einstein replied. "He pretends that we know perfectly well what the word 'observe' means, and that this exempts him from having to discriminate between 'objective' and 'subjective' phenomena. . . . I have a strong suspicion that, precisely because of the problems we have just been discussing, your theory will one day get you into hot water."

With the two men seemingly at an impasse, Heisenberg decided to change the subject. For several days, he had been struggling with a difficult professional decision. Heisenberg had spent seven productive months working with Bohr in Copenhagen a year earlier, shortly before his fateful trip to Heligoland. Now, Bohr had offered Heisenberg an opportunity to come to Copenhagen again, this time as Bohr's assistant. Heisenberg, naturally, had jumped at the opportunity. But, a few days later, he found himself in an incredibly fortunate dilemma. Heisenberg had been offered a tenured professorship in Leipzig—a permanent and prestigious position, and unheard of for someone so young. Unsure of what to do, he asked Einstein's advice. Einstein told him to go work with Bohr. Three days later, Heisenberg was on his way to Copenhagen, to once again sit at the feet of the quantum master himself.

Bohr and Einstein were friends—after their first meeting in 1920, Einstein wrote to Bohr that "seldom in my life has a person given me such pleasure by his mere presence as you have." Writing to his close friend Paul Ehrenfest, Einstein said Bohr "is like a sensitive child and walks about this world in a kind of hypnosis." Both Einstein and Bohr were great physicists of the same generation, and each had an enormous impact on the development of quantum physics. But the similarity mostly ends there. Unlike Einstein, Bohr continually worked with other physicists. Over the course of nearly half a century, Bohr took dozens of young physicists under his wing, mentoring them not only in physics but in all aspects of life. His enormous charisma and force of personality left a huge impression on all visitors to his institute in Copenhagen. "Even to the big shots, Bohr was the great God," as Richard Feynman, the American physicist, put it. To students and younger colleagues, Bohr was a father figure and sage of superhuman wisdom, who was the "wisest of living men," according to the American physicist David Frisch. John Wheeler, one of Bohr's most illustrious and influential students, compared Bohr's wisdom to "Confucius and Buddha, Jesus and Pericles, Erasmus and Lincoln." And, to many of Bohr's colleagues, he was a near-mystical figure, a font of unalloyed scientific truth. "We all look up to you as the profoundest thinker in science," wrote the English chemist Frederick Donnan in a letter to Bohr, "the Heaven-sent expounder of the real meaning of these modern advances. . . . I can and will think of you walking in your beautiful gardens, and stealing some moments of peace whilst the leaves and the flowers and the birds whisper their secrets to you."

Bohr's remarkable charisma was enhanced by his immense institutional power. The Danish government created and funded a research institute with the sole purpose of giving Bohr an environment to work in. The Danish Academy of Arts and Sciences chose Bohr to be the resident of the Carlsberg House of Honor, built and funded by the Carlsberg corporation, the great Danish beer-brewing company. The scion of a leading Danish intellectual family, Bohr regularly entertained at his home not only physicists but artists, politicians, and even the Danish royal family. For the young physicists who came to Copenhagen, "Bohr could provide intellectual stimulation and help in advancing careers, spiritual fulfillment and down-to-earth fun, material benefits and psychological counsel," as the historian of science Mara Beller put it. "He became a father figure who many young scientists were eager to honor and whose authority not many dared to challenge." Indeed, Bohr's influence on his students' lives often went well beyond the professional into the intensely personal: according to Victor Weisskopf, one of Bohr's most brilliant students, "any physicist working with Bohr was certain to be married after no more than two years."

Visiting the great sage of Copenhagen was intellectually and emotionally overwhelming, especially for young scientists. "Bohr had invited a number of us out to Carlsberg where, sipping our coffee after dinner, we sat close to him—some literally at his feet, on the floor—so as not to miss a word," wrote Otto Frisch, another of Bohr's students. "Here, I felt, was

Socrates come to life again, tossing us challenges in his gentle way, lifting each argument onto a higher plane, drawing wisdom out of us which we didn't know was in us (and which, of course, wasn't). Our conversations ranged from religion to genetics, from politics to art; and when I cycled home through the streets of Copenhagen, fragrant with lilac or wet with rain, I felt intoxicated with the heady spirit of Platonic dialogue."

But Bohr was a peculiar kind of sage—brilliant and insightful, yet plodding and obscure, sometimes infuriatingly so. "It is practically impossible to describe Niels Bohr to a person who has never worked with him," said George Gamow, a Russian physicist and former student of Bohr (who had a famously large personality himself). "Probably his most characteristic property was the slowness of his thinking and comprehension." Gamow then described the frustration that was watching a movie with the father of quantum physics:

> The only movies Bohr liked were those called The Gun Fight at the Lazy Gee Ranch or The Lone Ranger and a Sioux Girl. But it was hard to go with Bohr to the movies. He could not follow the plot, and was constantly asking us, to the great annoyance of the rest of the audience, questions like this: "Is that the sister of that cowboy who shot the Indian who tried to steal a herd of cattle belonging to her brother-in-law?" The same slowness of reaction was apparent at scientific meetings. Many a time, a visiting young physicist (most physicists visiting Copenhagen were young) would deliver a brilliant talk about his recent calculations on some intricate problem of the quantum theory. Everybody in the audience would understand the argument quite clearly, but Bohr wouldn't. So everybody would start to explain to Bohr the simple point he had missed, and in the resulting turmoil everybody would stop understanding anything. Finally, after a considerable period of time, Bohr would begin to understand, and it would turn out that what he understood about the problem presented by the visitor was quite different from what the visitor meant, and was correct, while the visitor's interpretation was wrong.

For his students and colleagues, the pull of Bohr's reputation, and his sheer force of personality, overcame the annoyances and peculiarities of working with him. If anything, those peculiarities endeared Bohr to his students more, for Bohr's quirks allowed them to see that it was not simply that they needed Bohr—he also needed them. Bohr's working style was slow, intense, and collaborative by nature. He was constantly wording and rewording his ideas and bouncing them off others. Writing was a painful process for Bohr, and nearly impossible for him to accomplish without help. In fact, in the years encompassing the crucial infancy of quantum theory, from 1922 to 1930, Bohr did not publish a single paper alone. And where Einstein's writing was clear and deceptively simple, Bohr's writing was tortuous and obscure, with famously long and convoluted sentences. Here, for example, is one of his shorter and more straightforward sentences, in which he is explaining that quantum "jumps" are the key difference between quantum physics and Newton's classical physics:

> Notwithstanding the difficulties which, hence, are involved in the formulation of the quantum theory, it seems, as we shall see, that its essence may be expressed in the so-called quantum postulate, which attributes to any atomic process an essential discontinuity, or rather individuality, completely foreign to the classical theories and symbolized by Planck's quantum of action.

Bohr was no more clear in speech than in writing. "At [a] 1932 conference, Bohr gave a fundamental report on the current difficulties of atomic theory," recalled his student Carl von Weizsäcker. "With an expression of suffering, his head held to one side, he stumbled over incomplete sentences." And Bohr's difficulties with expressing himself weren't limited to public talks. Describing a private conversation, Weizsäcker wrote that Bohr's "stumbling way of talking . . . would become less and less intelligible the more important the subject became." (Strangely, Bohr purportedly told his students to "never express yourself more clearly than you are able to think.") Yet this obscurity of thought merely added to Bohr's sagelike qualities. He could say a single word and leave his students puzzling over it for hours or days on end. And his obscurity did not diminish his students' feelings for him. Rudolf Peierls, one of Bohr's students (who later supervised the PhD of a young John Bell), said that "although often we could not understand [Bohr], we admired him almost without reservation and loved him without limits."

Three days after leaving Einstein in Berlin, Heisenberg arrived in Copenhagen. Since his previous stint at Bohr's institute, he had successfully defended his PhD, developed matrix mechanics, and been offered a faculty position. But, rather than returning victorious, Heisenberg was frustrated. His matrix mechanics was revolutionary—but his triumph had been short-lived. Six months after Heisenberg's work had first appeared in print, the Viennese physicist Erwin Schrödinger published a competing theory of quantum physics: wave mechanics.

Schrödinger had come up with wave mechanics while shacked up with his mistress in a resort in the Swiss Alps in December 1925. His theory was written in the relatively simple mathematical language of waves, with smoothly changing wave functions governed by the Schrödinger equation (as we saw in Chapter 1). Heisenberg was worried that Schrödinger's accomplishment would eclipse his own, and rightly so. The abstruse mathematics of Heisenberg's matrix mechanics was unfamiliar to most physicists at the time, and it had no obvious picture of the world to go along with it. Schrödinger's theory, meanwhile, used familiar mathematics with simple physical ideas; it was easy to handle and easy to think about. Schrödinger boasted that with his theory, physicists didn't have to "suppress intuition and to operate only with abstract concepts such as transition probabilities, energy levels, and the like." Much of the physics community agreed, even Heisenberg's erstwhile allies. Arnold Sommerfeld, Heisenberg's PhD adviser, said, "Although the truth of matrix mechanics is indubitable, its handling is extremely intricate and frighteningly abstract. Schrödinger has now come to our rescue." Born described Schrödinger's wave mechanics as "the deepest form of the quantum laws." Pauli, meanwhile, used Schrödinger's theory to do what he had been unable to accomplish with matrix mechanics alone—he managed to derive the brightness of the spectral lines in hydrogen, solving a problem that had been outstanding for more than seventy years.

Figure 2.1. Architects of the Copenhagen interpretation at the Niels Bohr Institute, 1936. Left to right: Bohr, Heisenberg, and Pauli.

Yet for all the successes of wave mechanics—and for all Schrödinger's bluster—it seemed that in the areas they overlapped, Schrödinger's wave mechanics gave the same results as Heisenberg's matrix mechanics. Schrödinger's theory, like Heisenberg's, reproduced the spectrum of the hydrogen atom perfectly: the different energy levels of Bohr's atom were, in Schrödinger's theory, associated with energy "eigenstates," special wave functions with constant energies. As Schrödinger soon discovered, matrix mechanics and wave mechanics were mathematically equivalent, using different tools to describe the same ideas: a single new theory of quantum mechanics. Problems like the brightness of spectral lines had been solved first with wave mechanics only because Schrödinger's equation was mathematically easier to handle than Heisenberg's matrices in most situations. But the two versions of quantum mechanics still differed radically in their interpretation. Schrödinger was sure that he could find a way to interpret all quantum phenomena as the smooth movement of the waves his equation described. Heisenberg was unconvinced. "The more I think about the physical portion of the Schrödinger theory, the more repulsive I find it," he wrote to Pauli. "What Schrödinger writes about the visualizability of his theory is 'probably not quite right,' in other words it's crap."

Yet Schrödinger's waves seemed more natural to most physicists than Heisenberg's matrices. Heisenberg, frustrated by the situation and fearful that Schrödinger's ideas would eclipse his own, wrote to his mentor Bohr. Bohr in turn wrote to Schrödinger, inviting him to visit Copenhagen to have "some discussions for the narrower circle of those who work here at the Institute, in which we can deal more deeply with the open questions of atomic theory." Schrödinger arrived by train on the first of October, 1926, and the debate started immediately, as Heisenberg later recalled:

> Bohr's discussions with Schrödinger began at the railway station and were continued daily from early morning until late at night. Schrödinger stayed in Bohr's house so that nothing would interrupt the conversations. And although Bohr was normally most considerate and friendly in his dealings with people, he now struck me as an almost remorseless fanatic, one who was not prepared to make the least concession or grant that he could ever be mistaken. It is hardly possible to convey just how passionate the discussions were, just how deeply rooted the convictions of each, a fact that marked their every utterance.

Schrödinger believed that his wave equation's success meant that all quantum phenomena could eventually be explained away as the behavior of continuous waves. Yet Bohr and Heisenberg pointed out that there were phenomena that seemed to demand quantum "jumps," like electrons jumping from one orbit to another in Bohr's atom, which could not be explained away by the smooth movement of waves. Schrödinger disagreed. "If all this damned quantum jumping were really here to stay, I should be sorry I ever got involved with quantum theory," he complained. Eventually, Schrödinger, worn out by Bohr's relentless questioning, caught a "feverish cold" in the damp dark Danish autumn and took to his bed in the Bohrs' house. While Margrethe, Bohr's wife, brought Schrödinger tea and cake, Bohr continued to press his advantage, sitting on the edge of Schrödinger's bed and saying in his quiet voice, "But you must surely admit that. . . ."

With neither side convinced by the other, Schrödinger went home. "No understanding could be expected, since, at the time, neither side was able to offer a complete and coherent interpretation of quantum mechanics," Heisenberg recalled. "For all that, we in Copenhagen felt convinced toward the end of Schrödinger's visit that we were on the right track." Fundamentally, the problem was that the meaning of Schrödinger's wave function was still not clear. But Max Born had discovered a piece of the puzzle that summer. He found that a particle's wave function in a location yields the probability of measuring the particle in that location—and that the wave function collapses once measurement happens. Born's insight ultimately won him a Nobel Prize, and rightly so. But Born's rule for handling wave functions also left physicists with new puzzles: What was a measurement? And why did wave functions behave differently when they were being measured, whatever that might mean? Born's idea and Schrödinger's mathematics had unlocked the quantum world, but at a price. The measurement problem had arrived.

Heisenberg wasn't particularly concerned with solving the measurement problem. He was more concerned with getting another offer for a tenured professorship. He was worried that Schrödinger's accomplishments had eclipsed his own and that he had made a mistake in returning to Copenhagen rather than accepting the permanent professional safety that had been offered by Leipzig. Hungry for another major insight to improve his chances on the job market—and to one-up Schrödinger—Heisenberg turned his attention to measurement, but not the measurement problem. Instead, he focused on something less difficult and more likely to yield results: the limitations on what we can learn about quantum objects. Combining Born's new idea with some of Einstein's suggestions from their meeting in Berlin, Heisenberg uncovered a pithy new truth that, he thought, put the lie to Schrödinger's idea of an orderly quantum world.

Heisenberg started thinking about what would happen if you tried to measure the position of a single particle, like an electron, to very high precision. He realized that you could do this the same way you'd look for a lost wallet in a dark field: shine a flashlight around until you've found what you're looking for. An ordinary flashlight wouldn't work for an electron, though—the wavelength of visible light is far too large for that. But Heisenberg knew you could find an electron using higher-energy light, with a shorter wavelength: gamma rays. Shine a gamma-ray flashlight around the room, and you'll find your electron. But gamma rays pack a punch—bounce one gamma-ray photon off an electron, and the electron will go careening off in some random direction. So you'll know where the electron was, but you won't know how fast it's going or where it's heading now.

Heisenberg wondered if this kind of trade-off between measuring an object's position and its momentum was unavoidable, or if it was just an artifact of his thought experiment. To his delight, he discovered that these limits on measurement were fundamental: buried in the mathematics of Schrödinger's wave mechanics, Heisenberg found a precise formulation of how much information you have to give up about an object's momentum in order to learn more about its position, and vice versa. You could know a lot about where an object was or a lot about how it was moving—but you couldn't know both at the same time.

At Bohr's urging, Heisenberg used the term "uncertainty principle" to describe this insight. Heisenberg's uncertainty paper paid off as he had hoped: the University of Leipzig again offered him a tenured professorship. He accepted, and in June 1927, Heisenberg, at twenty-five, became the youngest tenured professor in all of Germany.

Meanwhile, Bohr found that Heisenberg's uncertainty principle meshed well with his own new ideas about the true nature of the quantum world, which he called "complementarity." In typical Bohr fashion, his paper on complementarity became bogged down in a series of drafts filled with sentences that refused to end. But that September, Bohr ran out of time for rewrites. The International Physics Conference was meeting on the shores of alpine Lake Como, in northern Italy, and Bohr was scheduled to give the keynote address. Frantically revising his prepared statements up through the day of his talk, Bohr took the stage, speaking softly and haltingly.

Bohr started from the idea that "our usual description of physical phenomena is based entirely on the idea that the phenomena concerned may be observed without disturbing them appreciably." But, as Heisenberg's uncertainty principle made clear, "any observation of atomic phenomena will involve an interaction with the agency of observation not to be neglected." Therefore, Bohr continued, "an independent reality in the ordinary physical sense can neither be ascribed to the phenomena nor to the agencies of observation." In other words, one could not ask what was really happening inside of an atom when nobody looked—according to Bohr, the quantum world could only be considered real in conjunction with some kind of measurement apparatus to study that world. And the behavior of the objects in that world, as indicated by such an apparatus, would be best described as either particles or waves, but never both simultaneously. These descriptions are contradictory—a particle has a definite location, which waves don't; waves have frequencies and wavelengths, which particles don't—yet Bohr claimed that this "inevitable dilemma" was not a problem for quantum physics. "We are not dealing with contradictory but with complementary pictures of the phenomena," claimed Bohr, which are "indispensable for a description of experience."

This "wave-particle duality" shows up in all quantum phenomena. For example, in an old cathode-ray-tube TV, electrons shoot from the back of the TV toward the phosphorescent screen at the front of the TV, which lights up when an electron hits it. When an electron is shot out into the tube, its wave function obeys the Schrödinger equation, undulating and propagating outward like a wave. But when the electron hits the phosphorescent screen, it hits in one location, lighting up a particular spot on the screen, like a particle. So sometimes the electron behaves like a wave, and sometimes it behaves like a particle, but never both. According to Bohr, there cannot be a more complete description of an electron, or of anything—merely incomplete and incompatible analogies that never overlap. This, Bohr said, was the heart of complementarity, and it was inevitable and unavoidable. The new quantum theory had shown it was impossible to give a single consistent account of an electron that would work at all times.

Bohr pointed to the Heisenberg uncertainty principle as further justification for the inevitability of complementarity. Using Heisenberg's gamma-ray flashlight as an example, he pointed out that there was no way to avoid altering the momentum of an electron when observing its position, and vice versa. Bohr then echoed Mach, as Heisenberg had, and claimed the impossibility of measuring both properties of an electron simultaneously meant that it could not have both properties simultaneously. Position and momentum, like particles and waves, were complementary—never used at once but both needed for the complete description of a situation.

But Bohr was wrong. There was nothing inevitable or necessary about complementarity—other interpretations of quantum physics are possible. Indeed, the claim of inevitability is an awfully strong and strange claim to make about any interpretive issue in science, precisely because it is always possible to reinterpret any theory. Yet Bohr was convinced that complementarity was the deepest insight into nature found within the quantum theory.

Stranger still is Bohr's use of the gamma-ray flashlight to bolster his claims. It's certainly true that the thought experiment illustrates a world in which there are limits on our knowledge, but it's also a world where particles have well-defined positions and momenta at all times. Hitting an electron with a gamma ray can't alter the electron's momentum unless it has a momentum in the first place. We don't know what that momentum is—but that's certainly not the same thing as saying it doesn't exist.

As is always the case with Bohr, it's hard to be sure what he was actually trying to say, because his writing is so convoluted and obscure. But this is certainly how complementarity has often been understood. And as for Bohr's audience at Como, it's not clear what they understood at all. The reaction to his talk was muted. Many of the people in the audience were his students and colleagues—Heisenberg, Pauli, Born—and had spent much time in Copenhagen listening to Bohr expound on these ideas before. Many others were simply unimpressed. "[Complementarity] doesn't provide you with any equations which you didn't have before," said the English physicist Paul Dirac. (Dirac wasn't merely sniping—he had in fact discovered a new equation himself. He had skillfully fused quantum physics with special relativity, leading to a new theory of particle physics that came to be known as quantum field theory. Dirac's theory correctly predicted the existence of antimatter, a feat that would win him a Nobel Prize in 1933.) Eugene Wigner, the brilliant Hungarian mathematical physicist, agreed, stating that "Bohr's principle will not change the way we do physics." Schrödinger, of course, vehemently disagreed with Bohr—but Schrödinger wasn't there. He had just received a cushy appointment as professor of physics in Berlin and was still dealing with his move there from Switzerland. And there was nothing for Einstein to love in Bohr's ideas, but Einstein was also absent. Five years earlier, the fascist Benito Mussolini had taken control of Italy by marching on Rome with 30,000 Blackshirts, and Einstein had resolved to boycott all events in Italy as long as Mussolini and his thugs were in power. But, the next month, Bohr and many of the physicists at Como assembled again, for a prestigious invitation-only conference in Brussels—and this time, Einstein, Schrödinger, and more besides were all there. The stage for the quantum showdown was set.

3

Street Brawl

Ernst Solvay wanted to leave his mark on the world with his money. Like Alfred Nobel before him, he had profited through industrial applications of chemistry—though less explosively than Nobel, the father of dynamite—and, also like Nobel, he wanted to better the world by promoting scientific research. So, in 1911, Solvay used his money to organize a conference on the nascent quantum theory in his native Belgium. The conference was an enormous success, and Solvay decided to pour more money into organizing invitation-only conferences on subjects at the cutting edge of physics and chemistry. Solvay himself died in 1922, but his conferences continue to this day and are among the most rarefied of all scientific meetings. Yet the Fifth Solvay Conference, held in Brussels in October 1927, stands out from the rest. Seventeen of the twenty-nine attendees had won or would go on to win Nobel Prizes; one person there, Marie Curie, already had two Nobels to her name. In addition to Curie, Einstein, Planck, Schrödinger, Bohr, Heisenberg, Born, Dirac, and Pauli were there, and the conference photograph appears in many quantum physics textbooks. And, along with this picture, a historical fable has been handed down from generation to generation of physicists through an informal oral tradition, a sort of origin myth for quantum physics itself, that goes like this:

Once upon a time, a group of brilliant physicists discovered quantum physics. The new theory was wildly successful. Yet Einstein couldn't accept the radical new picture of nature revealed by quantum physics,

Figure 3.1. The Fifth Solvay Conference, Brussels, 1927. Front row: Einstein, center; Curie, third from left; Planck, second from left. Second row: Bohr, far right; Born, second from right; de Broglie, third from right. Back row: Heisenberg, third from right; Pauli, fourth from right; Schrödinger, center.

despite the pivotal role he had played in its early development (and despite the fact that older physicists had similarly argued against Einstein's theory of relativity a generation earlier). Famously protesting that "God does not play dice," Einstein had a series of informal debates with Bohr, beginning at Solvay in 1927, in which he repeatedly tried to find a way around Heisenberg's uncertainty principle. Ultimately, Bohr prevailed, and the rest of the physics community accepted that quantum physics was correct and that the Copenhagen interpretation was the correct way to understand it. But Einstein never accepted the new theory, and until the day he died he insisted that nature could not be fundamentally random. Thus, the fable concludes, even the greatest and most famous physicists can still be wrong.

Some of this story is true. It's true that Einstein and Bohr disagreed about quantum physics. It's true that they debated it at the Solvay conference in 1927 and afterward. And it's true that Einstein said that "God does not play dice," though he said it in a letter to Max Born in 1926, not in Brussels in 1927. But in almost every other important respect—Einstein's real problem with quantum physics, Bohr's defense of it, even the content of the Copenhagen interpretation and its general acceptance by the rest of the physics community after 1927—the truth is entirely different, and far more interesting, than the standard fable suggests.

Louis de Broglie, physicist and French nobleman, was among the first to speak at the Fifth Solvay Conference. De Broglie, who had defended his PhD thesis only three years prior, had been the first to suggest that all of the fundamental constituents of matter had both a particle and a wave aspect. He had borrowed much of his reasoning from Einstein; his adviser, Paul Langevin, had not been sure what to make of de Broglie's ideas, so Langevin wrote to Einstein to ask his opinion. Einstein replied enthusiastically, declaring that de Broglie had "lifted a corner of the great veil," and de Broglie got his doctorate.

Speaking to the assembled conference in Brussels, de Broglie presented a new idea. Skillfully manipulating the Schrödinger equation, he developed a novel picture of quantum physics with the same mathematics. Rather than particles and waves being incomplete, contradictory, "complementary" pictures of the quantum, de Broglie offered a quantum world where particles and waves lived in a peaceful coexistence, with particles surfing along "pilot waves" that govern their motion—anticipating Bohm's interpretation of quantum physics a quarter century later. De Broglie's particles moved in an entirely deterministic way, despite Born's statistical rule identifying the wave function as a tool for calculating probabilities. Yet the particles satisfied Heisenberg's uncertainty principle, because their paths were hidden from view—no experiment could reveal a particle's full trajectory, just as Heisenberg had said. De Broglie had found a way to restore determinism and causality to the quantum world without sacrificing the extraordinary match between theory and observation of the new quantum physics.

De Broglie's ideas were met with interest and vigorous debate. Wolfgang Pauli was quick to object. He claimed that de Broglie's theory contradicted existing theoretical work on particle collisions in quantum physics. De Broglie, floundering under Pauli's scrutiny, struggled to explain that Pauli was wrong. Pauli's objection was based on a deeply misleading analogy, which threw the French prince for a loop. While de Broglie's reply was largely accurate, Pauli remained unsatisfied.

Another, more serious objection to de Broglie's interpretation came from Hans Kramers, a Dutch physicist who had been one of Bohr's students. He pointed out that when a photon bounces off of a mirror, the mirror must recoil slightly from the impact. But de Broglie's theory, according to Kramers, could not explain the mirror's recoil. De Broglie admitted he could not answer this question. Unbeknownst to de Broglie or Kramers, it was in fact possible to explain the mirror's recoil using de Broglie's theory—it merely required treating both the photon and the mirror as quantum objects, not just the photon. But de Broglie, like most other physicists at the time, thought that quantum physics applied only to small objects, and this left him unable to reply to Kramers. Soon after the conference, de Broglie himself gave up on his ideas, for reasons related to Kramers's objection.

Born and Heisenberg spoke next, presenting their matrix-based formulation of quantum physics, in which irreducibly random quantum jumps played a central role. Toward the end of their presentation, they boldly claimed that quantum physics was "a closed theory, whose fundamental physical and mathematical assumptions are no longer susceptible of any modification." In other words, quantum physics was done, fully cooked: there was no further need to dig into the innards and find something more, either mathematically or interpretationally. Later on, Bohr spoke, mostly rehashing his Como lecture, emphasizing that the wave and particle descriptions of quantum phenomena were complementary rather than contradictory: both necessary for a complete description, but never to be used to describe the same object at the same time.

Finally, after several days of sitting and listening while saying almost nothing, Einstein rose to speak during a period of open discussion. He had been passing notes with his close friend Paul Ehrenfest, gently mocking the Copenhagen camp, and had waited to carefully formulate his thoughts before he replied to them. Everyone in the room knew that Einstein had serious reservations about Bohr and Heisenberg's ideas.

Now, all eyes were on him as he walked up to the chalkboard to sketch out a simple thought experiment that contained a devastating critique of the Copenhagen interpretation.

Why were Bohr, Heisenberg, and others so convinced that the quantum world couldn't be visualized? Why did they seem to think that things couldn't be real until they were observed? Why did they insist that the classical world obeyed fundamentally different rules from the quantum world? Why, in short, did they believe the strange assemblage of claims that came to be known as the Copenhagen interpretation?

Niels Bohr's force of personality is the most obvious answer. But the next question is why Bohr had these ideas—or even whether he did. Bohr's writing is so difficult and obscure that it is hard to say what Bohr's own positions were, which makes it even harder to divine the ideas that influenced Bohr. (Amazingly, Bohr's students and colleagues cited complementarity as a reason for this. According to his students, Bohr himself had said that "truth was complementary to clarity," and thus, they claimed, "Bohr was a very bad speaker, because he was too much concerned with truth"; similarly, "his sentences were long, involuted and opaque" because he "strove for precision.") But Bohr's obfuscatory writing hasn't thwarted attempts to trace the origins of his ideas: on the contrary, there's a cottage industry of theorizing about what was happening inside the head of one Niels Henrik David Bohr. Some have suggested that he was influenced primarily by Kant; others have pointed at his compatriot Søren Kierkegaard (buried just a few dozen yards away from Bohr in Copenhagen's Assistens Cemetery); still others have seen the influence of gnosticism in the contradictions of complementarity. Bohr's most loyal and vocal advocate, Léon Rosenfeld, saw a consistent strain of Marxism in Bohr's writing and thoughts—an opinion that surely had nothing whatsoever to do with the fact that Rosenfeld himself was an avowed Marxist. In short, the literature on Bohr is vast and inconclusive (though most seem to agree that Kant's writing really did have some influence on him).

Bohr's opaque writing and peculiar ability to inspire devotion in his students and colleagues aren't the whole explanation, though. Another part of the answer lies in the intellectual atmosphere of the period itself. For example, the antimaterialist culture of interwar Weimar Germany likely played a role. And Heisenberg and others were definitely influenced by Ernst Mach and his successors, the "Vienna Circle" of philosophers, who developed a school of thought called "logical positivism." Logical positivism picked up where Mach left off—according to them, any statement that made reference to something unobservable was not only bad science, it was literally meaningless. Thus, talking about what happens in quantum systems when nobody's looking is nonsensical.

The influence of the logical positivists on the founders of quantum physics was particularly personal in the case of Wolfgang Pauli. Pauli was born and raised in Vienna, and his godfather was Ernst Mach himself. Outspoken, quick-witted, and deeply talented, Pauli had enormous influence among the physicists of his day. Heisenberg and Bohr both craved his good opinion. But it was hard to come by—Pauli's scathing put-downs were legendary, earning him the nickname "the Wrath of God." "I do not mind if you think slowly, but I do object when you publish more quickly than you can think," he once told a colleague. He dismissively said of another physicist's paper that "it is not even wrong." Even his compliments were backhanded: after hearing Einstein lecture to a full house at the University of Munich, Pauli exclaimed, "You know, what Mr. Einstein said is not so stupid." And, when discussing matters of quantum interpretation, Pauli often sounded a positivist note. According to him, worrying about the position of an object before it's been measured was pointless. "One should no more rack one's brain about the problem of whether something one cannot know anything about exists all the same," he said, "than about the ancient question of how many angels are able to sit on the point of a needle."

Positivism influenced the rest of the Copenhagen camp, but to varying degrees. And they applied it in different ways, leading to inconsistent views among them. Bohr simply dismissed the idea of a quantum world altogether. "There is no quantum world," he said. "Isolated material particles are abstractions, their properties on the quantum theory being definable and observable only through their interaction with other systems." But Heisenberg thought there was a quantum world—just one that operated differently from our own. "The atoms or the elementary particles are not as real [as phenomena in daily life]; they form a world of potentialities or possibilities rather than one of things or facts." And Jordan thought that "observations not only disturb what has to be measured, they produce it"—he claimed that measuring an electron "compel[s] it *to assume a definite position*." But if there's no quantum world, as Bohr claimed, then measurements can't compel anything to happen there. And Pauli contradicted Bohr too. Pauli thought that observation introduced "indeterminable effects" that disturbed the systems being observed in uncontrollable ways. Yet observation certainly can't disturb the quantum world if there is no quantum world, as Bohr thought. Pauli may have even contradicted himself. He had dismissed the whole enterprise of talking about what happened when nobody was looking. But if it's meaningless to talk about things before they're observed, how could Pauli say observation disturbed anything at all? And Heisenberg and Jordan clearly contradicted Pauli. They had no qualms about making strong statements about unobserved systems. Thus, the myth that these physicists created a unified Copenhagen interpretation is just that—a myth.

Yet, despite their differences, Bohr, Heisenberg, and the rest of the Göttingen-Copenhagen group had a few things in common. They all agreed that it was pointless to talk about what was "really" happening in the quantum world. Making accurate predictions about the outcomes of measurements was, for them, enough. As Bohr put it years after Solvay, "It is wrong to think that the task of physics is to find out how nature is. Physics concerns what we can say about nature." Quantum physics, then, didn't have to present a coherent or consistent picture of how the world operated—indeed, according to Bohr's complementarity, such a picture was necessarily impossible. It was enough to merely describe measurable features of the world accurately, without talking about what was actually happening. Quantum physics, in short, shouldn't be taken seriously as a theory of the way the world actually is. Instead, quantum physics is a mere tool, an instrument for predicting the outcomes of measurements.

Yet, strangely, its unseriousness should be taken very seriously: in claiming their version of quantum physics as a "closed theory," Heisenberg and Born were ruling out the possibility of an explanation of the quantum world, independent of observation, even in principle.

This is where Einstein parted ways with Bohr, Heisenberg, and their ideological compatriots. "The programmatic aim of all physics," according to Einstein, was "the complete description of any (individual) real situation (as it supposedly exists irrespective of any act of observation or substantiation)." In this view, Einstein knew he was out of sync with the intellectual fashion of his day: "Whenever the positivistically inclined modern physicist hears such a formulation his reaction is that of a pitying smile." But Einstein found positivism wholly uncompelling, seeing it as a total rejection of the idea of a physical world, tantamount to claiming that reality exists only in our minds: "What I dislike in this kind of argumentation is the basic positivistic attitude, which from my point of view is untenable, and which seems to me to come to the same thing as [Irish philosopher George] Berkeley's principle, *esse est percipi* [to be is to be perceived]." Though Einstein had no doubts about the importance of the new quantum theory, he was convinced that Born and Heisenberg were wrong to claim that quantum physics was complete, and that Bohr's philosophy of complementarity was inadequate for understanding the true nature of the quantum world. His thought experiment was simple, elegant, and carefully designed to strike at the heart of this inadequacy.

Consider, Einstein said to the assembled Solvay conference, a stream of electrons passing through a very small hole in a screen (Figure 3.2). On the other side of the screen, there is a hemisphere of phosphorescent film that can register the impact of a single electron. According to quantum physics, the wave function for the stream of electrons should be uniform—the probability of an electron hitting the film should be the same at every location on the hemisphere. And that's fine—if quantum physics says you'll find ten electrons per square centimeter of film after conducting your experiment, then on average, that's what you'll

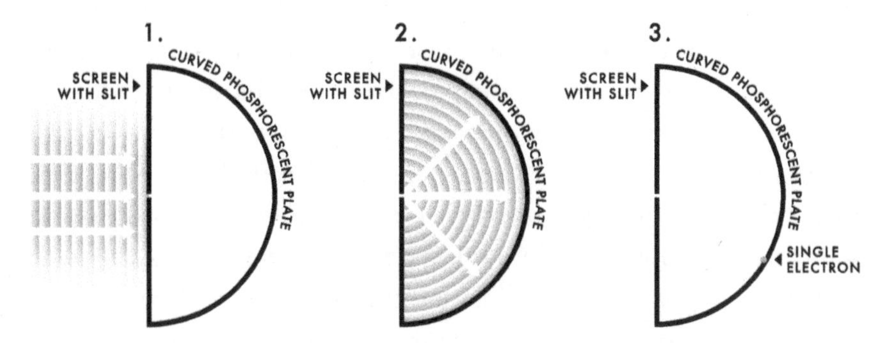

Figure 3.2. Einstein's thought experiment at Solvay. When the electron strikes the plate, how does the rest of the wave function "know" to collapse immediately? After the diagram in Bacciagaluppi and Valentini 2009, p. 486.

find. Quantum physics is great at describing the aggregate behavior of large groups of particles. But quantum physics can't do more than assign probabilities; it can't tell you exactly how many electrons will hit each part of the screen, it can only give you an average.

Now, Einstein asked his audience to consider the case in which a *single* electron is sent through the hole. Quantum physics still predicts that the electron is equally likely to hit at any location on the screen and cannot be more precise than that. That's fine, though—maybe it just means the theory is incomplete or limited in some way. But, Einstein reminded the group, Heisenberg and Born had claimed that quantum physics was closed, complete, perfect as it was. In that case, there cannot be anything that determines the particular location at which the electron hits the film. But this is a problem—and not because it introduces randomness into nature.

Instead, the problem is one of *locality:* the principle that something that happens in one location can't instantly influence an event that happens somewhere else. The wave function of our single electron is spread evenly across the hemisphere of film, and according to Heisenberg, Born, and Bohr, the electron itself isn't anywhere. The fact that the electron's wave function is evenly spread simply means that the film is equally likely to register the impact of an electron at any location. But, Einstein pointed out, what happens to the wave function when the film does register that impact in one particular spot? Born had shown that a particle's wave function was proportional to the probability of finding the particle in a particular place. But once the electron impacts the film in a particular spot, the probability of it hitting the film anywhere else immediately drops to zero. So, somehow, the wave function must instantaneously vanish, across the entire hemisphere, the moment that spot on the film indicates the electron hit. Anything less than instantaneous vanishing, and we run the risk of seeing a nonexistent second electron register on the film after the first, at some location where the wave function isn't yet zero. This "entirely peculiar mechanism of action at a distance," said Einstein, "implies to my mind a contradiction with the principle [i.e., the special theory] of relativity," which states very clearly that neither objects nor signals can travel faster than the speed of light. Thus, if quantum physics really is a complete description of nature, then it must violate relativity. To Einstein, the conclusion was obvious: the electron must have been in a particular location even before it hit the film, even though quantum physics could say nothing about where, exactly, it was. This was the only way Einstein saw to avoid invoking an instantaneous wave function collapse that would violate locality. Therefore, quantum physics was an incomplete description of nature, and more was needed to understand the true story of the quantum world. Specifically, to avoid contradictions with relativity, particles must have determinate locations at all times, in addition to wave functions. "I think that Mr. de Broglie is right to search in this direction," Einstein concluded.

The response to Einstein's thought experiment from the rest of the conference was muted incomprehension. Bohr, to his credit, admitted as much. "I feel myself in a very difficult position because I don't understand what precisely is the point which Einstein wants to [make]," he said. "No doubt it is my fault." Einstein's simple thought experiment offered a devastating critique of the Copenhagen position, but its simplicity may have paradoxically been a barrier to understanding: Einstein's presentation of it was quite brief and may have given the impression that he was confused about the nature of probability. Bohr, in particular, seems to have misunderstood Einstein rather badly: he recalled later that Einstein had doubts about Heisenberg's uncertainty principle, and introduced his thought experiment because he was looking for ways around

Figure 3.3. Einstein and Bohr, c. 1930.

it. Einstein's concerns about locality went unheard for the rest of the Solvay conference. But Einstein soon developed new thought experiments, doggedly pursuing the problems he saw with quantum physics.

At the next Solvay conference, in 1930, Einstein presented Bohr with another thought experiment, an imaginary contraption involving a spring scale with a box full of light hanging from it, timed with an accurate clock. Bohr, again, thought that Einstein was trying to get around a quantum uncertainty principle. After thinking for a while, Bohr revealed that Einstein's thought experiment "failed" because he (Einstein) had forgotten to take his own theory of general relativity into account.

This episode has become a legend in the history of quantum physics—Einstein hoisted by his own petard. But in reality, the problem was with Bohr. Einstein's thought experiment in 1930 was never intended to get around any kind of uncertainty principle—his complaint, once again, was locality, just as it had been at Solvay three years earlier. According to his friend Paul Ehrenfest, Einstein "absolutely no longer doubted the uncertainty relations" and had developed this thought experiment "for a totally different purpose." Bohr had once again missed the point.

Several years later, Einstein came back with yet another thought experiment to demonstrate his concerns about locality, one that would resound for decades. Einstein and two colleagues, Boris Podolsky and Nathan Rosen, published a paper in 1935 with the provocative title "Can Quantum Mechanical Description of Physical Reality Be Considered Complete?" This paper, known as the EPR paper after its authors' initials, is sometimes depicted as Einstein's last desperate move in his showdown with Bohr. The truth is far messier—and much more intriguing.

The EPR paper, on the face of it, isn't about locality—ironically, it appears to be a way around the Heisenberg uncertainty principle. But instead of devising a way to measure the momentum and position of a single particle directly at a single time, as Einstein had purportedly done in his earlier thought experiments, the EPR paper goes about it indirectly. The thought experiment at the heart of the paper imagines a pair of particles, A and B, which collide head-on, interact in a very specific and delicate way, then fly off in two opposite directions. Momentum is always conserved—it's a basic law of nature—so the total momentum of the particles is fixed over time. And, because of the way the particles interact, the distance between them at any time is easy to calculate.

In Newton's physics, this would be like two identical billiard balls hitting each other head-on, then bouncing off to opposite ends of an enormous pool table. Because the total momentum of the two balls has to be zero, knowing one ball's speed and direction would instantly tell you the other ball is moving at the same speed in the opposite direction. Similarly, finding one ball would allow you to calculate the location of the other ball, if you know the time and location of the collision.

In quantum physics, the situation is a little trickier. According to the Copenhagen interpretation, particles don't have properties like position or momentum (or anything else) until those properties are measured. But, EPR argued, measurements made on one particle couldn't instantly affect another particles far away. So, to get around the uncertainty principle, just wait until particles A and B are very far apart, then find the momentum of A. Measuring A's momentum lets you infer B's momentum without disturbing B at all. Then simply measure the position of B.

Now you know B's position and momentum, to arbitrary precision, at the same time. Therefore, argued EPR, a particle can have a definite position and momentum at the same time. But because quantum physics doesn't let you simultaneously predict the position and momentum of a single particle, EPR argued that it must be *incomplete*—there must be features of the world that quantum physics doesn't account for. The EPR paper closes by holding out hope for a better theory that can account for those things: "While we have thus shown that the wave function does not provide a complete description of the physical reality, we left open the question of whether or not such a description exists. We believe, however, that such a theory is possible."

The most famous scientist in the world blasting away at a well-known (if not well-understood) theory in such harsh terms made for a media frenzy, of course—especially when Podolsky leaked the story to the press early. "EINSTEIN ATTACKS QUANTUM THEORY" blared the *New York Times* on May 4, 1935, several days before the EPR paper was published. "Scientist and Two Colleagues Find It Is Not 'Complete' Even Though 'Correct.'" Einstein, furious, sent a statement to the newspaper in reply: "Any information upon which the article 'Einstein Attacks Quantum Theory' . . . is based was given to you without my authority. It is my invariable practice to discuss scientific matters only in the appropriate forum and I deprecate advance publication of any announcement in regard to such matters in the secular press."

Podolsky's leak wasn't the only reason Einstein was upset. Despite the fact that his name was on the EPR paper, Einstein hadn't actually written it himself—and he wasn't happy with it either. Shortly after its publication, Einstein told Schrödinger that the EPR paper "was written by Podolsky after much discussion. Still, it did not come out as well as I had originally wanted; rather the essential thing was, so to speak, smothered by the [mathematics]." Later in the same letter, he said that he "couldn't care less" about the uncertainty principle; his real problem with quantum physics had nothing to do with that.

For Einstein, the crucial bit of the EPR thought experiment once again had to do with locality. If you measure A's momentum, you know B's momentum too. But because B is far away from A, then, assuming locality, there's no way that making a measurement on A could have affected B immediately. B's momentum must have been set when A and B collided, just like billiard balls.

But quantum physics doesn't let you calculate the momenta of A and B once they collide. Instead, the quantum wave function connects A and B in a strange way. Because of their collision, A and B share a single wave function, rather than having their own individual wave functions. But that shared wave function doesn't say what the particles' momenta are before a measurement is made. It simply ensures that, once A's momentum is measured, B's momentum will always be equal and opposite.

According to the Copenhagen interpretation, particles don't have definite properties until those properties are measured. So if A and B have definite momenta before they're measured, then the Copenhagen interpretation is wrong and quantum physics is an incomplete description of nature. But if A and B *don't* have definite momenta before they're measured, then the very act of measuring A's momentum must affect B instantly, in order to ensure that its momentum is equal and opposite to A's—even if A is in New York City and B is on the Moon. And that violates locality. In short, quantum physics is either incomplete or nonlocal. This forced choice was what had been "smothered" in the EPR paper, according to Einstein.

Einstein rejected any violation of locality, calling it "spooky action at a distance" in a letter to Max Born. He pointed out that there was no reason to assume any weird connection of this sort—the facts at hand could easily be explained by the incompleteness of quantum theory:

> When I consider the physical phenomena known to me, and especially those which are being so successfully encompassed by quantum mechanics, I still cannot find any fact anywhere which would make it appear likely that [locality] will have to be abandoned. I am therefore inclined to believe that the description of quantum mechanics in the sense of [the Copenhagen interpretation] has to be regarded as an incomplete and indirect description of reality, to be replaced at some later date by a more complete and direct one.

Meanwhile, the rest of the physics community was shocked by the EPR paper. "Now we have to start all over again, because Einstein proved

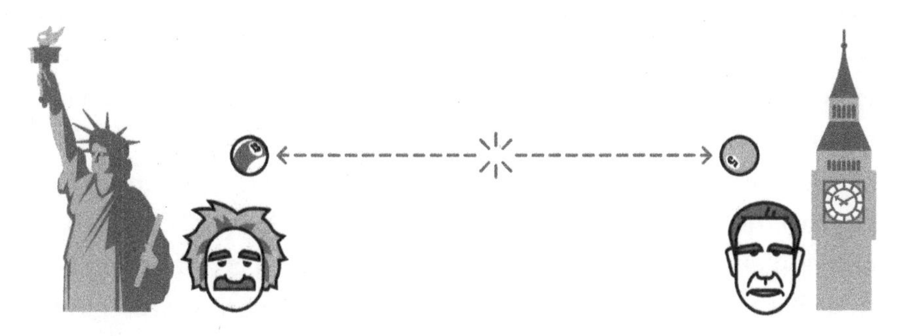

Figure 3.4. The EPR experiment. Two billiard balls collide and fly off in opposite directions. When Albert measures the momentum of his billiard ball, he instantly infers the momentum of Niels's billiard ball, even though he is in New York and Niels is in London. Either Niels's billiard ball already had that momentum in London before Albert made his measurement in New York, or there is "spooky action at a distance," instantly connecting the two balls across the Atlantic.

that it does not work," wailed Dirac. In a fury, Pauli wrote a letter to Heisenberg, calling Einstein's behavior a "disaster" and asking Heisenberg to write a reply for publication. When Heisenberg heard that Bohr was working on his own reply, though, he shelved his draft and let the master himself respond to Einstein's newest heresy.

"This onslaught came down upon us as a bolt from the blue. Its effect on Bohr was remarkable," said Léon Rosenfeld. "As soon as Bohr had heard my report of Einstein's argument, everything else was abandoned; we had to clear up such a misunderstanding at once." Bohr immediately began crafting a reply with Rosenfeld's help. Normally a painfully slow writer, Bohr managed to crank out a paper replying to EPR in six weeks—"an astonishing speed" for Bohr, according to Rosenfeld—and sent it off to *Physical Review,* the same journal where the EPR paper had appeared.

Bohr considered the EPR thought experiment carefully in his reply. He agreed that measuring A's momentum could not "mechanically" disturb B—there was "no question" of that. But, he maintained, there was still "the question of *an influence on the very conditions which define the possible types of predictions regarding the future behavior of the system.*" Unfortunately, it's not clear what distinction Bohr was trying to draw between "mechanical disturbance[s]" on one hand and "influences" on the other. Was he saying that measuring A could immediately affect B?

Maybe. Did he think that quantum physics therefore had to be nonlocal? Maybe again. There has been an enormous amount of ink spilled trying to decipher Bohr's reply to EPR; there is no clear consensus on what he meant, or whether he thought quantum physics was nonlocal.

Bohr himself later apologized for the quality of his writing. Looking back nearly fifteen years later, he wrote that he was "deeply aware of the inefficiency of expression" in the crucial part of his reply to EPR. But he didn't elaborate on his response, except to say that, in the quantum world, it's impossible to draw a sharp distinction between the behavior of the things you want to measure and their interaction with the devices you use to measure them. It's unclear how this relates to the EPR argument, and it definitely doesn't address Einstein's concerns about locality.

Despite Bohr's muddled writing, the mere fact of his reply to EPR assuaged the concerns of most of the rest of the physics community—though most physicists agreed with Max Born that Bohr's writing was "frequently nebulous and obscure." Few actually read what Bohr had written. But whether Bohr himself thought the Copenhagen interpretation was nonlocal, most other physicists didn't. As far as they were concerned, Bohr's reply simply meant that the Copenhagen interpretation was alive and well, and EPR's accusations of incompleteness could be safely ignored.

But Schrödinger was still unconvinced by the Copenhagen interpretation. Writing to Einstein after reading the EPR paper, Schrödinger said, "I am very pleased that in the [EPR paper] you have publicly called the dogmatic quantum mechanics to account."

Schrödinger also pointed out something surprising about the EPR thought experiment. The strange connection between particles A and B, causing them to share a wave function, was not unusual. Writing about this to Einstein, and in several papers written later that year, Schrödinger dubbed this connection "entanglement."

Entanglement, Schrödinger found, is pervasive in quantum physics. When any two subatomic particles collide, they almost always become entangled. When a group of objects forms some larger object, like subatomic particles in an atom or atoms in a molecule, they become entangled. In fact, nearly any interaction between any particles would cause them to become entangled, sharing a single wave function in the same way as the particles in the EPR thought experiment.

Schrödinger's observation that entanglement shows up throughout quantum physics only deepened the problem for the Copenhagen interpretation. For any entangled system, Einstein's choice applied: either the system is nonlocal, or quantum physics can't fully describe all the features of that system. And Schrödinger had just shown that nearly any quantum interaction would result in an entangled system. Thus, the challenge the EPR paper posed wasn't limited to some tiny corner of quantum physics—it was deeply embedded in the fundamental structure of the theory.

But Einstein's fears that the forced choice between nonlocality and incompleteness had been smothered in the EPR paper were sadly justified. In a letter to Einstein, Schrödinger vented his frustration over how badly other physicists had missed the point: "It is as if one person said, 'It is bitter cold in Chicago'; and another answered, 'That is a fallacy, it is very hot in Florida.'" Einstein himself received many letters from other physicists vigorously defending the Copenhagen interpretation and pointing out where the EPR paper had gone wrong—but he was amused to find that the letters all disagreed about where exactly that was! Many were under the impression that the EPR argument, and all of Einstein's problems with quantum theory, were based on a desire for a clockwork deterministic universe, like the universe of Newton's physics. They might have been misled by Einstein's famous exclamation about a dice-playing God. But Einstein's concerns had little, if anything, to do with determinism—they were about the importance of locality and a physical reality that exists independently of anyone observing it. Quantum physics, said Einstein, "avoids reality and reason." In his view, physics had been led astray by following Bohr. Writing to Schrödinger, Einstein described Bohr as a "talmudic philosopher [who] doesn't give a hoot for 'reality,' which he regards as a hobgoblin of the naive."

Yet, in the eyes of most contemporary physicists, Einstein's concerns were irrelevant at best and misguided at worst. The English physicist Charles Darwin (named after his famous grandfather) said, "It is a part of my doctrine that the details of a physicist's philosophy do not matter much." Darwin had once been Bohr's student, like many of the physicists at the forefront of quantum physics research—and almost none had ever worked with Einstein. Thus, in the clash between the two on matters of quantum philosophy, most physicists were inclined to follow "Bohr's Sunday word of worship," as the physicist Alfred Landé put it, while pursuing their own research on more down-to-earth subjects in quantum physics. After all, quantum physics worked, so why worry? The new theory allowed physicists to calculate and predict an enormous variety of phenomena with unprecedented accuracy, most of which had little if anything to do with the mysteries of entanglement. Other mysteries, more amenable to experimental exploration, beckoned—particularly the dark and powerful ones that lay in the atomic nucleus. Less than four years after EPR was published, those mysteries were revealed—and the world went to war.

4

Copenhagen in Manhattan

In the winter of 1955, Werner Heisenberg gave a series of lectures at the University of St. Andrews in Scotland. The Cold War was in full swing; in the space of the previous decade, Heisenberg had gone from an enemy alien of the UK to a citizen of a trusted ally. Yet he was uneasy about his reputation among his fellow physicists and took the opportunity of his talk in Scotland to bolster it.

First, Heisenberg preached the familiar gospel of Copenhagen. "The idea of an objective real world whose smallest parts exist objectively in the same sense as stones or trees exist, independently of whether or not we observe them," Heisenberg said, "is impossible." How, then, does our world of stones and trees emerge from the world of atoms and molecules? "The transition from the 'possible' to the 'actual' takes place during the act of observation," said Heisenberg. And what happens when we're not looking? According to Heisenberg, that question can't even be asked. "If we want to describe what happens in an atomic event, we have to realize that the word 'happens' can apply only to the observation, not to the state of affairs between two observations." And what about the measurement problem? What makes observation so special? Whatever it is, it is "physical," not "psychical," said Heisenberg. "The transition from the 'possible' to the 'actual' takes place as soon as the interaction of the object with the measuring device, and thereby with the rest of the world, has come into play; it is not connected with the act of registration of the result by the mind of the observer." Yet on the question of what constituted a "measuring device" and why it obeyed different rules than the quantum world, Heisenberg was infuriatingly unclear. Nowhere in his lectures did he propose anything like a solution to the measurement problem.

But Heisenberg also made sure to leave very little daylight between his own views and Bohr's in his talk. "Since the spring of 1927 one has had a consistent interpretation of quantum theory, which is frequently called the 'Copenhagen interpretation.'" It was an exaggeration, at best, to claim that there had been a single consistent interpretation of quantum physics since 1927—and it was certainly not true that anything whatsoever was "frequently" called the Copenhagen interpretation at the time. In fact, Heisenberg himself had coined the phrase just a few months earlier, in an essay he had written for Bohr's seventieth birthday. In both his essay and his lecture, Heisenberg depicted the Copenhagen interpretation as a unified body of work that was developed by Bohr, himself, and a handful of others in 1927—and in both essay and lecture, Heisenberg took it upon himself to defend the Copenhagen interpretation against its enemies. "Many attempts have been made to criticize the Copenhagen interpretation and to replace it by one more in line with the concepts of classical physics or materialistic philosophy," Heisenberg warned his audience in Scotland. But, he claimed, this was simply impossible—ruled out entirely by the astonishing success of quantum physics, which could only be interpreted in One True Way, the Copenhagen interpretation.

The phrase "Copenhagen interpretation" was new, but this was hardly the first time that someone who had worked in Copenhagen claimed there was only one way to interpret quantum physics. Yet Heisenberg had extra reason to paint himself as an architect and defender of the quantum orthodoxy now. A common enemy had mended relations between the UK and Germany, and Heisenberg was likely hoping to perform the same kind of maneuver. Heisenberg's appalling activities during the war had nearly destroyed his relationships with Bohr and the rest of his colleagues. But the crucible of war had also radically reshaped physics itself—and luckily for Heisenberg and his precious reputation, these dizzying changes had made physicists much more receptive to the Copenhagen interpretation.

On May 16, 1933, Max Planck, the physicist whose black-body radiation law had set off the quantum revolution, met with Adolf Hitler. As the head of the Kaiser-Wilhelm-Gesellschaft, the foremost scientific institute in Germany, it was customary for Planck to meet with new heads of state. Hitler had been chancellor for less than four months, and he had already seized dictatorial power over the young Weimar Republic, using the threat of domestic terrorism in the aftermath of the Reichstag fire as an excuse. Now, Hitler had passed a law decreeing that anyone not of "pure Aryan" descent was barred from holding any civil service jobs, including public university professorships. For Planck, this was simply a step too far. Surely "there are different kinds of Jews, some valuable for mankind and others worthless," Planck told Hitler, "and that distinction must be made."

"That is not right," replied Hitler. "A Jew is a Jew. All Jews stick together, like leeches."

Planck tried a different tack. "It would be self-mutilation to make valuable Jews emigrate, since we need their scientific work."

At the suggestion that he could ever need the help of a Jew, Hitler snapped. "If the dismissal of Jewish scientists means the annihilation of contemporary German science, then we shall do without science for a few years!" Speaking faster and faster, Planck recalled later, the Führer "whipped himself into such a frenzy that I had no choice except to fall silent and leave." Jews no longer had a place in German science, and there was nothing Planck could do about it.

Germany's universities, all of which were public, had been at the heart of European intellectual life for over a century. Now, 1,600 scholars were out of a job. The burden fell disproportionately on the sciences: since nineteenth-century German idealist philosophy had sneered at the sciences as "materialistic" and therefore inferior, there had been fewer impediments to Jews advancing in the sciences. Now, over a hundred German physicists were unemployed—a full quarter of all physicists in what had been the unrivaled center of the physics world. With a single act, physics in Germany had been destroyed.

Einstein would have been first among the newly unemployed—but he had seen Germany's fate coming. He and his wife Elsa had left their home in Berlin to tour the United States months before Hitler came to power. "Take a very good look at it. You will never see it again," Einstein had told Elsa as they left. Once the Nazis took over, Einstein, the most famous Jew in the world, was a marked man. Einstein's stepdaughter managed to smuggle his papers safely out of his Berlin apartment before the Nazis could destroy them; Hitler's goons ransacked the apartment four times in three days, but Einstein's family and papers all made it out of the country. Einstein met his family and collected his belongings in Belgium, renounced his German citizenship, then returned to the United States, where he took up a post at the newly minted Institute for Advanced Study in Princeton. He remained in America for the rest of his life.

Physicists who lacked Einstein's foresight but shared his Jewish heritage fled Nazi Germany after the Civil Service Act. They mostly landed in the United States and the UK, drastically shifting the center of the physics world (and changing the international language of physics from German to English). Max Born was unceremoniously suspended from his post at Göttingen. "All I had built up in Göttingen, during twelve years' hard work, was shattered," he wrote. "It seemed to me like the end of the world." He and his family went to Cambridge for a time, then India, and ultimately resettled in Scotland for the duration of the war.

As Hitler expanded his control beyond Germany's borders during the 1930s, more Jews who had the means to escape fled. By the Anschluss of March 1938, which united Germany with Hitler's native Austria, many of the great Jewish intellectuals of Viennese culture had already left: Ludwig Wittgenstein was teaching in Cambridge, Karl Popper was a lecturer at the University of New Zealand, and Billy Wilder was writing lines for Greta Garbo in Hollywood. The best-known physicist in Austria, Erwin Schrödinger, wasn't Jewish, but his wife was. Schrödinger had been at the University of Berlin in 1933 but had quit in protest when Hitler came to power. After Hitler invaded Austria, Schrödinger publicly recanted his anti-Nazi views, but this wasn't enough for the new regime. Dismissed from his university post for "political unreliability," Schrödinger fled to Ireland with his wife. Once there, he wrote a letter to Einstein, apologizing profusely for his "great duplicity."

Italy's Jews started to feel the pressure of the Nazi anti-Semitic policies after Hitler's visit to Mussolini's Fascist Italy in the summer of 1938. "The racial campaign . . . acquired momentum at an amazingly fast pace," wrote Laura Fermi. "We at once decided to leave Italy as soon as possible." Her husband, Enrico, was the pride of Italian physics, one of the foremost experts on both theoretical and experimental nuclear physics in the world. But with Italy unsafe for the family of a Catholic man and a Jewish woman, Enrico and Laura quietly made plans to leave. Their plans were complicated by Mussolini's fascist economic policy, which made it illegal to take more than pocket change out of Italy. Then Niels Bohr intervened. When Fermi came to Copenhagen for a conference that summer, Bohr took him aside and—breaking an unwritten rule of the physics community—told Fermi that his name was in the running for a Nobel Prize that year. Would the prize, which came with a $1 million cash award and an excuse to travel abroad, be useful this year, Bohr asked? Or would another time be more convenient, given the political situation? Fermi told Bohr that this year would be a particularly good time for the prize. Returning home, Fermi found that the Italian government had confiscated all Jewish passports, including Laura's; pulling a few strings, he managed to get her passport back in time to attend the Nobel ceremony in Stockholm. After Stockholm, the Fermis went to Copenhagen to visit Bohr, and, finding the wheels of American immigration greased by the words "Nobel Prize winner," they set sail for Manhattan just before Christmas, arriving on January 2, 1939.

Well-established physicists like Einstein, Born, and Fermi were often able to secure new jobs in their new countries before they had even arrived. But the lives of students and younger researchers were far more thoroughly disrupted. "My heart aches at the thought of the young ones," Einstein wrote to Born in 1933. Einstein was soon involved with a British-led effort to help the academic victims of the Nazi regime, which met with some success. By the time Hitler invaded Poland and started World War II on September 1, 1939, more than a hundred physicists had emigrated from the European continent to the United States and the UK—some of the younger ones simply fleeing, refugees without the promise of a job in their new country, coming with a single small bag across the Channel or the Atlantic. Some came with nothing. Some never came at all.

John von Neumann, like Einstein, had made it out of Germany early. He and his friend and fellow Hungarian Eugene Wigner were both offered positions at Princeton in 1930. Knowing that the two men were unlikely to simply pack up and leave Europe, Princeton offered them half-time appointments: visit Princeton for half the year, and then go back to their jobs at the University of Berlin, where they could lounge in coffee shops with Einstein and Schrödinger, for the other half. Both men accepted the generous offer, but they had different opinions of the New World. Von Neumann immediately took to the States, holding dinner parties with his wife almost nightly, always in impeccable attire (von Neumann once rode a mule into the Grand Canyon while wearing a three-piece pinstriped suit). Wigner was more reluctant to leave Europe behind. Yet it was clear to him that he wouldn't be able to return to Berlin indefinitely. "There was no question in the mind of any person that the days of foreigners [in Germany], particularly with Jewish ancestry, were numbered," Wigner recalled. "It was so obvious that you didn't have to be perceptive. . . . It's like, 'Well, it will be colder in December.' Yes, it will be. We know it will." When Hitler came to power, Wigner and von Neumann simply didn't return to Berlin—both men had been fired from their German posts, being of Jewish descent.

Von Neumann and Wigner were two of a brilliantly gifted group of Hungarian Jewish scientists of their generation. Their astounding mathematical abilities and diverse scientific talents led their colleagues to jokingly suggest that Hungary was merely a cover story obscuring their true origin. "These people were really visitors from Mars," said their colleague Otto Frisch. "For them . . . it was difficult to speak without an accent that would give them away and therefore they chose to pretend to be Hungarians whose inability to speak any language without accent is well known; except Hungarian, and [these] brilliant men all lived elsewhere." Von Neumann in particular seemed almost inhuman in his brilliance. His colleagues at Princeton said that he "was indeed a demigod, but he had made a thorough, detailed study of human beings and could imitate them perfectly." Von Neumann and the Martians did often think about things differently from their colleagues—including the foundations of quantum physics.

Not long after he started visiting Princeton, von Neumann completed a textbook on quantum physics that became an instant classic. Other textbooks on the subject had been written before, but von Neumann breezily dismissed the best known and most technically sophisticated of them in the introduction to his own, claiming (accurately) that it "in no way satisfies the requirements of mathematical rigor." The book contained von Neumann's subtly flawed "impossibility proof," but this errant result was a (nearly invisible) blemish on what was otherwise an impressive technical achievement. Von Neumann articulated quantum physics in a mathematics as formal as his clothes, deriving well-known results from a handful of fundamental postulates. Among those postulates was one that von Neumann knew to be essential to the theory as it was understood at the time: he stated that wave functions normally obey the Schrödinger equation but collapse on measurement. "We therefore have two fundamentally different types of interventions which can occur in a system," von Neumann wrote. When an object remains undisturbed, the Schrödinger equation "describes how the system changes continuously and causally in the course of time." But, once a measurement is made, the smooth regularity of the Schrödinger equation goes out the window. "The arbitrary changes by measurements," said von Neumann, are "discontinuous, non-causal, and instantaneously acting."

Here, von Neumann dissented from Bohr's view. Bohr held that measurement devices and other large objects must be described in the language of classical physics, and that this somehow accounted for the results of quantum experiments without invoking any kind of wave function collapse. Exactly how this worked was something Bohr and his followers were all unclear about—and this lack of clarity was unacceptable to von Neumann in his quest to make quantum physics more mathematically rigorous. Instead, he held that quantum physics applied to large objects as well as small ones. Quantum physics, in von Neumann's view, was a theory of the whole world. But this made the measurement problem far more stark. If normal objects are subject to the laws of quantum physics in the same way that atoms are, then normal objects can't collapse wave functions, since wave function collapse violates the Schrödinger equation. And if normal objects don't collapse wave functions, that would lead straight to the paradox of Schrödinger's cat. The punk-rock particles from the Introduction were in two seemingly contradictory states—a bizarre situation known as a *superposition*—and because their wave function never collapsed, they ultimately forced Schrödinger's cat to be in a superposition as well, somehow both dead and alive. Yet we only ever see living cats or dead cats, not a superposition of them (whatever that might mean). Von Neumann wanted to avoid this problem, which is why he had been so explicit about the existence of wave function collapse in his book. But this still left the problem of how and why that collapse occurred.

Von Neumann's solution was to make the observer—whoever was looking—responsible for wave function collapse. "We must always divide the world into two parts, the one being the observed system, the other the observer," Von Neumann said. "Quantum mechanics describes the events which occur in the observed portion of the world, so long as they do not interact with the observing portion, with the aid of the [Schrödinger equation], but as soon as such an interaction occurs, i.e. a measurement, it requires the [collapse of the wave function]."

It's not entirely clear what von Neumann meant by this. Some took him to be saying that consciousness itself causes the collapse of the wave function; this was a view promoted by physicists Fritz London and Edmond Bauer in a book they wrote several years later, heavily influenced by von Neumann's work. Wigner also adopted this view later. But this is a weird view. Stating that consciousness collapses wave functions does arguably solve the measurement problem but only at the price of introducing new problems. How could consciousness cause wave function collapse? Since wave function collapse violates the Schrödinger equation, does that mean that consciousness has the ability to temporarily suspend or alter the laws of nature? How could this be true? And what is consciousness anyhow? Who has it? Can a chimp collapse a wave function? How about a dog? A flea? "Solving" the measurement problem by opening the Pandora's box of paradoxes associated with consciousness is a desperate move, albeit one that seemed reasonable at the time, in the absence of other fully developed solutions to the measurement problem.

Strange as it was, von Neumann may have also held the view that consciousness is responsible for wave function collapse. But he side-stepped this question in his textbook; there, he claimed that conscious observers held no special status in the theory. "The boundary between the [observer and observed] is arbitrary to a very large extent," he wrote. Sounding a positivist note, he claimed that "experience only makes statements of this type: an observer has made a certain (subjective) observation; and never any like this: a physical quantity has a certain value." He also claimed that Bohr's work supported this "dual description" of nature. Yet von Neumann's view on quantum interpretation certainly didn't line up with Bohr's. Indeed, there was a wide gap between Bohr and the "Martians" not only on wave function collapse and the application of quantum theory to measurement devices, but also on complementarity. Wigner had disparaged complementarity when Bohr had first unveiled the idea at Como in 1927, and von Neumann had made little use of the idea in his textbook. Now that von Neumann and others were questioning the Copenhagen orthodoxy on several fronts, a confrontation over the foundations of quantum theory seemed likely.

But by the late 1930s, Bohr, von Neumann, and Wigner had little time to think about the foundations of quantum physics. War was obviously on the horizon, and new developments in more practical branches of physics overtook concerns about the philosophical underpinnings of the field. In January 1939, Bohr and his assistant Léon Rosenfeld took a steamer across the Atlantic, bearing to Manhattan the latest news from the Continent: the German physicist Otto Hahn had split the atom. Bohr immediately tackled the problem. With the help of his former student John Wheeler, the father of quantum physics set about unveiling the mysteries of uranium.

The phenomenal power of atomic bombs ultimately derives from the delicate balancing act that is performed in the nucleus of every atom. The cloud of electrons surrounding an atomic nucleus is bound to the nucleus by the electric attraction between the negatively charged electrons and the positively charged protons in the nucleus. But that same electric force also tries to rip the nucleus apart—like charges repel, and the closer together they are, the more they repel. A typical atomic nucleus is 100,000 times smaller than the surrounding electron cloud, which is itself a million times smaller than the width of a human hair. At such close quarters, the electrical repulsion between the protons in the nucleus, left unchecked, would send them flying off at nearly the speed of light. Instead, atomic nuclei are held together by an even stronger force, unimaginatively dubbed the "strong nuclear force." The strong force binds together the protons and neutrons in atomic nuclei. Neutrons are electrically neutral—hence the name—but they feel the strong force just like protons. They play a crucial role in the nuclear tug-of-war between electrical repulsion and strong force attraction, aiding the latter without affecting the former. While the strong force isn't quite strong enough to keep two protons together by itself, adding a neutron to the mix increases the "stickiness" of the strong force without adding any electrical charge, creating a stable atomic nucleus of two protons and one neutron (helium-3).

The nuclear struggle between the sticky strong force and the repellent electrical force ultimately depends on the size of the nucleus. For small nuclei, the strong force wins out easily, and adding more protons and neutrons generally just makes it stronger. But the strong force can only act over very short distances, comparable to the size of a proton itself—anything much larger than a trillionth of a millimeter (a distance known as one fermi, after Enrico) is too much for it. After a certain point, the nucleus gets too big, the electric force starts to win the tug-of-war, and nuclei become weaker as more protons and neutrons are added. Specifically, that point is around nickel (28 protons and 34 neutrons) and iron (26 protons and 30–32 neutrons). Bigger nuclei than that are less stable, and beyond a certain size—namely lead, which has 82 protons and over 100 neutrons—there are no stable nuclei at all.

Uranium is far past that point. With 92 protons, it doesn't matter how many neutrons you add to uranium—it will eventually decay. But there are two forms of uranium nuclei that will stick around for billions of years before they do: uranium-235 and uranium-238. The numbers refer to the total number of protons and neutrons in the nuclei: U-235 has 143 neutrons and 92 protons, for a total of 235. U-238 has 3 more neutrons, which makes it slightly heavier. But they're both uranium: the chemical identity of an atomic nucleus is determined solely by the number of protons that it has. Chemistry is all about electromagnetic interactions between atoms. The chemical properties of an atom are entirely determined by the number of electrons it has—and the number of electrons that surround a particular atomic nucleus is determined in turn by the number of protons in that nucleus. Nuclei with the same number of protons but different numbers of neutrons are different *isotopes* of the same element—they differ in weight but not in their chemical properties.

Bohr and Wheeler, building on the work of refugee physicists Lise Meitner and her nephew Otto Frisch, found that the two isotopes of uranium have very different nuclear properties. Specifically, hitting a U-235 nucleus with a neutron leads the nucleus to *fission:* it splits into two smaller nuclei, releasing a fabulous quantity of energy, along with a few free-floating neutrons. With enough U-235—a critical mass—the neutrons left over from fission will hit more U-235 nuclei, which will split in turn, releasing even more neutrons and starting a chain reaction. Left uncontrolled in 120 pounds of pure U-235—a small sphere of the dense metal, less than twenty centimeters across—a nuclear chain reaction would explode with the power of 15,000 tons of TNT, enough to instantly level a small city. Controlling the reaction by absorbing some of the excess neutrons would allow you to power a small city instead, for days on end, with the same 120 pounds of U-235.

U-238 is a different story. Those three extra neutrons give it a little more stability, so hitting it with a neutron won't split it as easily. This makes it impossible to build a bomb out of U-238. And fortunately, about 99.3 percent of uranium in nature is U-238. To build an atomic bomb, you would need to separate the tiny quantity of U-235 from the enormous bulk of U-238—and since they're chemically identical, the only way to separate them is to take advantage of the fact that U-238 is 1.3 percent heavier than U-235. This guaranteed that nuclear power would be phenomenally difficult to achieve, requiring enormous quantities of uranium and city-sized industrial diffusion and centrifuge facilities. "It can never be done unless you turn the United States into one huge factory," concluded Bohr.

Yet the risk of not pursuing nuclear power was too high. If Nazi Germany were to build an atomic bomb, the war would be over. The Einsteins and Fermis and Borns of the world would never be able to escape Hitler's Reich. "A little bomb like that," said Fermi, cupping his hands as he looked out over Manhattan, "and it would all disappear."

"Can you guess where I found out about [fission]? In . . . the infirmary." Eugene Wigner had jaundice. "I was in the infirmary for six weeks. It was a wonderful period, because jaundice doesn't hurt really," Wigner recalled. "They fed you on potatoes, beans, and everything boiled in water, and the food was not good. But the rest and the detachment were wonderful." Wigner shared the news of uranium fission with his visiting friend Leo Szilard, another Hungarian refugee physicist who had realized the enormous possibilities of a nuclear chain reaction years earlier. "Szilard was in Princeton, and he came to visit me every day, and we discussed fission problems and this and that. Well, the theory of Bohr and Wheeler of course occupied us very much. . . . Szilard came to me one morning and said, 'Wigner, now I think there will be a chain reaction.'"

Debating what to do next, the two Hungarians enlisted a third: Edward Teller, who had settled in Washington, DC. Over the summer of 1939, this "Hungarian conspiracy" developed a plan to alert the US government to the fact that "Hitler's success could depend on [nuclear fission]," as Szilard put it. Putting their plan into motion, they recruited a fourth conspirator: Albert Einstein. The Hungarians hoped that a letter from the most famous scientist in the world would get the attention of President Roosevelt. Spending several weekends with Einstein at his vacation house on Long Island, Szilard, with help from Teller and Wigner, crafted a letter to pass along to FDR. The plan worked, to a point: the letter did get FDR's attention, but he appointed Lyman Briggs, the ineffectual leader of the Bureau of Standards, to head a Uranium Committee. Briggs and his committee did little, and the project stalled for over a year while Hitler occupied Denmark, captured Paris, and relentlessly bombed London.

When the US government finally started to seriously investigate atomic power in the fall of 1941, Wigner met with Arthur Compton, an American physicist who was preparing a report for FDR's Top Policy Group on the feasibility of developing atomic bombs. "[Wigner] urged me, almost in tears, to help get the atomic program rolling," wrote Compton. "His lively fear that the Nazis would make the bomb first was the more impressive because from his life in Europe he knew them so well."

Several months after the attack on Pearl Harbor, the American atomic bomb project was turned over to the military. It was assigned to General Leslie Groves, an administrator with the Army Corps of Engineers. Groves had just completed directing the building of the Pentagon (then the world's largest building) and protested his assignment—he wanted to be sent to the front. But once he learned more about the potential outcome of the work, he warmed to it. Groves selected the Berkeley physicist Robert Oppenheimer as the scientific director of the top-secret project, code-named Manhattan. Under the "peculiar sovereignty" of the Manhattan Project, Fermi, Wigner, and other refugee European physicists joined their American colleagues at Los Alamos, high in the New Mexico desert, to build a bomb, racing against their German rivals.

Many physicists at Los Alamos thought the Nazis had a head start on nuclear power, and they had good reason to think so. Germany had been the center of the physics world for generations, and America had long been considered a scientific backwater. Fission, after all, had been first discovered in Germany. Germany had also been in the war longer, and thanks to Hitler's invasion of Czechoslovakia, it had access to vast uranium resources. And, despite Hitler's racist civil service laws, many good physicists remained in the country. Otto Hahn, the nuclear chemist who discovered fission, stayed in Germany, though he wanted nothing to do with the Nazis. Instead, he quietly continued his research, protecting his Jewish colleagues where he could and corresponding with the ones in exile, like Meitner and Frisch. Hahn's friend, the Nobel Prize–winning physicist Max von Laue, took his opposition one step further and risked his life by repeatedly and publicly condemning Hitler's regime from within. But most German physicists didn't follow Hahn's example, and almost none took von Laue's principled stance. And some, like Pascual Jordan, eagerly aided Hitler's regime. Finding the Nazi ideology appealing for aesthetic reasons—and in line with his idealist stance on the philosophy of science—Jordan joined not only the Nazi Party in 1933, but also the Brownshirts, Hitler's paramilitary storm troopers. And other physicists, like Johannes Stark and Philipp Lenard, had been Nazis even before Hitler came to power, and applied Hitler's racial "philosophy" to physics, declaring relativity and quantum theory to be "Jewish physics."

Werner Heisenberg fell somewhere between the extremes of von Laue's conscientious objection and Jordan's full embrace of the Nazi philosophy. Heisenberg condemned the bewilderingly idiotic "Deutsche Physik" of Stark and Lenard, and helped von Laue in successfully ending the campaign against quantum physics and relativity in Germany. But Heisenberg also remained in Hitler's Reich largely out of a sense of obligation and patriotism, and morally compromised himself by collaborating with the Nazis while simultaneously hiding behind the "apolitical" nature of science. He had been offered many positions across the United States and the UK in the six years between Hitler's rise to power and the outbreak of war—most recently, during his tour of the States in the summer of 1939. Heisenberg turned them all down, insisting only that "Germany needs me." While Heisenberg was not a Nazi, he left very little doubt about the depths of his loyalty to Germany, no matter who was leading the country—leaving a physics summer school in Michigan early, he said he had to return to Germany for "machine gun practice in the Bavarian Alps."

Shortly after the war started, Heisenberg was (unsurprisingly) chosen as one of the leaders of the German nuclear program. The project floundered almost from the start. Heisenberg's grasp on experimental physics had been poor since his student days in Munich, and he made simple errors in calculating the relevant quantities—"though a brilliant theoretician [Heisenberg] was always very casual about numbers," his former colleague Rudolf Peierls recalled. Communication breakdowns and clerical errors plagued the project; interference from the Nazi scientific bureaucracy forced personnel decisions based on political beliefs rather than scientific talent. And a crucial realization—that purified graphite could be used to moderate and control nuclear chain reactions—escaped the attention of Heisenberg and his colleagues. After dismissing the possibilities of impure graphite, they instead focused their efforts on a much more rare and expensive moderator, heavy water, which slowed their progress even further. By 1942, just as the American bomb program was building up steam, the German program had come to a nearly total halt. In an Army Ordnance conference in Berlin in 1942, Heisenberg told his Nazi superiors that although it was unlikely that a bomb could be completed before the end of the war, a nuclear reactor held promise as a new power source for the Reich's war engine. Shortly thereafter, Heisenberg was made the de facto head of the nuclear program in Germany, despite the fact that he had never led an experimental team before in his life. Heisenberg's team worked on creating a controlled nuclear chain reaction up until the end of the war in 1945, unaware that Fermi had already done it in Chicago in 1942—and unable to prevent a nuclear meltdown should they succeed and lose control of the reaction. Indeed, Heisenberg had much less control than he thought: he hoped to "make use of warfare for physics" by doing interesting nuclear physics research with the blood money he received from the Nazi regime—even if that research handed Hitler nuclear power. Heisenberg "had agreed to sup with the devil," wrote Peierls years later, "and perhaps he found that there was not a long enough spoon."

By December 1944, it was clear to Heisenberg that Germany was near defeat—speaking with his colleague Gregor Wentzel at a dinner party in Switzerland, he wistfully sighed that "it would have been so beautiful if we had won." He returned to his fission lab in Hechingen and made one last push to complete the reactor, but he was out of time. In April 1945, with the Allies closing in on Germany from all sides, Heisenberg was forced to flee from his research. He bicycled 250 kilometers in seventy-two hours, moving only at night so he wouldn't be shot by Allied aircraft, and made his way to his family in Urfeld. Days later, he was apprehended there by an American military task force, Operation Alsos, that had been sent deep into Europe to capture and interrogate the German nuclear physicists.

The Alsos team whisked off Heisenberg, Hahn, von Laue, and several other German physicists to Farm Hall, an English manor house that had been converted into a military intelligence post. The house was supplied with sporting equipment, chalkboards, a radio, and plenty of food—more comforts than the average English family, grumbled one of their military minders. The average English family certainly didn't have Farm Hall's full complement of hidden microphones in each room. "I wonder whether there are microphones installed here?" asked Kurt Diebner, one of the German physicists, several days after their arrival. "Microphones installed?" replied Heisenberg, laughing. "Oh no, they're not as cute as all that. I don't think they know the real Gestapo methods; they're a bit old fashioned in that respect." Reassured, the German physicists freely discussed physics, politics, and current events, eagerly reading the newspapers that their British minders had supplied in the express hope of provoking discussion.

Heisenberg and the others also discussed the mystery of their continued detainment—inquiring, they were told only that they were being held "at His Majesty's pleasure." Thinking themselves the world's foremost experts on nuclear physics—and convinced that any American bomb effort could not possibly have surpassed the German one, since German physics was superior—they hatched wild schemes to alert the press of their plight, to escape to Cambridge and see their colleagues there who (they presumed) were desperate to consult them on their knowledge of nuclear matters. They even spoke matter-of-factly about their fates being decided personally by the "Big Three," Truman, Churchill, and Stalin, who were meeting in Potsdam at the time. Some managed to convince themselves that their association with the Nazis would not be held against them personally, and that they would be able to use their status as elite physicists to escape to Argentina and start new lives.

Finally, after weeks in luxurious captivity, the bubble burst. On the evening of August 6, 1945, just before dinner, Major Rittner, the British military intelligence officer in charge of Farm Hall, quietly took Otto Hahn aside and told him that the Americans had dropped an atomic bomb on Hiroshima. "Hahn was completely shattered by the news," wrote Rittner:

> He felt personally responsible for the deaths of hundreds of thousands of people, as it was his original discovery which had made the bomb possible. He told me that he had originally contemplated suicide when he realized the terrible potentialities of his discovery. . . . With the help of considerable alcoholic stimulant he was calmed down and we went down to dinner where he announced the news to the assembled guests. As was to be expected, the announcement was greeted with incredulity.

"I don't believe a word of the whole thing," said Heisenberg upon hearing the news. "I don't believe it has anything to do with uranium." Hahn jeered, "If the Americans have a uranium bomb then you're all second raters. Poor old Heisenberg." After they heard the BBC report the news in great detail later that night, Heisenberg and the others accepted the truth: they had been beaten.

Over the next few days, Heisenberg attempted to work out how his project had fallen so far behind; his fumbling calculations show that he had never really understood how to even build a bomb in the first place, though he had certainly thought he'd understood it. And the bickering of the other scientists at Farm Hall confirmed what documents captured by Alsos had already suggested: the Nazi bomb program, unlike the Manhattan Project, was a disorganized mess, with vital information compartmentalized and no clear vision of how to proceed. Yet, in those same few days, the Farm Hall transcripts make it clear that Heisenberg and his student, Carl von Weizsäcker, purposefully constructed a revisionist narrative of their wartime activities. According to them, while the Americans had built a weapon of death and destruction on unprecedented scales, they, the Germans, had deliberately pursued only a nuclear reactor, being unwilling to build a massive new weapon for Hitler's Reich—thereby placing the responsibility for their failure on their supposed moral clarity, rather than their sheer incompetence.

While Heisenberg had been working to such honorable ends during the war, his mentor had been nearly killed. Bohr had returned to Copenhagen after his visit to America in 1939, arriving home several months before the outbreak of war that September. Germany invaded Denmark before sunrise on the morning of April 9 the following year; two hours later, the Danish government surrendered. Hitler was determined to make Denmark a "model protectorate," demonstrating his peaceful ways to the rest of the world. He managed to curb his bloodlust and refrained from imposing anti-Semitic laws on the Danes for over three years. Finally, in October 1943, the SS arrived in the streets of Copenhagen. They planned to round up the city's Jews during Rosh Hashanah, the Jewish New Year and one of the holiest days on the Jewish calendar. But when they went from door to door, they found that nearly all the Jews in the city had simply vanished. A German diplomat, Georg Duckwitz, had warned the leaders of Denmark's Jewish community days earlier, and most of the country's Jews had gone into hiding by the time the SS arrived. One of them was Niels Bohr, who had been ferried to safety in a fishing boat along with his family across the Øresund strait and into neutral Sweden three days before the Nazis arrived to arrest him at his institute. In Stockholm, Bohr met with King Christian X and pled his case, asking the king to offer Danish Jews asylum in Sweden. Swedish radio announced the asylum offer that evening, and over the next two months, the Danish resistance and Swedish coast guard arranged passage for hundreds of small fishing boats, rowboats, and canoes, each boat ferrying two, three, four Jewish Danes to safety. More than 7,000 Jews—95 percent of the Jews that lived in Denmark at the time—managed to evade the Nazis.

Stockholm, crawling with Nazi agents, wasn't safe for Bohr—and the Allies decided that Bohr himself was too important to remain in Sweden anyhow. The British Royal Air Force dispatched a high-altitude Mosquito bomber, a small plane designed to fly higher than antiaircraft fire, to bring Bohr to the UK. The small aircraft's bomb bay was specially equipped to carry Bohr, with an oxygen mask and a pair of headphones to allow the pilot to communicate with his precious cargo. But the headphones were too small for Bohr's enormous head. Unable to hear the command to turn on his oxygen, Bohr passed out. The pilot realized there was a problem and flew the plane low over the North Sea; Bohr survived. After a briefing in England, Bohr flew to the United States, where he was whisked to the headquarters of the Manhattan Project in Los Alamos, traveling under the alias Nicholas Baker. Guiding "Nicholas" around the facilities, Teller was looking forward to showing Bohr that his pessimism about nuclear power had been misguided. "But before I could open my mouth, [Bohr] said, 'You see, I told you it couldn't be done without turning the whole country into a factory. You have done just that.'"

Bohr was more right than he knew. By the end of the war, the Manhattan Project had cost the nation nearly $25 billion, employing 125,000 people at thirty-one different locations across the United States and Canada. Hundreds of physicists were called away from their everyday laboratory work to satisfy Manhattan's relentless hunger for people and materials. After the war ended, physics research in the United States never returned to what it was before the war. Damned by their success in building the bomb, military research dollars poured into physics. In 1938, before the war, total spending on physics research in the United States was about $17 million, and nearly none of it came from the government. Less than a decade after the war, in 1953, physics research funding was just shy of $400 million—an increase by a factor of twenty-five in just fifteen years. And, by 1954, 98 percent of the money for basic research in the physical sciences in the United States was coming from the military or defense-oriented government agencies, like the Atomic Energy Commission, successor to the Manhattan Project.

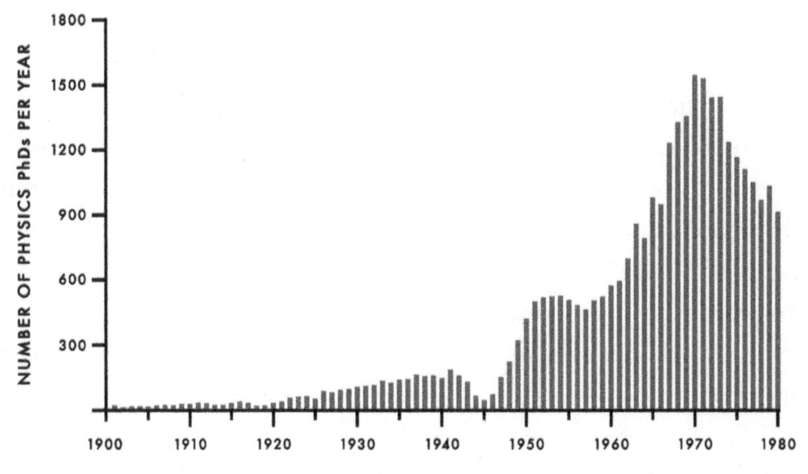

Figure 4.1. Physics PhDs granted by US institutions per year, 1900–1980.

Along with that money came people. As the war ended under the shadows of two mushroom clouds, young veterans, funded by the GI Bill, flocked to universities to learn about the new physics. "My interest in physics," wrote a Harvard physics PhD student in 1948, "was aroused while I was working on the Atomic Bomb in New Mexico while in the Army." Another wrote that he had a "feeling that the work [physics] was important as a result of the war"; still another wrote that "the war introduced me to the scientific life." Physics departments were inundated with students. In 1941, about 170 US graduate students earned their PhDs in physics. By 1951, that number was over 500 and climbing, a far faster rate of growth than any other academic field over the same span of time (Figure 4.1). And by 1953, half of all physics PhDs were under the age of thirty. Educating physicists was no longer seen as a matter of mere scientific necessity but an essential investment in military infrastructure. Henry Smyth, commissioner of the Atomic Energy Commission and former chair of the Princeton Physics Department, spoke of the "stockpiling and rationing of scientific manpower" in an address to the American Association for the Advancement of Science in 1950. Scientists, he said, "have become a major war asset. It is important that they be used to the greatest advantage. . . . I am speaking of scientists not as men who enrich our culture but as tools of war needed for the preservation of our freedom."

Plenty of physicists were alarmed and unhappy about this new state of affairs. "The hot and cold wars have so changed my profession that I can hardly recognize it anymore," complained the Dutch American physicist Samuel Goudsmit. "We physicists are among the maladjusted veterans of the Second World War." Goudsmit, one of the handful of Jewish European physicists who immigrated to the United States well before Hitler came to power, pined for the "string-and-sealing-wax days" before the war, when physics was done on a paltry budget with spare parts lying around. Less than a decade after the end of the war, the flood of people and money had radically altered the day-to-day work of physics:

> It's been a shock. We've got marvelous laboratories for basic research, which is the real love of any self-respecting physicist, but somehow

> we don't have the same tender affection for them that we would have had years ago, when acquiring a three-hundred-dollar spectroscope was reason enough for throwing a party. Today we're given a multi-million-dollar piece of equipment, and the minute the dedications ceremonies are over, we're poring over plans for an even more powerful one. In the old days, physicists gave themselves up wholly to a single-minded study of the fundamental laws of the universe. Now we feel called upon to do things of a sort we'd never even imagined we'd be doing—thoroughly unscientific things. We sit down with the Defense Secretary to help him figure out his next year's budget. We brief the President of the United States on the nation's nuclear stockpile. . . . Some of us are in industry, designing electronic equipment, and some of us are attached to the American embassy staffs in England, France, and Germany. Colleagues of mine who never even bothered to vote before Hiroshima now sit at the elbows of our United Nations representatives when the subject of atomic energy is on the agenda.

Goudsmit himself had his own taste of these unscientific activities during the war—he was the civilian leader of the Alsos mission, which had gathered up Heisenberg and other top German nuclear physicists and sent them to Farm Hall. He had also done work on radar at MIT (another major wartime physics research program, employing thousands and costing millions) and consulted for the Royal Air Force. Before the war, he had worked at the University of Michigan and had planned to retire from research and devote himself to teaching full-time. After the war, he changed his mind. "I felt caught up in the violent upsurge of everything associated with physics that had followed Hiroshima," Goudsmit recalled, "and I wanted to be more closely associated with it than seemed possible on a university campus." Goudsmit became the head of the physics department at Brookhaven National Laboratory, one of the newly minted government-run pure research labs. Yet, despite his administrative post in the new edifice of "Big Science," Goudsmit remained uneasy about the changes in his field. "The conditions we work under today certainly aren't hastening [a] breakthrough," he said in 1953.

> A quarter of a century ago we could exchange ideas in Bohr's study with no government secrets, weapons programs, or spy cases to bother us. . . . None of us were distracted by offers to become college presidents or big wheels in industry, and governments didn't give a hoot about physicists. There was no trying to elbow one's way to power, for the simple reason that there wasn't any place to exercise power. No huge laboratories, no military projects. . . . We all felt that we belonged to a sort of lodge, with a worldwide membership of only four hundred or so, and everyone knew everyone else well—or at least knew what everyone else was doing. Now four times that number will turn up for a meeting of just American physicists, and most of them will be strangers to each other.

Research into the meaning of quantum physics was one of the casualties of the war. With all these new students crowding classrooms around the country, professors found it impossible to teach the philosophical questions at the foundations of quantum physics. Before the war, courses in quantum physics on both sides of the Atlantic, like Heisenberg's in Leipzig and Oppenheimer's in Berkeley, spent a great deal of time on conceptual issues. Textbooks and exams from the prewar period asked students to write detailed essays on the nature of the uncertainty principle and the role of the observer in the quantum world. But, with ballooning class sizes, detailed discussion of philosophy became all but impossible. "With these subjects [such as uncertainty, complementarity, and causality] lecturing is of little avail," complained a physics professor at the University of Pittsburgh in 1956. "The baffled student hardly knows what to write down, and what notes he does take are almost certain to horrify the instructor." Smaller quantum physics classes in smaller departments afforded foundational questions more time—about five times as much as larger classes—but, as enrollments surged, few small physics classes were left. The larger classes focused on "efficient, repeatable means of calculation," rather than focusing on foundations. And textbooks nearly dropped questions about foundations altogether, as a new generation of reviewers in physics periodicals praised a new batch of texts for "avoiding philosophical discussion" and "philosophically tainted questions." Textbooks that bucked the trend were condemned for spending too much time on the "musty atavistic to-do about position and momentum." The era of Big Science had arrived—and it had no patience for puzzling over the meaning of quantum physics.

Heisenberg repeated his story about the German bomb program to anyone who would listen for the rest of his life. Goudsmit, who had access to the Farm Hall reports and had seen the pathetic remnants of the Nazi nuclear program firsthand, knew Heisenberg's story was a fabrication. But, with the existence of the Farm Hall transcripts itself classified, Goudsmit could state only that Heisenberg was lying, without explaining how he knew. The first popular account of the Manhattan Project, *Brighter Than a Thousand Suns,* written by the Swiss journalist Robert Jungk in 1958, repeated Heisenberg's story almost verbatim. So did *The Virus House,* the first book dedicated solely to the history of the German bomb program, which relied heavily on interviews from Heisenberg and his fellow former Farm Hall detainees. (The author, David Irving, was later revealed to be a Holocaust denier.)

Despite Heisenberg's public-relations campaign, there was a cloud of suspicion over his head for the rest of his days. His relationship with Bohr, in particular, was never the same—after Jungk's book was published, Bohr drafted an angry letter to him over details of Heisenberg's account of a meeting they had in 1942. But, in typical Bohr style, he drafted several versions of the letter and never sent it. Nonetheless, Bohr and Heisenberg did speak again and even met several times after the war. (Heisenberg's lies were, after all, rather minor compared to some. Pascual Jordan maintained that he had never truly supported the Nazi cause, despite his publications extolling the virtues of a National Socialist approach to science. He even had the audacity to send a letter to Max Born, his mentor who had been forced out by Hitler's racist policies, explaining that he hadn't really been a Nazi and asking for a character reference for his "de-Nazification." Born replied with a list of his friends and family members murdered by the Nazis.) Given the damage done to his reputation by his activities during the war, Heisenberg's invention of a single unified Copenhagen interpretation may have been an attempt to rework the history of quantum physics to his benefit. The Copenhagen interpretation wasn't a pure fabrication—there were certainly similarities in the positions that Bohr and his students and colleagues took—but the differences between Heisenberg's lecture and the writings of Bohr himself should have been enough to tip off anyone who was paying attention that no such beast had ever really existed.

Yet the idea of a single settled interpretation of quantum physics, associated with the giants Bohr and Heisenberg, went over well in the post-Manhattan world of Big Science. Most physicists were perfectly happy with the jumble of ideas that purportedly constituted the Copenhagen interpretation itself, since questions about the meaning of quantum physics had little bearing on their work. The mathematical formalism of the theory continued to work remarkably well in a wide variety of postwar applications of physics to the military-industrial complex, which turned most physicists to work in nuclear physics or solid-state physics (the branch of physics that, shortly after the war, led to the development of the silicon transistor, as well as many of the other materials later responsible for the shrinking size and ballooning importance of computers). Questions of interpretation, while vital for the progress of science in the long term, were immaterial when it came to the hard-nosed applications of quantum theory that were so suddenly and desperately prized. The Copenhagen interpretation's promise of a complete yet obscure answer to the quantum mysteries allowed the new army of postwar physicists to calculate answers without worrying about the meaning of the theory. The shift of physicists to the United States aided in this as well—in contrast to the great theorists of Europe, physicists in the States had always had an experimental and pragmatic bent. The questions at the foundations of quantum physics that had seemed so vital to Einstein and Bohr were dismissed by the new crop of American physicists as dreamy trifles, hardly suitable as subjects of inquiry to be funded by the rivers of money flowing from the Pentagon.

But not all American physicists were pragmatic enough to swallow the Copenhagen interpretation. "Bohr's principle [complementarity] puts nature on the fence and leaves it there," grumbled Henry Margenau, a philosophically minded physicist working at Yale. "It relieves its advocates of the need to bridge a chasm in understanding by declaring that chasm to be unbridgeable and perennial; it legislates a difficulty into a norm." One American physicist in particular was destined to cause serious trouble for the Copenhagen interpretation. He had worked with Oppenheimer at Berkeley during the war and was recruited by Princeton shortly thereafter. In 1947, David Bohm arrived at Princeton, a freshly minted assistant professor. He had accepted the Copenhagen interpretation in his career to date, but soon he would be irritated by pesky doubts. Within five years, those doubts ballooned into a full-scale personal rebellion against the quantum orthodoxy. David Bohm was about to do the impossible: defy von Neumann's proof, shock John Bell out of the uneasy peace he had made with Copenhagen—and change quantum physics forever.

Part II

Quantum Dissidents

We emphasize not only that our view is that of a minority but also that current interest in such questions is small. The typical physicist feels that they have long been answered, and that he will fully understand just how if ever he can spare 20 minutes to think about it.

—John Bell and Michael Nauenberg, 1966

5

Physics in Exile

Max Dresden entered the crowded seminar room, all eyes on him as he took his place alone in front of the chalkboard. Dresden was a physicist at the University of Kansas, and, during his visit to Princeton's Institute for Advanced Study in 1952, he had volunteered to give a talk on the fascinating new work by David Bohm. Dresden was eager to hear what his audience thought of Bohm's work: the "Princetitute" was home to some of the finest minds in all of physics, including Einstein himself—though, as Dresden looked out at the room, Einstein's unruly white hair was nowhere to be seen.

Dresden's students had brought Bohm's paper to his attention, and he had initially dismissed their questions by citing von Neumann's famous proof that the Copenhagen interpretation was the only way of understanding quantum physics. But, after repeated pestering, Dresden finally looked at Bohm's paper, and he was surprised by what he found. Bohm had discovered a totally new way to interpret quantum physics. Rather than refusing to answer questions about the quantum world, as the Copenhagen interpretation did, Bohm's interpretation depicted a world of subatomic particles that existed whether or not anyone was looking at them, particles with definite positions at all times. These particles, in turn, each had "pilot waves" that determined their motion, which also behaved in an orderly and predictable fashion. Somehow, Bohm had found a way to tame the chaotic and unknowable world of the quantum—and he had done it without sacrificing accuracy, because Bohm's theory was mathematically equivalent to "normal" quantum physics.

Over the course of his talk, Dresden presented Bohm's ideas and mathematics to his audience. When he finished, the moment he dreaded arrived: the floor was opened to questions from the room of luminaries. Dresden had offered to give this talk on less than a week's notice and desperately hoped that he was prepared for what was sure to be a high-level technical discussion of someone else's ideas.

Instead, to Dresden's horror, the room erupted in vitriol. One person called Bohm a "public nuisance." Another called him a traitor, still another said he was a Trotskyite. As for Bohm's ideas, they were dismissed as mere "juvenile deviationism," and several people implied that Dresden himself was at fault as a physicist to have taken Bohm seriously. Finally, Robert Oppenheimer, the director of the institute, spoke up. Oppenheimer was one of the most influential and famous physicists alive; he had led the Manhattan Project to success during the war, and he had mentored a blazingly brilliant team of physicists at Berkeley before that, including Bohm himself. Dresden watched in shock as Oppenheimer suggested to the room that "if we cannot disprove Bohm, then we must agree to ignore him."

Bohm wasn't there to defend his own ideas. He had been on the Princeton University faculty just a few months earlier, but now he was trapped in Brazil, exiled and blacklisted from his native country, while his former colleagues dismissed his new theory out of hand.

This story—Dresden's discovery of Bohm's paper, his visit to Princeton, the astonishingly obtuse responses of the physicists there—might be accurate. It's certainly one of the stories that people tell about Bohm and the reception of his ideas; Oppenheimer's supposed quote about ignoring Bohm has become particularly notorious. But there are a lot of stories that people tell about Bohm, many of which are poorly sourced or lacking sources entirely. These stories exist because, a quarter century after his death, David Bohm remains an intensely polarizing figure. He is written off as a kook, a deluded mystic, a hopelessly conservative throwback who wanted to return to the physics of Isaac Newton. He is also hailed as a visionary, the patron saint of heretics in the One True Church of Copenhagen.

One of the problems with writing about David Bohm is that he really was persecuted and was forced to flee across the globe at some of the most crucial points in his life, which means that many of his most interesting personal papers were lost or destroyed. Furthermore, the people who disagreed with Bohm are the people who won—and, being victors, they went about doing what victors do with history, which makes it even more difficult to separate myth from fact. And, to make things even worse, there's pushback from Bohm's defenders that takes things too far, overcompensating for the revisionist history of the orthodox camp. There's a biography of Bohm, written by his friend and colleague David Peat; he paints Bohm as a sort of secular saint with improbably clear vision about the nature of reality. Moreover, the biography is riddled with factual errors, and also takes some quotes out of context and produces others with no clear evidence that they were ever said. Finally, interest in Bohm's work increased markedly shortly after he died and shows no signs of stopping, leading to a slew of new questions that would have been easy to answer had anyone bothered to ask them of Bohm before he died in 1992. This convoluted situation has produced a remarkable number of myths and legends about a relatively obscure physicist who died well within living memory.

Those legends are important. They tell us something about the role Bohm plays in the culture of quantum physics. They also tell us about the reactions that Bohm's ideas provoked. Behind those legends is a remarkably simple theory about how the quantum world works—and the remarkably complex life of a single unfortunate and brilliant man.

We know for a fact that David Joseph Bohm was born on December 20, 1917, in Wilkes-Barre, Pennsylvania. Bohm's father, Samuel, was a Jewish immigrant from Hungary who came to Pennsylvania alone at the age of nineteen; there, he met and married Frieda Popky, a Lithuanian Jew who had come over to the United States with her family years earlier. Samuel Bohm was a down-to-earth man who owned a furniture store in town, and he was known to the locals as a wheeler-dealer and a skirt chaser. Frieda Bohm, in contrast, was a shy homemaker—she had been quiet and withdrawn ever since her family left Europe—who went through violent mood swings. Her erratic behavior worsened as Bohm grew older; she heard voices, broke a neighbor's nose, and threatened to kill her husband, ultimately ending up in a psychiatric institution. Though David was close with his mother, her frightening behavior forced him to seek refuge in books. Upon discovering science fiction, Bohm was hooked, and his interests steered toward science. Bohm's father had little patience for the "scientism" of his son—when David informed him of the existence of other planets orbiting the Sun, Samuel merely dismissed the fact as irrelevant to the world of human affairs—but he nonetheless paid for Bohm's college education, sending him to Penn State (which was a small rural college at the time, not the enormous state university it is today).

At Penn State, Bohm's brilliance was obvious to his friends and professors alike, as were his personal quirks. Bohm had a "talent for getting people to want to take care of him," according to his friend Melba Phillips—and "a talent for being unhappy." Bohm was continually concerned with his health and suffered from terrible stomachaches from his Penn State days onward. Despite all this, he worked hard and won a spot in the physics PhD program at Caltech upon his graduation from Penn State in 1939. The Pennsylvania immigrants' son had made good: he was at one of the leading centers for physics in the world. But after one semester at Caltech, he became dissatisfied with the course work and research options available. Bohm thought the research being done at Caltech was incremental rather than fundamental, and he found the environment too competitive for his liking. "I wasn't really happy there at Caltech," he recalled later. "They're not interested in science. They were more interested in competition and getting ahead and mastering techniques and so on." Discouraged and uncertain about his future, he went home to Wilkes-Barre for a summer. When he returned to Pasadena in the fall, he became even more unhappy. "In general I was getting a little bit, not exactly depressed, but probably a little low." At a friend's suggestion, Bohm approached a charismatic young visiting professor to ask whether there were any spots open in his research group at Berkeley. By the start of the next semester, Bohm had moved up the California coast to work with his new mentor, J. Robert Oppenheimer.

In Oppenheimer, Bohm found a kindred spirit: a Jew from the East Coast who wanted to tackle the largest outstanding problems in theoretical physics, and who was also interested in a wide range of intellectual pursuits beyond physics. But there were also profound differences between Bohm and Oppenheimer: most notably, while Bohm's family was solidly working-class, Oppenheimer came from a wealthy, well-connected family in the Manhattan social scene. Despite the anti-Semitic "Jewish quotas" that existed at the time, Oppenheimer had managed to go to Harvard for his undergraduate degree. After graduating summa cum laude in three years, he went off to Europe and earned his PhD under Max Born. Later, Oppenheimer spent time working with Pauli in Switzerland, and, though he never studied in Copenhagen, Oppenheimer met Bohr and came to know him quite well. When Oppenheimer—or "Oppie," as he was known to his friends and students—came back to the United States, he set to work turning Berkeley into the first great department of theoretical physics in the country. By the time Bohm showed up in 1941, physicists in Berkeley knew that "Bohr was God and Oppie was his prophet," as Joe Weinberg, one of Oppie's other graduate students, put it. When Bohm arrived, Weinberg set about converting the new student. "With Weinberg I had intense discussions of Bohr on complementarity," Bohm recalled later. "At that time, I was convinced that Bohr's approach was the right approach and for many years I continued with Bohr's approach. . . . I was carried away with it because Weinberg was a very intense, convincing person and since Oppenheimer was also behind it that gave it a lot of weight in my mind."

Quantum physics wasn't the only thing that occupied Bohm's mind in Berkeley. He was also keeping an eye on the war raging in Europe—and what he saw there made communism look good. "Until, say in 1940 or '41, I wouldn't have had much sympathy with the Communist Party," Bohm recalled later. "The thing that deeply impressed me was the collapse of Europe in the face of the Nazis, which I felt was due to the lack of will to resist. . . . I thought the Nazis were a total threat to civilization. . . .

It seemed that the Russians were the only ones who were really fighting them. That was the main thing. Then I began to listen to what they said more sympathetically." Bohm joined the Berkeley campus chapter of the Communist Party in November 1942. But he found the reality of the party less appealing than the idea. "I began to feel that they did nothing but talk about things of no significance, about trying to organize protests of affairs on the campus, and so on. . . . The meetings were interminable." Bohm left the party after several months, but he remained a Marxist in his political convictions for many years afterward.

Bohm's politics presented a problem for him when it came time to defend his PhD. Despite the fact that Oppenheimer had personally requested that Bohm be transferred to Los Alamos, Bohm had been denied security clearance by the Army. Army security lied to Oppenheimer, telling him that Bohm was a security risk because he still had relatives in Europe that could be used against him; in reality, Bohm had been denied clearance because of his association with Weinberg, who was also a member of the Communist Party. But Bohm's thesis research, on interactions between atomic nuclei, was nonetheless very relevant for the work going on at Los Alamos—so much so that it was immediately classified beyond Bohm's clearance. His notes and calculations were seized by the Army, and he was forbidden from writing his own thesis. Oppenheimer swooped in to the rescue, assuring the UC Berkeley administration that Bohm deserved his PhD.

Bohm continued on at Berkeley for a couple of years after the war, publishing papers on various recondite areas of quantum physics. In 1947, on the basis of that work and a favorable interview report from John Wheeler, Princeton's physics department hired Bohm as an assistant professor. "Bohm has been recommended to us as one of the ablest young theoretical physicists that Oppenheimer has turned out," wrote Henry Smyth, the chair of the department. (Several years later, Smyth would write of "stockpiling scientific manpower.")

Bohm found Princeton's campus and climate a disappointing change after his time in Berkeley, and he thought the faculty there were "very status conscious." But Bohm quickly settled in. He started teaching quantum physics classes using Oppenheimer's old notes and began research collaborations with several promising graduate students. He formed a small group of close friends and even struck up a relationship with Hanna Loewy, the stepdaughter of a professor at the Institute of Advanced Study. Bohm's relationship with Loewy soon turned serious, and there was talk of marriage. Loewy brought Bohm home to meet her mother, Alice, and her stepfather, Erich Kahler. Bohm also met one of Kahler's closest friends: Albert Einstein.

On Wednesday, May 25, 1949, David Bohm appeared before the House Un-American Activities Committee (HUAC). Sitting opposite six congressmen—including one Representative Richard M. Nixon—and another half-dozen congressional staffers, Bohm was asked about the extent of his association with the Communist Party. "I can't answer that question," he replied, "because it might tend to incriminate and degrade me, and, also, I think it infringes on my rights as guaranteed by the First Amendment." The committee asked him to repeat himself, then asked him dozens more questions, asking Bohm to implicate several of his former colleagues and friends from his time at Berkeley, including Joe Weinberg. Bohm refused. Then he went home and didn't think much more about it for over a year. "Apparently the whole issue was dying away," he recalled later.

Bohm had other troubles on his mind. He was assembling his quantum physics course materials into a textbook, taking great care to explain and defend the Copenhagen interpretation. But doubts had crept in, and, as the book neared completion in the summer of 1950, those doubts grew. "When I finished the book, I wasn't quite satisfied that I really understood it," Bohm said. Then, on December 4, 1950, a US marshal walked into Bohm's office and arrested him.

Bohm was taken down to a federal courthouse in Trenton and indicted for contempt of Congress, for refusing to testify to HUAC. Loewy, along with Bohm's student Sam Schweber, drove to Trenton and bailed Bohm out. When they returned to Princeton, they found that Harold Dodds, the president of the university, had already suspended Bohm

Figure 5.1. David Bohm after testifying to HUAC in May 1949.

from research and teaching duties, and forbidden him to set foot on campus. Bohm was blacklisted.

In February 1951, while he waited for his day in court, Bohm had a small party to celebrate the publication of his new textbook, *Quantum Theory*. Bohm's book laid out quantum physics in a simple and straightforward way, with an emphasis on concepts rather than equations. It devoted an entire section to the measurement problem, in which Bohm diligently defended the Copenhagen interpretation. "I wrote this book which I had hoped to be from Bohr's view," he recalled later. "I tried to understand it as best I could. I taught the thing [quantum physics] for three years and put out notes and then finally a book." Bohm's book was released to generally favorable reviews. Bohm even received a "very enthusiastic" response to his book from the notoriously harsh Wolfgang Pauli, who said he'd enjoyed Bohm's approach to the subject.

Shortly after his book came out, Bohm received a phone call that changed the course of his life. "Einstein telephoned," Bohm said. "I was staying at a house with some friends of his, and he wanted to see me." Einstein had read Bohm's book and wanted to talk with him about it. "I went to see him and we discussed the book," Bohm recalled. "He [Einstein] thought that I had done as well as you could for explaining this theory, but he still was not satisfied that it was adequate. Basically, his objections were that the theory was conceptually incomplete, that this wave function was not a complete description of the reality and there was more to it than that. That was his basic objection." Einstein was harping on the same problem he had identified twenty-five years earlier: quantum physics, for all its successes, was stubbornly mute on the question of what was real. "We discussed it and he felt that one needs a theory in which one could discuss some reality which was existing and would stand by itself and did not always have to be referred to an observer," Bohm recalled. "He really felt quite definite that the quantum theory was not doing this. Therefore, though he accepted that it was giving the right results . . . he felt that it was incomplete."

Bohm walked out of Einstein's office with one thought ringing in his head: "Can I make another way of looking at it?" Was there another way to interpret the strange mathematics of quantum physics? Or was the Copenhagen interpretation the only way to think about the theory? "It seemed that Einstein was right and I already felt dissatisfied," Bohm recalled. "I began to wonder, does [the wave function] give a complete description of reality?" Einstein was sure that it didn't. Bohm took that idea and ran with it. In a matter of weeks, he discovered there was a simple way to rewrite the fundamental equations of quantum theory. The predictions and results remained the same—the new version was mathematically equivalent to the old—but the picture of the world suggested by the math, the story it told, was radically different from the Copenhagen interpretation.

Bohm was amazed at what he had found. He wrote up his ideas and sent them off in a pair of papers to be published in *Physical Review*, the most prominent research journal in physics. In the meantime, Bohm had more good news. On May 31, he appeared in federal district court in Washington, DC, where he was cleared of all charges. But the next month, under tremendous pressure from President Dodds, the Princeton physics department announced that they would not be renewing Bohm's contract, leaving him out of a job. Einstein wrote several letters of recommendation for Bohm but to no avail. Despite his legal innocence, Bohm remained on the blacklist.

Toward the end of that summer, Bohm (with help from Einstein and Oppenheimer) found a job at the University of São Paulo, in Brazil.

Bohm had never been outside the United States and he didn't speak a word of Portuguese. But he had no other options—and he also suspected that he was under FBI surveillance. He left for Brazil in October.

Throughout his ordeal, Bohm remained hopeful that when his papers appeared in print the upcoming January, his new view of quantum theory would inspire debate and gain him recognition among his fellow physicists. "It is hard to predict the reception of my article," he wrote to a friend back at Princeton not long after arriving in Brazil, "but I am happy that in the long run it will have a big effect." What he really feared, he continued, was "that the big-shots will treat my article with a conspiracy of silence; perhaps implying privately to the smaller shots that while there is nothing demonstrably illogical about the article, it really is just a philosophical point, of no practical interest." Bohm attempted to learn Portuguese, adjusted to yet another new (and to him, unpleasant) climate—and waited for his ideas to finally meet the world.

In Bohm's interpretation of quantum physics, much of the mystery of the quantum world simply falls away. Objects have definite positions at all times, whether or not anyone is looking at them. Particles have a wave nature, but there's nothing "complementary" about it—particles are just particles, and their motions are guided by pilot waves. Particles surf along these waves, guided by the waves' motion (hence the name). Heisenberg's uncertainty principle still holds—the more we know about a particle's position, the less we know about its momentum, and vice versa—but according to Bohm, this is simply a limitation on the information that the quantum world is willing to yield to us. We may not know where a given electron is, but in Bohm's universe, it's always somewhere.

This simple idea allowed Bohm to cut through the thicket of quantum paradoxes. The Copenhagen interpretation doesn't let you ask what's happening to Schrödinger's cat before you look in the box, insisting only that it's meaningless to talk about the unobservable. But, in Bohm's pilot-wave interpretation, not only can you ask but there's an

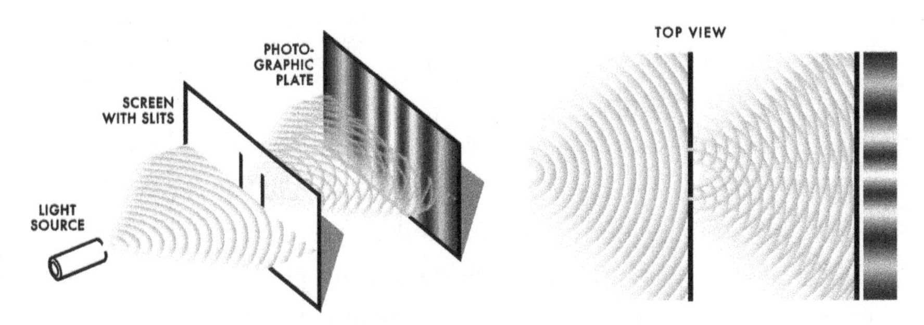

Figure 5.2. Waves in the double-slit experiment interfere with each other.

answer: before you look in the box, the cat is either dead or alive, and opening the box merely reveals which is true. The act of observation has nothing to do with the condition of the cat.

At first blush, this seems far too easy. If, in Bohm's theory, there is nothing strange about particle positions or Schrödinger's cat, how can it possibly hope to reproduce all the bizarre results of quantum physics? But the guarantee is in the mathematics: Bohm's theory is mathematically equivalent to the Schrödinger equation, the central equation of quantum physics, and therefore it must make the same predictions as any other interpretation. This is technically true, but it doesn't give a feel for how Bohm's interpretation actually works. To do that, we'll need to look at one of the strangest experiments in all of quantum physics: the double slit.

The great physicist Richard Feynman famously said that the double-slit experiment "has in it the heart of quantum mechanics," and that "in reality, it contains the *only* mystery." Yet, for all that hype, it's a surprisingly simple experiment. Set up a screen in front of a photographic plate, and place two narrow closely spaced slits in the screen. Then shine a light on it. The light waves will interfere with each other on the other side of the slits, creating a pattern of light and dark bands on the photographic plate (Figure 5.2). There's nothing particularly quantum about this—waves create interference patterns all the time, whether they're overlapping waves from two stones thrown in a pond or sound waves coming from two stereo speakers. Wave interference isn't mysterious: in spots where the peaks of one wave line up with the valleys of another, they cancel out and the waves vanish; when the peaks of both waves line up with each other, they're amplified. This creates the patterns of dark and light bands in Figure 5.2.

The weirdness really begins when you shine a much dimmer light on the double slit. Rather than shining a flashlight on the double slit, send the minimum amount of light possible: one photon at a time. Now, each photon faces a choice, like our nanometer Hamlet from the Introduction: go through the left slit or the right slit? Once a photon goes through a slit, it hits the photographic plate behind the slit, leaving a dot on impact. Repeat this over and over again, and you might expect to see two groups of dots, one lined up behind each slit (Figure 5.3a). After all, photons are particles—little tennis balls of light. If you threw tennis balls through a (much larger) double slit, you'd expect them to mostly hit the back wall in two clusters, one behind each slit. But photons aren't really tennis balls of light, and they do something extraordinary instead: though each one hits the plate in a single location, their impacts collectively form an interference pattern on the plate (Figure 5.3b). Even though each photon went through the double slit individually, they still somehow "knew" where to arrive on the photographic plate in order to form an interference pattern. Something was interfering with each photon as it went through, despite the fact that particles don't interfere with each other, and there was only one particle in the double slit at a time anyhow.

Puzzled by the results of your experiment, you repeat it, but with a twist. This time, you attach a little photon detector to each slit, in an attempt to determine which slit each photon goes through, so you can figure out how the interference pattern on the plate is formed. The results convince you of what you had already suspected but hadn't dared to believe: the photons are deliberately messing with you. Now that you're watching them so closely, they refuse to form an interference pattern at all and instead form exactly the two groups of dots that you had expected before (Figure 5.3a). What gives? How can the photons behave differently just because you're watching them? How do they know you're watching them at all?

The Copenhagen interpretation, true to form, gives a mystical pseudo-answer steeped in the language of Bohr's philosophy of complementarity.

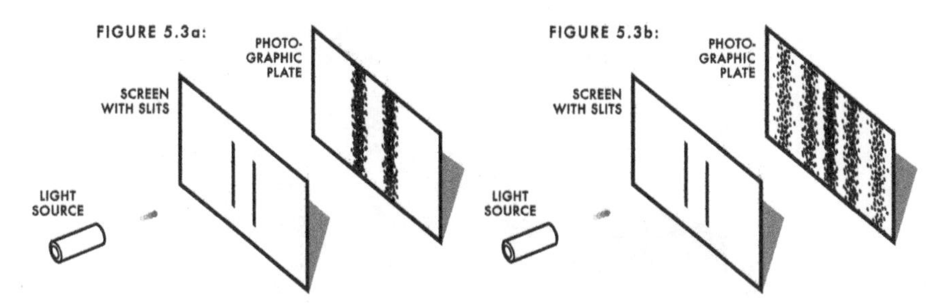

Figure 5.3. (a) We wouldn't expect individual photons passing one at a time through the double slit to produce an interference pattern. (b) Somehow, individual photons passing through the double slit do manage to interfere with themselves.

The idea of particles, Copenhagen claims, is complementary to the idea of waves. The ideas are contradictory—photons cannot be both particles and waves—but both are necessary, in alternation, for describing this experiment. When you aren't measuring the position of a photon, it is a wave. Thus, photons can interfere with themselves as they pass through the double slit. But measuring the location of a photon forces it to behave as a particle: when the photon hits the screen behind the double slit, it must strike in only one spot. Similarly, putting photon detectors on each slit causes the photon to behave as a particle as it passes through the double slit: the detectors force each photon to pass through only one slit, and thus not interfere with themselves, when before they were free to behave as waves and pass through both slits. But asking where the photon was before the measurement is meaningless: waves have no singular location. The property measured was created by the measurement itself, and to ask about its location beforehand is mere sophistry. Any attempt to picture how this is possible, any attempt to give an account of how the quantum world behaves between measurements, is doomed to fail, because, as Bohr said, there is no quantum world.

Bohm accounted for the strange results of the double-slit experiment by doing exactly what the Copenhagen interpretation said was impossible: he gave a detailed account of what happens in the quantum world whether or not anyone is looking. Photons, according to Bohm, are

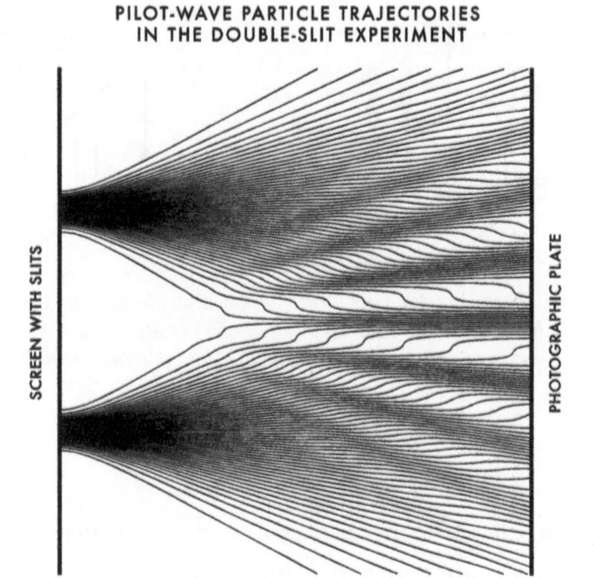

Figure 5.4. Particle trajectories guided by pilot waves in the double-slit experiment (top-down view). Figure produced with Mathematica code graciously provided by Professor Charles Sebens of UCSD.

particles surfing on waves. While a particle can only pass through one slit, its pilot wave passes through both and interferes with itself. That self-interference, in turn, affects the motion of the particle, because it is guided by the wave. The wave pushes the particle onto a path ensuring the appearance of an interference pattern on the photographic plate after enough photons have been sent through the double slit (Figure 5.4). Putting photon detectors on each slit affects each photon's pilot wave—no matter how ingenious the design, any photon detector must alter a photon's pilot wave, as ensured by Heisenberg's uncertainty principle, which in Bohm's interpretation places limits on how much measuring devices can avoid interfering with the things they attempt to measure. The effect of these measurements on the photons' pilot waves alters their trajectories, causing them to form a pair of clusters on the photographic plate rather than an interference pattern. In Bohm's account, although measurement can influence a particle's motion, all particles have definite positions whether or not anyone is looking at them.

Bohm's interpretation is a lot like de Broglie's old interpretation, presented at the 1927 Solvay conference. The mathematics of the two interpretations are essentially identical, differing only in their emphasis of certain ideas over others, and the key physical insight is the same: a quantum world composed of particles guided by waves. But Bohm succeeded where de Broglie failed. Bohm handily solved the problems raised by Pauli, Kramers, and others a quarter century earlier at Solvay, by insisting that everything be treated in a quantum way—both the things being measured and the devices doing the measuring. This was a truly radical idea: taking quantum physics seriously as a way of accounting for the entire world. In Bohm's pilot-wave interpretation, strange quantum behaviors are minimized for larger objects, which is why we don't see them in the everyday world. But every object, big and small, is ultimately governed by the same set of quantum equations.

The Copenhagen interpretation, in contrast, did not see quantum physics as a way to account for the whole world—and especially not experimental equipment involved in making measurements, like photographic plates or double slits. According to Bohr, one of the fundamental features of quantum physics was "the necessity of accounting for the functions of the measuring instruments in purely classical terms, excluding in principle any regard to the quantum." Quantum mechanics was a physics of the small, not of the large, and never the twain shall meet: when Bohr's student George Gamow wrote about a fantasy world in which quantum effects appeared at large scales, purely as a way of explaining to nonscientists how quantum physics worked, Bohr "was irritated rather than amused." Quantum physics, according to Copenhagen, was not to be taken seriously as a theory of the entire world. It was, instead, a theory about how we interfaced with the world of the extremely tiny, a pragmatic invention, a means for predicting the outcomes of experiments and nothing more. And this was as it should be, according to Bohr: he claimed that the job of physicists was "not to disclose the real essence" of the world around us, but simply to find "methods for ordering and surveying human experience."

Is Bohr right? Is it wrong to say that physicists should attempt to figure out how the world actually is? Is it enough to come up with theories that make accurate predictions about the outcomes of experiments? And if Bohm's theory gives the same predictions as "regular" quantum physics (whatever that is), then what's the point of it? How can there be any important difference between two competing theories that make the same predictions?

These questions point to difficult issues in the philosophy of science (some of which we'll meet again in Chapter 8). The short answer is that no, Bohr is not correct, at least not straightforwardly. The picture of the world that comes along with a physical theory is an important component of that theory. Two theories that are identical in their predictions can have wildly different pictures of the world—like putting the Earth at the center of the universe rather than the Sun—and those pictures, in turn, determine a lot about the daily practice of science. If you think that the Sun is at the center of the solar system, rather than the Earth, you're likely to conclude that there's nothing special about Earth, or our solar system, and that there could easily be planets around other stars, even though both astronomical theories give the same predictions about how different lights will move across the sky here on Earth. The story that comes along with a scientific theory influences the experiments that scientists choose to perform, the way new evidence is evaluated, and ultimately guides the search for new theories as well.

In his 1952 papers outlining his new interpretation, Bohm made exactly this point. "The purpose of a theory is not only to correlate the results of observations that we already know how to make," he wrote in the conclusion to his second paper, "but also to suggest the need for new kinds of observations and to predict their results." Bohm laid part of the blame for the Copenhagen interpretation at the feet of logical positivism, the philosophy of science inspired by Mach (which we first encountered in Chapter 3). The Copenhagen interpretation, in Bohm's view, was "guided to a considerable extent" by the idea that objects that can't be seen aren't real, an idea Bohm ascribed to positivism. Yet, as Bohm pointed out, "the history of scientific research is full of examples in which it was very fruitful indeed to assume that certain objects or elements might be real, long before any procedures were known which would permit them to be observed directly." Bohm then gave the example of atoms, the existence of which Mach resisted to the end, despite the overwhelming evidence to support them, because they couldn't be seen. Bohm made this point again shortly after he arrived in Brazil, in a letter he wrote to his friend and fellow physicist Arthur Wightman:

> Tentative concepts are needed, even before empirical evidence is available, to guide the choice and design of experiments, as well as to aid in their interpretation. . . . Very often, the actual empirical evidence for a new idea comes from surprising quarters (witness Brownean [*sic*] movement, the first evidence for existence of atoms, discovered by a biologist). However, such evidence can be appreciated only by people who are alert to the possibilities. For this reason, I would argue in favor of the widest possible diffusion of knowledge of *all* the possibilities among physicists. At a time like this, physicists ought to know of all the possibilities, and to feel that while they do not know which of these are correct, they must be ready, if necessary, to abandon even what seemed most secure and beautiful in the old point of view, in favor of what may seem arbitrary and ugly in the new point of view, if this should help explain something.

Yet, as Bohm pointed out in his 1952 papers, "[Positivism's] reflection still remains in the philosophical point of view implicitly adopted by a large number of modern theoretical physicists." To positivistically inclined physicists, it wasn't just that there was no need for a new interpretation of quantum physics—according to them, there was no need for any interpretation whatsoever. Quantum physics perfectly correlated and predicted observations, and that's all a scientific theory needed to do on a strictly positivist account of science. Any ideas about what nature was actually like that came along with a theory were just extra baggage. This was the logic behind Bohr's "rhetoric of inevitability," as the historian of science Mara Beller called it. Bohr and his followers said the Copenhagen interpretation was not just the correct way of understanding quantum physics—they said it was the only way to do it, the necessary and inevitable conclusion of the quantum revolution. "Every feature of [the Copenhagen interpretation]," claimed Léon Rosenfeld, one of Bohr's closest colleagues, "has been forced upon us as the only way to avoid the ambiguities which would essentially affect any attempt at an analysis in classical terms of typical quantum phenomena." Thus, according to Bohr's camp, the search for another interpretation wasn't just unnecessary, it was a waste of time. By the time Bohm's papers appeared, seven years after the end of World War II and the changes it brought to the culture of physics, this view was prevalent among physicists.

Bohm, of course, had put the lie to the rhetoric of inevitability by creating a live alternative to the Copenhagen interpretation. But recognition that Bohm had achieved anything at all with his theory was hard to come by. Bohm had anticipated that he might be ignored or disparaged, but when word reached him of his work's reception in Princeton, he was, understandably, slightly unhappy.

"As for . . . the Princetitute, what those little farts think is of no consequence to me. . . . I am convinced that I am on the right track." Bohm, in his Brazilian isolation, could vent his frustration only through letters to his friends. And those same letters were his only indication of what was happening in the wider world of physics. Weeks after arriving in October 1951, Bohm was summoned to the US Consulate in São Paulo. Once there, his passport was confiscated and stamped valid for return only to the United States. But Bohm was afraid of what would happen to him if he did return to his homeland. "The best possible interpretation is that they simply do not want me to leave Brazil," Bohm wrote to Einstein, "and the worst is that they are planning to carry me back because perhaps they are reopening this whole dirty business again." Bohm had hoped to travel to Europe, to meet with other leading physicists there and defend his ideas. "It is really necessary for me to give talks, in Europe if possible, and perhaps even in U.S., if Europe is not possible; or else nobody will take the trouble to read [my paper]," he wrote to a friend. Without a passport, Bohm would have to mount his defense remotely. It didn't go well.

Before his papers appeared in print, Bohm had sent drafts to several of the founding fathers of quantum physics (some of whom had written to Bohm just months earlier with praise for his textbook). De Broglie wrote back, pointing out that he had thought of similar ideas twenty-five years earlier, but Pauli and others had set him straight by raising important problems with the pilot-wave theory. Pauli wrote back next, throwing those same problems at Bohm. But Bohm managed to handle them with style and aplomb through his brilliant insight that measurement devices themselves must be incorporated in his quantum descriptions. Pauli, after a lengthy and heated exchange of letters over the next few months, finally conceded that Bohm's theory was consistent, though he still maintained that, since there was no way to test it against "normal" quantum physics, it remained "a check that cannot be cashed." Ultimately, Pauli thought that Bohm's ideas were simply "artificial metaphysics."

Niels Bohr himself never wrote back to Bohm. But Bohm did receive a report from his friend, Art Wightman, who was visiting Bohr's institute at the time. According to Wightman, Bohr thought that Bohm's theory was "very foolish," and didn't say much else. Von Neumann, meanwhile, was less dismissive; he thought Bohm's ideas were "consistent," and even "very elegant," but suspected that Bohm would have difficulties extending his theory to encompass the quantum phenomenon of spin—a suspicion that ultimately proved to be wrong.

Von Neumann's suspicion was likely born of his own "impossibility" proof demonstrating the necessity of the Copenhagen interpretation. Bohm knew that his own theory showed that something was wrong with that proof—or at least that it was a less powerful proof than commonly supposed by other physicists. Bohm discussed how his theory evaded von Neumann's proof toward the end of his second paper laying out his pilot-wave theory. But his analysis of von Neumann's proof was somewhat unclear at best and simply incorrect at worst. And without a clear and pithy explanation of what had gone wrong in von Neumann's proof, many physicists assumed the flaw was with Bohm's theory instead—that it simply couldn't be correct, because von Neumann had shown such theories were impossible.

There were a few physicists who did come around to supporting Bohm's view: most notably, Louis de Broglie, who took up his old interpretation and started a priority dispute with Bohm over the work. Bohm resisted acknowledging de Broglie's contributions at first. "If one man finds a diamond and then throws it away because he falsely concludes it is a valueless stone, and if this stone is later found by another man who recognizes its true value, would you not say that the stone belongs to the second man?" Nonetheless, the dispute was short-lived and resolved amicably. When Bohm wrote a book on his new interpretation several years later, de Broglie wrote a glowing introduction, in which he described Bohm's work as "elegant and suggestive." De Broglie's institute in Paris became one of the few places in the world where dissent against the Copenhagen interpretation was the norm.

Bohm also hoped for support from Soviet physicists and other Communists. His interpretation made quantum physics explicitly about stuff existing in the world, rather than an abstract statement of what physicists can say about experimental outcomes. This lined up well with the emphasis on "materialism" and rejection of positivism that was a common thread in many strains of Marxist thought. Mach's positivism, in particular, was a common Marxist punching bag. Even Lenin himself had condemned Mach; in *Materialism and Empirio-criticism*, Lenin called Mach's philosophy "reactionary" and "solipsist." There were some Soviet physicists who took up this sort of charge against the Copenhagen interpretation, such as Dmitrii Blokhintsev and Yakov Terletsky, whose work Bohm encountered after developing his own interpretation.

Bohm's theory had also appeared during the height of Zhdanovism, an ideological campaign by Stalin's USSR to stamp out any intellectual work that had even the faintest whiff of a conflict with the ideals of Soviet communism. Though there were certainly versions of the Copenhagen interpretation that could be compatible with Soviet state ideology, the aura of positivism that hung around the Copenhagen interpretation was enough to keep most Soviet physicists from defending Bohr's ideas publicly under Stalin's rule. This led to the "age of banishment of complementarity" in the USSR, according to historian of science Loren Graham.

Some of Bohm's fellow Marxists did respond positively to his work: several of de Broglie's students (most notably Jean-Pierre Vigier) found both Marxism and pilot waves appealing. Yet many Marxist physicists didn't back Bohm's ideas. While Blokhintsev and Terletsky were critical of Bohr's principle of complementarity and the other trappings of the Copenhagen interpretation—sometimes vocally so—they did not support Bohm's interpretation, instead pursuing their own alternatives to the quantum orthodoxy. Indeed, Bohm suspected that Zhdanovism may have simply kept most physicists behind the Iron Curtain from discussing questions of quantum interpretation at all. "I ask myself the question, 'Why in 25 years didn't someone in USSR find a materialist interpretation of quantum theory?' It wasn't really very hard," he wrote to his friend Miriam Yevick. "In USSR, there has been much criticism of quantum theory on ideological grounds, but it produced no results, because it may have scared people away from these problems, rather than stimulate them."

In any event, the policy of Zhdanovism died with Stalin in 1953, leading to the (relative!) relaxing of ideological strictures in the Soviet Union under Khrushchev. This freed Russian physicists who had sat at Bohr's feet to be more vocal in their support of the Copenhagen interpretation; one of them, Vladimir Fock, campaigned for Bohr's ideas throughout the Soviet physics education system, referring to the pilot-wave interpretation as the "Bohm-Vigier illness." And Bohm conjectured that others were hesitant to criticize the Copenhagen interpretation not only out of allegiance to Bohr but also out of fear that they would appear to be doing so for ideological reasons. The USSR had already produced Lysenkoism, a bogus alternative to Darwinian evolution based on a "proper Marxist" understanding of biology. Soviet biology and agriculture took decades to recover from the damage done by Lysenko and his pseudo-scientific cronies. The last thing that good physicists in the USSR wanted was a similar fiasco in quantum physics.

One Marxist in particular had it in for Bohm: Léon Rosenfeld, Bohr's right-hand man in Copenhagen. Rosenfeld's twinned devotions to complementarity and Marxism led Pauli to nickname him "square root of Bohr × Trotsky." Rosenfeld took it upon himself to defend the One True Quantum Physics from Bohm. "I certainly shall not enter into any controversy with you or anybody else on the subject of complementarity," he wrote to Bohm, "for the simple reason that there is not the slightest controversial point about it." Rosenfeld spent much of his time participating in this nonexistent controversy, devoting remarkable effort to preventing the spread of Bohm's ideas. Rosenfeld successfully prevented Bohm from publishing a paper in *Nature*; he also prevented a translation of a

Russian paper critical of complementarity from appearing in *Nature* by convincing the translator to withdraw it. He even managed to prevent the publication of an English translation of a book by de Broglie on pilot waves. And when Bohm published his book on the pilot-wave interpretation several years later, Rosenfeld wrote a scathing review, claiming Bohm had hopelessly misunderstood quantum physics: "It is understandable that a pioneer advancing in unknown territory does not find the best path at the outset; it is less understandable that a tourist still becomes lost after that territory has been surveyed and mapped down to one part in twenty thousand." Rosenfeld's views were largely shared by the physics community, as one of his friends commented in a letter to him: "I was much amused by the onslaught on David Bohm. . . . Half a dozen of the most eminent scientists have got their knife into him. Great honor for somebody so young."

Those eminent scientists included not only Rosenfeld and Pauli but also Werner Heisenberg, who dismissed Bohm's theory as "a kind of 'ideological superstructure,' which has little to do with immediate physical reality," and Max Born, who said that Pauli "slays Bohm not only philosophically but physically as well." Yet there were private ideological divisions among Heisenberg, Born, Pauli, Rosenfeld, and the rest of the old guard. Rosenfeld thought Heisenberg was flirting with idealism—a devastating insult from a Marxist—while Pauli and Born thought Rosenfeld was too politically motivated in his science. But the founders of the Copenhagen interpretation closed ranks against Bohm despite their disagreements.

And it wasn't just the old guard who found Bohm's ideas distasteful. Younger physicists, insofar as they paid attention to Bohm at all, were also dismissive of him. In particular, many were troubled by an unavoidable fact of Bohm's theory: it was nonlocal, allowing particles to influence each other instantaneously at long distances. A single particle, wandering the universe on its own without bumping into anything, is guided in its path by its own pilot wave and is perfectly local. But introduce a second particle that interacts in any way with the first, and suddenly they are linked—entangled—and the pilot wave of one particle will change depending on the precise location of the other particle, no matter how distant it may be. This kind of "spooky action at a distance" also appeared in the Copenhagen interpretation—it was exactly what Einstein had argued against in the EPR paper. But many physicists were still unaware of the EPR argument, and most that were aware of it profoundly misunderstood it. To them, the manifest action at a distance in Bohm's theory was another strike against it when compared to Copenhagen.

There was also the question of whether Bohm's ideas would actually lead to new research insights. In particular, because Bohm's theory involved faster-than-light connections between particles, it appeared difficult to extend Bohm's ideas to incorporate special relativity. Relativistic quantum theory, known as quantum field theory (QFT), was already an active and productive area of research in the United States and Europe. QFT was originally pioneered by Dirac and was being led at the time by people like Feynman, Julian Schwinger, Sin-Itiro Tomonaga, and Freeman Dyson. QFT met with great success: Dirac had used it to predict the existence of antimatter, which won him a Nobel Prize; others had used it to prove deep connections between seemingly unrelated quantum properties and to explain the increasingly complex results in high-energy particle physics that were pouring in from the growing number of particle accelerators around the world. And nonrelativistic quantum theory was also being used to great success in other areas, such as solid-state physics. According to Sam Schweber, Bohm was still highly regarded for his other work in physics—but nobody could see how to apply his new ideas about quantum theory to the wide variety of interesting problems at hand. "So much was happening, both in [solid-state] physics and high-energy physics, that people weren't that much concerned with foundations," recalled Schweber. Bohm's interpretation of quantum theory, he said, "is not generative. It's very difficult to see how to do Bohmian quantum mechanics when you want to generalize it to quantum field theory. It lay on the side."

Bohm's theory would have to account for the successes of QFT and connect to other areas of already active research if it were to survive. But Bohm, stuck in Brazil, found progress slow. "I alone am supposed in a year or two to produce a scientific revolution comparable to that of Newton, Einstein, Schrödinger, and Dirac all rolled into one," he complained to a friend. Bohm's exile also made it difficult for him to keep in touch with the latest developments in quantum physics; he dismissed his friend Richard Feynman's latest work in QFT, which would eventually garner Feynman a Nobel Prize, as "long and dreary calculations on a theory that is known to be of no use." Bohm's geographical and ideological isolation was taking a serious toll on his scientific work.

Even staunch opponents of the Copenhagen orthodoxy, like Erwin Schrödinger, didn't lend Bohm support. Schrödinger still had massive problems with the Copenhagen interpretation, even after a quarter century, and continued to fight it until his death. "The impudence with which you assert time and again that the Copenhagen interpretation is practically universally accepted, assert it without reservations, even before an audience of the laity—who are completely at your mercy—it's at the limit of the estimable," he wrote to Max Born in 1960. "Have you no anxiety about the verdict of history?" Yet when Bohm wrote to Schrödinger about the pilot-wave interpretation, he received only a reply from his secretary, who said that Schrödinger wasn't interested in his work. "Schrödinger does not deign to write me himself, but he deigned to let his secretary tell me that His Eminence feels that it is irrelevant that mechanical models can be found for the quantum theory," grumbled Bohm. "Of course, his Eminence did not find it necessary to read my papers. . . . In Portuguese, I would call Schrödinger 'um burro,' and leave it for you to guess the translation." Schrödinger was preoccupied with his own attempt to interpret quantum physics, a picture of the quantum world in which there was only a wave function, with no particles at all. Particles guided by pilot waves were wholly uninteresting to him.

Most disappointing of all, though, was Einstein's reaction to Bohm's work. Einstein was certainly sympathetic to Bohm's motivations—it was Einstein's advice that had given Bohm the courage to develop his ideas in the first place—but he was not at all pleased with the answer that Bohm had arrived at. "Have you noticed that Bohm believes (as de Broglie did, by the way, 25 years ago) that he is able to interpret the quantum theory in deterministic terms?" Einstein wrote to his old friend Max Born. "That way seems too cheap to me."

Einstein's letter didn't go on to explain what, exactly, was "too cheap" about Bohm's ideas. But the pilot-wave interpretation did have a few features that were clearly unacceptable to Einstein. Objects could move in strange ways or fail to move at all when it seemed like they should. Einstein pointed out that in Bohm's theory, a particle trapped in a box could be motionless despite having an enormous amount of kinetic energy (energy of motion). This contradicted the principle that quantum physics should agree with classical physics for large objects. In reply, Bohm pointed out that in such a situation, if you opened the box, the walls of the box would interact with the particle, ensuring the previously motionless particle would go shooting out of the box at high speed—a high speed corresponding to the kinetic energy it had before the box was opened in the first place. Strange, certainly, but any theory would have to be strange to reproduce the counterintuitive results of quantum physics. (To Einstein's credit, he arranged for Bohm's reply to his criticism to be published alongside his own views.)

Einstein was also unhappy with the idea of nonlocality. He knew the Copenhagen interpretation was nonlocal, so this feature of Bohm's theory wasn't any worse than the usual view. But Einstein couldn't see any physical reason to give up locality—the EPR argument made it clear that quantum physics was either nonlocal or incomplete, and Einstein's money was on the latter. Writing to Born, he said that "when I consider the physical phenomena known to me, and especially those which are being so successfully encompassed by quantum mechanics, I still cannot find any fact anywhere which would make it appear likely that [locality] will have to be abandoned."

Einstein also wanted to find other ways to describe what was going on at the quantum level altogether. The Copenhagen interpretation and Bohr insisted on the necessity of using classical concepts, along with classical descriptions of measuring apparatuses. Bohm's ideas broke with both, but not as thoroughly as Einstein had hoped for. Einstein wanted a new way of looking at nature, a theory underlying quantum physics that would reveal some previously unknown truth, rather than a new way of interpreting the existing quantum theory. Einstein hoped to find such a picture in a unified field theory, something that would unite his theory of general relativity with the deeper reality he was convinced lay underneath the mathematics of quantum physics. Writing of Einstein after his death, Born said that "his ideas were more radical [than Bohm's], but 'music of the future.'"

History was repeating itself: just as in Solvay twenty-five years earlier, the defenders of the Copenhagen interpretation presented a united front despite their private disagreements, while the rebellion, unable to agree on a single position, fizzled out.

After two years, Bohm was itching to leave Brazil. His theory was being alternately ignored or disparaged, and he couldn't travel to speak in its defense. He turned to Einstein for help. Despite his distaste for the pilot-wave interpretation, Einstein was still generally supportive of Bohm. Now, Einstein pulled some strings to set Bohm free. He contacted Nathan Rosen, his former assistant and coauthor of the EPR paper, and asked him whether he could hire Bohm at his new physics department in his new country: Israel. Bohm, as a talented physicist and Jewish political refugee, seemed like a natural fit for Israel; Einstein, as the most famous Jew in the world, had considerable pull there. Rosen arranged for a job for Bohm, but without a passport, Bohm was still trapped in Brazil. With his job offer in hand, Bohm attempted to get Israeli citizenship; when that failed, Einstein suggested getting Brazilian citizenship and traveling on that passport. Bohm's Brazilian contacts smoothed the wheels of government for him, and Bohm became a Brazilian citizen on December 20, 1954. Several months later, Bohm finally left Brazil, after nearly four years there.

Bohm took well to life in Israel. He met a fellow immigrant, Sarah Woolfson, and they were soon married. He published a book on his version of quantum physics. He traveled to Europe to meet and work with other physicists. He even went to Bohr's institute in Copenhagen a couple of times, though he worked solely on plasma physics there, and there is no record that he ever spoke with Niels Bohr about the pilot-wave interpretation. Bohm also worked with a particularly talented student in Tel Aviv, Yakir Aharonov. Wishing to avoid tainting Aharonov with the scent of his own heresies, he made a pact with him at the outset of their collaboration: they would do all of their work in "normal" quantum physics, rather than Bohm's new version. Together, they found a surprising new consequence of quantum physics that would be Bohm's best-known work in "normal" physics: the Aharonov-Bohm effect, an unusual feature of the behavior of electrons and other charged particles traveling near electromagnetic fields.

Bohm, meanwhile, convinced himself that he had been wrong about the pilot-wave interpretation. After writing his book on the subject, Bohm decided that he had been mistaken and that his interpretation didn't work after all—though he still didn't think the orthodox Copenhagen view could be right either. He abandoned his interpretation for a variety of reasons: he couldn't see how to make it work with special relativity, he was discouraged by the lack of interest from the wider physics community, and he couldn't see a way forward with the ideas from his own theory. "Because I did not see clearly, at the time, how to proceed further," he said years later, "my interests began to turn in other directions." This change came at roughly the same time as another, related, major intellectual shift for Bohm: in the wake of the brutal suppression of the Hungarian Uprising in 1956, Bohm abandoned his Marxism. This change in philosophy altered Bohm's thinking about the nature of the quantum world, which further motivated him to abandon his old ideas.

As Bohm searched for a new approach to quantum physics, he finally found some stability in his professional life. He left Israel in 1957, taking a temporary appointment at the University of Bristol in the UK. Several years later, he found a permanent position at Birkbeck College at the University of London. And eventually, he was offered two permanent positions in the United States: one at the newly formed Brandeis University in Boston, and another several years later at the New Mexico Institute of Mining and Technology. But when he attempted to take these jobs, he faced a new problem: the US government, upon learning of his Brazilian citizenship, had stripped him of his American citizenship. And with his ties to communism still fresh in the minds of State Department officials, Bohm's application to regain his citizenship in his native land was not looked upon kindly. He could return and become an American again, they told him—if he publicly renounced communism. Although Bohm was no longer a Marxist of any stripe, he considered it unethical to publicly renounce his former political views simply as a means to some other practical end. "I feel it wrong to say it [criticize communism] in order to regain American citizenship. For then, I am saying something not mainly because I think it is true, but rather, for some ulterior purpose. It's rather like writing a scientific article in order to impress one's superior, so as to get a better job." Unwilling to compromise his integrity, Bohm remained at Birkbeck.

Meanwhile, the pilot-wave interpretation sank into obscurity as Bohm searched for a new way to understand the quantum world. But back at Princeton, where all his troubles began, a new alternative had already been found.

6

It Came from Another World!

Albert Einstein gave the last lecture of his life in Princeton, New Jersey, on April 14, 1954. It was a guest lecture in John Wheeler's graduate seminar on relativity, but the subject inevitably turned to the role of the observer in quantum physics. ("On quantum theory I use up more brain grease than on relativity," Einstein once told his friend Otto Stern.) Einstein outlined his objections to quantum physics; afterward, the students asked questions, attempting to defend Bohr's views as they had been taught by Wheeler. Einstein handled them with patience, asking in return, with a slight smile, "When a mouse observes, does that change the state of the universe?"

One of the first-year graduate students in the room that day made note of Einstein's pithy challenge to the Copenhagen interpretation. A year later, Einstein was dead, and the student, Hugh Everett, was putting Einstein's words to good use in defense of his own new interpretation of quantum physics. Unlike Einstein—and like Bohm—Everett attempted to resolve the problems of quantum physics through the mathematics of quantum physics itself, rather than attempting to find an entirely new theory. But, unlike Bohm, Everett's solution didn't involve pilot waves. Everett's answer to the questions that lurked at the foundations of quantum physics was wholly original—and far stranger than anything Bohm or Einstein had ever proposed.

Hugh Everett III was born on November 11, 1930, into a family with Virginia roots that stretched back generations on his father's side; his paternal great-grandfather had fought for the Confederacy in the Civil War. Everett's father, Hugh Everett Jr., was a military engineer and logistics officer whose existence revolved around the Army. His mother, Katharine, was a free-spirited writer and pacifist. She and Hugh Jr. were a bad match temperamentally and philosophically, and they divorced (scandalous, for the time) several years after Hugh III was born. Hugh grew up in Bethesda, Maryland, with his father and stepmother. His family dubbed him "Pudge," because of his somewhat stocky build; Everett hated the nickname, but it stuck with him for the rest of his life.

With his nose usually buried in a science-fiction book, Hugh showed early signs of academic talent—and a taste for paradoxes. At twelve, he wrote a letter to Einstein, claiming to have solved the question of what would happen if an immovable object met an unstoppable force. The letter is lost, but Einstein replied, saying that while unstoppable forces and immovable objects weren't real, "there seems to be a very stubborn boy who has forced his way victoriously through strange difficulties created by himself for this purpose."

A year later, Everett won a scholarship to St John's, a Catholic military prep academy in Washington, DC. There, he excelled in nearly all of his classes, even the required religious instruction, despite the fact that his vocal atheism had already earned him a new nickname: "the heretic." Everett graduated with honors in 1948 and went on to study at Catholic University, also in DC, where he studied chemical engineering and mathematics, quickly impressing his professors and classmates with his remarkable facility for math and logic.

Unsurprisingly for a talented student of logic, Everett's taste for paradoxes was still lively. His patience for his required religious instruction wearing thin, Everett challenged one of his devout professors at Catholic with a "proof" of the nonexistence of God. The professor was purportedly sent into a state of serious religious doubt and despair, to Everett's dismay. Everett wasn't particularly interested in actually convincing anyone to fundamentally shift their worldview—he just wanted to have fun. And for Everett, fun entailed gaming out the logical consequences of a statement and winning the argument at hand. Sending someone into a tailspin of religious faith wasn't the goal at all. Everett resolved not to show his proof to anyone devout again—but it was a resolution he couldn't keep. He showed the proof to religious friends on and off again for the rest of his life, unable to keep from delighting in the absurd.

Everett graduated from Catholic in 1953 and won a place in Princeton's physics PhD program. He had applied six weeks late, but it didn't matter, as the Princeton faculty were eager to meet such an extraordinarily talented student. Everett had scored in the 99th percentile on the brand-new physics GRE, and he had stellar letters of recommendation: "This is a once-in-a-lifetime recommendation. . . . Everett is by far the best student I have had at Princeton, Rutgers, or Catholic University. Everett has a better knowledge of mathematics than most of the graduate students at Catholic University and probably *no* graduate student is his equal in native ability." The National Science Foundation (NSF), also brand new, was similarly impressed, and paid for Everett's graduate tuition and stipend.

At Princeton, Everett became particularly close with three of his classmates, and they shared an apartment later on. "Everett was a lot of fun. He enjoyed needling people," recalled Hale Trotter, one of the three. "He was very competitive at whatever it was, if it was a poker game or it was ping pong," recalled another, Harvey Arnold. "[Everett] always wanted to go away the winner and he would make you stay there until he succeeded." Charles Misner, the final member of the trio, agreed, calling Everett "a brilliant oddball . . . [whose] favorite sport was one-upmanship," though Misner was hasty to point out that "it was always friendly competition" with Everett.

Everett's Princeton friends were also impressed with his brilliance. "It surprised me after I got to know him that he was as brilliant as he was," recalled Arnold. "It didn't come across until you got close to him. And then you would recognize that this guy would be on top of the world. He was smart in a very broad way. I mean, to go from chemical engineering to mathematics to physics and spending most of the time buried in a science fiction book. I mean, this is talent."

In his early days at Princeton, Everett put that talent to work in a mathematically rigorous field befitting someone with his sense of competition: the mathematical theory of games. Everett's interest was practical as well as personal; game theory was the language spoken by military strategists and operations researchers in the Pentagon, where Everett already had ambitions to work after earning his PhD. Princeton was among the best places in the world to study game theory at the time. Von Neumann, one of the founders of the field, was just down the street at the Institute for Advanced Study, and other game theory giants like Oskar Morgenstern and Albert Tucker were at the university itself. There was also a weekly game theory seminar with lectures from Princeton faculty and visiting luminaries, including John Nash. Everett attended the seminar regularly during his first year, and ultimately wrote and presented a short paper there that went on to be a classic in the field.

When he wasn't occupied by his game theory habit, Everett's attention was increasingly captured by quantum physics. Most graduate courses on quantum physics in the United States at the time hardly discussed the puzzles at the heart of the subject; the course Everett took in his first year at Princeton was no exception. But reading both von Neumann's classic textbook and Bohm's newer one, Everett saw there was a problem lurking at the heart of the theory. Von Neumann's textbook made it clear that wave function collapse was separate from the Schrödinger equation, something extra added to make sense of the theory. But where did it come from? Bohm's book, with its valiant attempt to defend the Copenhagen interpretation, made it clear to Everett that the usual way of thinking about quantum physics couldn't answer that question. Bohm's pilot-wave papers, meanwhile, provided a concrete alternative to the standard view. Working on that kind of research problem was disreputable—independently of the fact that Bohm himself was politically radioactive at the time—but Everett didn't particularly care about what was or wasn't reputable. And Einstein's disparaging attitude toward the Copenhagen interpretation made it easier to think about challenging it—as did the fact that several other experts on the foundations of quantum physics in Princeton at the time, like Wigner and von Neumann, didn't always see eye to eye with Bohr.

Meanwhile, one of Everett's professors, John Wheeler, was obsessed with his own disreputable problem, general relativity. Despite the theory's universal acceptance, it wasn't seen as a reasonable field of research

Figure 6.1. Bohr at Princeton in 1954. From left to right: Misner, Trotter, Bohr, Everett, and David Harrison.

at the time. Wheeler was interested in the same problem Einstein was trying to solve: marrying general relativity to quantum physics in a single theory of quantum gravity, with the ultimate goal of describing the entire universe, including its origin, in the still more disreputable nascent field of quantum cosmology. He enlisted Charlie Misner, Everett's friend, in this work. "Everyone talking with Wheeler at that time was likely to be encouraged to think about quantum gravity," recalled Misner. Everett's interest in fundamental problems in quantum theory—and his obvious talent—made Wheeler a natural choice to be Everett's advisor.

But the influence of Wheeler and his own taste for paradoxes weren't the only reasons Everett was interested in the measurement problem. Everett's competitive nature was also at play—and his opponent this time was the assistant of Niels Bohr himself. In the fall of 1954, during Everett's second year at Princeton, Bohr came to the Princetitute for four months. He brought with him his assistant, Aage Petersen, a Danish physicist only a few years older than Everett. Everett became friends with Petersen, and, through him, he gained access to Bohr. That fall, Arnold saw Everett wandering the Princeton campus with Petersen and Bohr, lost in conversation. When Bohr lectured on campus, Everett and Misner attended. They heard the old quantum master dismiss the idea of a "quantum theory of measurement" as wrongheaded.

Around the same time, Everett passed his qualifying exams and started to think seriously about his PhD thesis. Everett wanted to do a short and fun thesis, but he needed a suitable subject. It occurred to him over drinks. "One night at the Graduate College after a slosh or two of sherry," Everett recalled, talking with Misner years later, "you and Aage were starting to say some ridiculous things about the implications of quantum mechanics and I was having a little fun, joshing you and telling you some of the outrageous implications of what you said, and, ah, as we had a little more sherry and got a little more potted in the conversation—don't you remember, Charlie? You were there!" Misner didn't remember, which Everett chalked up to "too much sherry," and continued:

> *Everett:* Well, anyway, the whole business started with those discussions, and my impression is I went to Wheeler then later and said, "Hey, how about this, this is the thing to do." . . . [T]his obvious inconsistency in the [quantum] theory or whatever I thought of it then. . . .
>
> *Misner:* It is strange that he would be so interested in it—all in all, because it certainly went against the normal tenets of his great master, Bohr.
>
> *Everett:* Well, he still feels that way a little bit.

At the time, Wheeler "was preaching this idea that you ought to just look at the equations and obey the fundamentals of physics while you follow their conclusions and give them a serious hearing," according to Misner. For his PhD thesis, Everett took Wheeler's advice. He looked at the outrageous implications of quantum physics and gave them a serious hearing. What he found was far more astonishing than anything in his beloved science-fiction stories.

We met the measurement problem back in Chapter 1. The problem, in a nutshell, is this: Quantum wave functions move along nice and smoothly, always obeying one simple and deterministic law, the Schrödinger equation—except when they don't. When a measurement happens, wave functions collapse. How and why wave function collapse happens—and what constitutes a "measurement" anyway—is the measurement problem, the central puzzle of quantum physics.

Everett thought that measurement, as presented in von Neumann's textbook, was "a 'magic' process in which something quite drastic [occurs] (collapse of the wave function), while in all other times systems [are] assumed to obey perfectly natural continuous laws." Measurement shouldn't be fundamentally different from other physical processes. And even worse, according to Everett, von Neumann's approach doesn't even tell you what measurements are. If a measurement only happens when someone looks at a system, who, in particular, has to look? Everett argued that this line of reasoning leads inevitably to solipsism—the idea that you are the only being in the universe, and everyone else is somehow illusory or secondary, existing in states of indeterminate reality until you, the High Arbiter of Wave Function Collapse, deign to observe them. In his thesis, Everett admitted that this is an internally consistent view, but that "one must feel uneasy when, for example, writing textbooks on quantum mechanics, describing [wave function collapse], for the consumption of other persons to whom it does not apply."

Bohr's idea that the quantum world of the small obeyed entirely different rules from those that governed the classical world of the large offered a possible way out of this dilemma—but at the cost of a unified picture of the world free of contradiction, a price Everett was (justifiably) unwilling to pay. "The Copenhagen interpretation is hopelessly incomplete because of its a priori reliance on classical physics (excluding *in principle* any deduction of classical physics from quantum theory, or any adequate investigation of the measuring process)," Everett complained, "as well as a philosophic monstrosity with a 'reality' concept for the macroscopic world and denial of the same for the microcosm." Writing to Petersen, Everett stated his intentions quite clearly. "The time has come . . . to treat [quantum physics] in its own right as a fundamental theory without any dependence on classical physics, and to derive classical physics from it." Like Bohm before him, Everett wanted to take quantum physics seriously as a theory of the entire world.

Rejecting both von Neumann and Bohr, Everett came up with his own solution to the measurement problem. Rather than explaining wave function collapse, Everett stated that wave functions never collapse at all. This in itself was not new; Bohm said the same thing. But Bohm had also added particles with definite positions into the theory, which accounted for the outcomes of measurements. Everett didn't add particles—he didn't think he needed them. Instead, he insisted that a single universal wave function was all there was: a massive mathematical object describing the quantum states of all objects in the entire universe. This universal wave function, according to Everett, obeyed the Schrödinger equation at all times, never collapsing, but splitting instead. Each experiment, each quantum event, spun off new branches of the universal wave function, creating a multitude of universes in which that one event had every possible outcome. Everett's shocking idea came to be known as the "many-worlds" interpretation of quantum physics.

The many-worlds interpretation sounds absurd at first blush, and probably at second blush too. We live in one world, not a multitude of them. If every quantum event—and in a fully quantum world, that's every event of any kind—leads the universe to split, where are these other universes? How could there possibly be so many of them without any indication that they're there? And, for that matter, how can any single event—one photon going through one double-slit apparatus, for example—cause the entire universe to split? To understand how the many-worlds interpretation accounts for these problems, let's take a second look at a simple quantum experiment, even simpler than the double slit: Schrödinger's cat.

Way back in the Introduction, we met Schrödinger's thought experiment, which has given the ASPCA nightmares for over eighty years. Put a cat in a box, along with a vial of poison and a lump of weakly radioactive metal; set up a Geiger counter (radiation detector) and a hammer so that the hammer will smash the vial if the detector detects any radiation. Then leave the cat in the box long enough that there's a 50-50 chance that the metal has emitted radiation. Now what? Is the cat dead or alive? According to the Copenhagen interpretation, the question is meaningless—you can't ask what happened before you open the box, because it's unobservable. According to Bohm and the pilot-wave interpretation, the question is quite meaningful, we just don't know the answer to it. The cat is either dead or alive, and we'll find out which when we open the box.

But what does the mathematics say? What does Schrödinger's equation say about Schrödinger's cat? Well, the wave function of the lump of metal is half "radiation was emitted" and half "no radiation was emitted." That interacts with the wave function of the detector, which means they get entangled. So now instead of two wave functions, one for the lump of metal and one for the detector, you have one for the both of them, and now that's in a weird state: half "radiation was emitted and the detector detected it" and half "no radiation was emitted and the detector didn't see anything." As the quantum Rube Goldberg machine continues on its merry way, wave functions keep getting entangled: the wave function of the hammer entangles with the wave function of the detector and metal; the wave function of the vial entangles with the wave function of the hammer and detector and metal, and so on, eventually including the cat itself. The whole system—cat, box, metal, poison, and all—ultimately ends up sharing a single wave function, and that wave function, again, has two equal parts: one part in which radiation was emitted and the cat is dead, and another in which no radiation was emitted and the cat is alive.

So far, so good. Now what happens when you open the box? The usual answer—the answer of the Copenhagen interpretation and the answer given by von Neumann's famous textbook—is that a measurement causes the wave function to collapse. But what if it doesn't? What if we treat you the same way we treated everything in the box? Well, in that case, when you look in the box, you're interacting with it—which means you get entangled with the shared wave function of the box and everything in it. So now we have an even bigger wave function, still with two parts: one where you see a dead cat and a smashed vial of poison, and one where you see a happy cat and an intact vial. Which part of

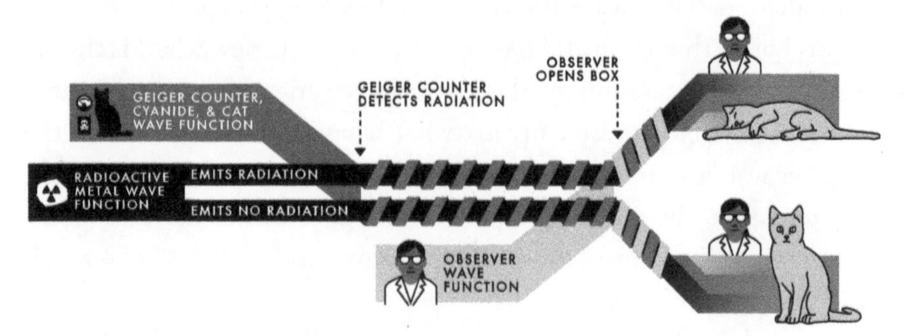

Figure 6.2. Branching in the many-worlds interpretation.

the wave function is real? Everett, remembering Wheeler's advice to take the consequences of physical laws seriously, answered that both are real. There's no way to pick one as more real than the other; the Schrödinger equation treats them equally. So when you perform this experiment, Everett said, both outcomes occur—and you split in two.

Of course, it doesn't seem like we split in two when we perform experiments—or anytime at all for that matter. But Everett had an answer ready for this as well. If I ask the "you" that sees the living cat how many cats you see, you'll answer "just one." And if I ask the same question of the "you" in the other branch of the wave function, the one with the dead cat, the answer will be the same (though your tone of voice will probably be quite different). The same thing happens, Everett pointed out, if I ask each copy of you how many selves you see. There is only one copy of you in each branch of the wave function, and, even if you repeat the experiment, this will still be true—there will be more branches, but each branch will still only have one copy of you. And the Schrödinger equation dictates that each branch will carry on independently of the others, with hardly any interaction between branches.

Strange enough, but we're not done. As you interact with the things in your environment, they get entangled with you, and then other things get entangled with them, and so on. Eventually, we have a single complicated and messy wave function for the entire universe—the universal wave function. And as more events happen, that universal wave function splits into more and more noninteracting parts, each merrily marching along to the deterministic beat of the Schrödinger equation.

These are the many worlds of Everett's interpretation. They may seem absurd on the face of it: there is, after all, only one world that we experience. But if that's your objection, Everett's reply is that you're hardly alone: to each person in each branch of the universal wave function, their world appears to be the only world, just as there appeared to be only one cat in the box and one you looking at it. This is a hallmark of the many-worlds interpretation: the appearance of a single world, despite the true existence of many.

When Everett finished a draft of his thesis in January 1956, Wheeler was the first to see it. Wheeler had enormous respect for Everett's abilities. Writing to the NSF, Wheeler said that Everett "originated an apparent paradox in the interpretation of the measurement problem in quantum theory. . . . In discussions of this paradox with graduate students and staff members here at Princeton, and with Niels Bohr, Everett brought to light new features of the problem that make it in and of itself an appropriate subject for an outstanding thesis when further developed. . . . [Everett] really is an original man."

But Wheeler was caught between several competing interests. He wanted to support a brilliant student's work, and he also wanted to find a way forward with quantum cosmology; supporting Everett's idea of a "universal wave function" would further both of those interests. But Wheeler also wanted to remain loyal to his mentor and friend, Bohr. Indeed, Wheeler idolized Bohr, writing that "nothing has done more to convince me that there once existed friends of mankind with the human wisdom of Confucius and Buddha, Jesus and Pericles, Erasmus and Lincoln, than walks and talks under the beech trees of Klampenborg Forest with Niels Bohr."

Wheeler was a political animal; he knew how to work with others and how to keep other people happy with his ideas, in exactly the way that Einstein could not. He knew that supporting Everett at the cost of his relationship with Bohr was a bad career move. "John Wheeler got along with everybody," recalled Misner. "But in Hugh's case, Wheeler had a very difficult time applying his usual tactics because he couldn't just

Figure 6.3. Wheeler (right), with Einstein (left) and Nobel Prize winner Hideki Yukawa in Princeton, 1954.

encourage Hugh to follow his ideas and present them as powerfully as possible, since they ran contrary to Bohr's ideas." But Wheeler was also unwilling to give up Everett's theory of a universal wave function—he saw it as a possible path forward for quantum gravity, an opportunity too good to pass up. This left Wheeler with only one option: he would attempt to secure Bohr's own blessing for Everett's work before endorsing it himself.

In mid-1956, Wheeler had his chance. He was invited to a visiting post at Leiden University, in the Netherlands, for several months. Once Wheeler had settled in there, he sent Everett's thesis draft, appropriately titled *Wave Mechanics Without Probability*, to Bohr, along with an introductory note. Tripping over himself to excuse any possible perception of Everett as contradicting Bohr, Wheeler wrote that "the title itself . . . like so many of the ideas in it, need further analysis and rephrasing." Wheeler himself soon followed in person and discussed Everett's thesis with Bohr, Petersen, and others in Copenhagen for several days.

Writing to Everett after his Copenhagen visit, Wheeler at first sounded hopeful and clear-eyed about the work yet to be done. "[Bohr and Petersen and I] had three long and strong discussions about it. . . . Stating conclusions briefly, your beautiful wave function formalism of course remains unshaken; but all of us feel that the real issue is the words that are to be attached to the quantities of the formalism." Wheeler implored his student to come to Copenhagen himself and resolve these problems, and offered to pay half of Everett's steamship fare to make it happen. "[Bohr] would welcome very much a several weeks' visit from you to thrash this out. . . . Unless and until you have fought out the issues of interpretation one by one with Bohr, I won't feel happy about the conclusions to be drawn from a piece of work as far reaching as yours. Please go (and see me too each way if you can!). So in one way your thesis is all done; in another way, the hardest part of the work is just beginning. . . . How soon can you come?" The last few sentences in the letter must have been unpleasant for Everett, who had already lined up a job in operations research at the Pentagon starting three weeks later, on the assumption that (as Wheeler had previously told him) his thesis would be accepted and his degree granted by the end of the summer.

But Bohr, Petersen, and others in Copenhagen were less enthusiastic about Everett's ideas than Wheeler thought. "There are some notions of Everett's that seem to lack meaningful content, as, for example, his universal wave function," wrote Alexander Stern, an American physicist studying with Bohr at the time, who had taken on the task of giving a seminar on Everett's work in front of Bohr and the rest of the institute. Stern's letter gives a sense of the attitude in Copenhagen toward Everett's ideas. "The basic shortcoming in his method of approach of his erudite, but inconclusive and indefinite paper is his lack of an adequate understanding of the measuring process. Everett does not seem to appreciate the FUNDAMENTALLY irreversible character and the FINALITY of a macroscopic measurement. . . . [I]t is an INDEFINABLE interaction." Stern went on to claim, without further explanation, that there was no contradiction between the Schrödinger equation and wave function collapse—that the measurement problem was not a problem at all—and that Everett's claim that there was such a contradiction was "not tenable." Ultimately, he dismissed Everett's ideas as either "a matter of theology" or "metaphysics," since the extra worlds postulated by Everett could never be seen or perceived directly in any way.

Despite the dim view that Copenhagen took of Everett's work, Wheeler still wanted the universal wave function and its promise of a quantum cosmology. To get Bohr's imprimatur, then, the words that went along with the universal wave function would have to change to better match the Copenhagen interpretation. Wheeler hoped for a way to retain what he liked in Everett's ideas while still using Copenhagen's language. His next letter to Everett made that clear—and also showed how much his estimate of the work involved had already changed, in light of the reaction from Copenhagen:

> [Resolving the issues with Bohr] is going to take a lot of time, a lot of heavy arguments with a practical tough-minded man like Bohr, and a lot of writing and rewriting. The combination of qualities, to accept corrections in a humble spirit, but to insist on the soundness of certain fundamental principles, is one that is rare but indispensable; and you have it. But it won't do much good unless you go and fight with the greatest fighter. Frankly, I feel about 2 more months of nearly solid day by day argument are needed to get the bugs out of the words, not out of the formalism [i.e. the idea of the universal wave function].

Wheeler also wrote back to Stern, vigorously defending the universal wave function—while eagerly displaying his own support for Bohr and the Copenhagen interpretation. Even more astonishing, he claimed that Everett supported the Copenhagen interpretation as well:

> I would not have imposed upon my friends the burden of analyzing Everett's ideas . . . if I did not feel that the concept of "universal wave function" offers an illuminating and satisfactory way to present the content of quantum theory. I do not in any way question the self consistency and correctness of the present quantum mechanical formalism when I say this. On the contrary, I have vigorously supported and expect to support in the future the current and inescapable approach to the measurement problem. To be sure, Everett may have felt some questions on this point in the past, but I do not. Moreover, I think I may say that this very fine and able and independently thinking young man has gradually come to accept the present approach

> to the measurement problem as correct and self consistent, despite a few traces that remain in the present thesis, draft of a past dubious attitude. So, to avoid any possible misunderstanding, let me say that Everett's thesis is not meant to question the present approach to the measurement problem, but to accept it and generalize it.

Several days later, Wheeler followed up with another letter to Everett, enclosing Stern's letter and his own reply to Stern. This letter suggests he had become still more concerned about the difficulty of reconciling Everett's ideas with Bohr's. "Your thesis must receive heavy revision of words and discussion, very little of mathematics, before I can rightfully take the responsibility to recommend it for acceptance. Moreover, I think it will be humanly impossible to come to agreement on all issues unless you and I are in the same place for several weeks, or unless you and Bohr and associates are in the same place for several weeks, or both." Wheeler went on to say to Everett that he was "sure that [your work] will receive discussion of a scope comparable to what has attended Bohm's publications," a backhanded compliment at best. Unsurprisingly, later in the same letter, Wheeler felt the need to assure Everett that he was his "'promoter,' and one actively interested in your reputation and promising future."

Despite Wheeler's insistence that Everett go to Copenhagen immediately, Everett didn't go. Part of the reason was a message from Petersen, informing Everett that Bohr was out of town until the fall—and that Bohr and his circle wanted Everett to do more work before he visited. "I think it would be very helpful to us if, as a background of your criticism, you gave a thorough treatment of the attitude behind the complementary mode of description [i.e., the Copenhagen interpretation] and as clearly as possible stated the points where you think that this approach is insufficient." "While I am doing [that], you might do the same for my work," Everett shot back. "I believe that a number of misunderstandings will evaporate when it has been read more carefully (say 2 or 3 times)." Nonetheless, Everett still wanted to go to Copenhagen, but the schedule Petersen proposed was another problem; Everett was supposed to start a new job at the Weapons Systems Evaluation Group (WSEG) at the Pentagon in less than a month, designing war games and doing operations research on nuclear strikes for the military. Spending time on a detailed reply to Rosenfeld and Bohr, in addition to the new work Wheeler required of Everett for his thesis, all on top of the day-to-day work of his new job, was just not possible—nor was a two-month-long trip to Copenhagen in the fall (as Petersen had proposed) for work totally unrelated to WSEG.

Wheeler couldn't make Everett go to Copenhagen, but he could make him slave over revisions to his PhD thesis. When Wheeler returned to the United States at the end of the summer of 1956, that's exactly what happened. "Hugh and I worked long hours at night in my office to revise the draft," Wheeler recalled later. Wheeler told his friend and colleague Bryce DeWitt that "I sat down with Everett and told him what to say." Finally, six months later, Everett submitted the radically revised and shortened version of his thesis—with a new title, "'Relative State' Formulation of Quantum Mechanics"—which emphasized the mathematical formalism of the universal wave function and downplayed the "splitting" into many worlds. With Wheeler's approval, Everett finally received his PhD in physics from Princeton in April 1957. His shortened thesis was judged "very good" and published in *Reviews of Modern Physics*. It appeared with a short companion paper by Wheeler, where he claimed that Everett's interpretation "does not seek to supplant the [Copenhagen interpretation], but to give a new and independent foundation for that [interpretation]."

Nonetheless, the physicists in Copenhagen still didn't agree with Wheeler. Everett has "some confusion as regards the observational problem," wrote Bohr to Wheeler after Wheeler sent him a copy of Everett's shortened thesis. Bohr, true to form, said he didn't have time to write down all of his ideas on the subject and promised that Petersen would write with a more detailed reply to Everett. Petersen's comments were indeed more extensive, and more devastating. "I think that most of us here [in Copenhagen] look differently upon the problems and don't feel those difficulties in quantum mechanics which your paper sets out to remove," Petersen wrote. "The very idea of observation belongs to the frame of classical concepts." In other words, Petersen and the others in Copenhagen thought that the process of observation had to be classical—that it was impossible in principle to explain observations using quantum physics. Instead, the world must be split in two: the classical and the quantum, and quantum physics could never be used to describe classical events like observations and measurements. But several sentences later in the same letter, Petersen contradicts himself—he says that there are quantum effects in measurement devices, but they can be safely ignored because the devices are large. Astonishingly, Petersen uses this to justify the split between the classical and quantum, the same split that supposedly makes it impossible to give a quantum description of a measurement apparatus in the first place! "There is no arbitrary distinction between the use of classical concepts and the [quantum] formalism since the large mass of the [measurement] apparatus compared with that of the individual atomic objects permits that neglect of quantum effects," Petersen wrote. Everett spotted the contradiction immediately and called Petersen on it in his reply. "You talk of the massiveness of macrosystems allowing one to neglect further quantum effects . . . but never give any justification for this flatly asserted dogma," wrote Everett. "It most certainly does *not* follow from [the Schrödinger equation] which leads to quite strange superposition states [like Schrödinger's cat] even for macrosystems when applied to any measuring processes!" Everett also pointed out that applying the Heisenberg uncertainty principle to measurement devices, as Petersen had done in his reply—and as Bohr had done in his replies to Einstein thirty years earlier—violated the Copenhagen interpretation's strict interdiction against using quantum physics to describe measurements. Yet Petersen and the rest of the Copenhagen camp did not address this point and continued to ignore the criticisms of the Copenhagen interpretation that Everett had laid out in his thesis.

And aside from Bohr's circle in Copenhagen, few people saw Everett's work. Wheeler sent Everett's thesis to a handful of other physicists, such as Schrödinger, Oppenheimer, and Wigner. Many did not even bother to write back. Some who did write back did so simply to argue against it, as Bohr, Petersen, and Stern had. Wheeler tentatively promoted the universal wave function at a 1957 quantum gravity conference in Chapel Hill, but the concept met with a similar fate there. Richard Feynman, who was at the conference (and who had once been a student of Wheeler's himself), simply found Everett's ideas too preposterous to accept. "The concept of a 'universal wave function' has serious difficulties," he told the assembled conference, because it forces you "to believe in the equal reality of an infinity of possible worlds"—a bridge too far, even for a rebel like Feynman.

Not everyone dismissed Everett's new interpretation out of hand. Norbert Weiner, the father of cybernetics and a giant in game theory, was on Wheeler's short mailing list for the thesis; he told Wheeler and Everett that he was "sympathetic to [their] point of view." Wheeler also sent Everett's thesis to Henry Margenau, at Yale, a notable dissenter from the Copenhagen orthodoxy who had been complaining about the measurement problem for years, calling wave function collapse "a mathematical fiction" and a "grotesque claim," and protesting that "measurement should not . . . be given sacramental unction and expected to perform a redemptive act." Unsurprisingly, he approved of Everett's ideas, though he admitted that he had not had time to read the thesis carefully.

Bryce DeWitt, Wheeler's colleague, fellow quantum cosmologist, and co-organizer of the Chapel Hill quantum gravity conference, was skeptical of Everett's thesis at first. "I am afraid that it is precisely at the most crucial point in Everett's argument where many people, including myself, will be unable to swallow your implication. . . . What I am *not* prepared to accept" is the branching of worlds required by Everett's theory, wrote DeWitt to Wheeler. "I can testify to this from personal introspection, as can you. I simply do *not* branch." Wheeler passed DeWitt's reply on to Everett; in his reply, Everett drew an analogy between DeWitt's objection and early objections to the Copernican Sun-centered model of the solar system, tinged with his usual irony:

> One of the basic criticisms leveled against the Copernican theory was that "the mobility of the earth as a real physical fact is incompatible with the common sense interpretation of nature." In other words, as any fool can plainly see the earth doesn't *really* move because we don't experience any motion. However, a theory which involves the motion of the earth is not difficult to swallow if it is a complete enough theory that one can also deduce that no motion will be felt by the earth's inhabitants (as was possible with Newtonian physics). Thus, in order to decide whether or not a theory contradicts our experience, it is necessary to see what the theory itself predicts our experience will be.

> Now in your letter you say, ". . . I simply do *not* branch." I can't resist asking: Do you feel the motion of the earth?

DeWitt, amazed, could only laugh and say "touché." He was fully convinced—and for the time being, Everett's sole disciple.

Everett, finally armed with his PhD, continued to work at WSEG and other parts of the Cold War military-industrial complex for the rest of his life, never returning to academia. It's tempting to conclude that he left academia because of his poor treatment at the hands of Wheeler and Bohr's circle, but the reality is that Everett never wanted to be an academic. He had planned to leave academia long before Wheeler's disastrous visit to Copenhagen; after all, he already had the position at WSEG worked out by the time Wheeler wrote to him after his visit to Bohr's institute. In those letters he sent from Leiden, Wheeler implored Everett to pursue an academic career. But as with his pleas for Everett to visit Copenhagen, Wheeler's wishes for Everett's career fell on deaf ears. Everett was keenly interested in fundamental physics, but it was hardly his only interest, professional or otherwise. Everett cared about fine food, cocktails, cigarettes, travel—and women. He wanted a *Mad Men* lifestyle. A career as an academic couldn't give him that, but a career as a Cold War technocrat could. By 1958, Everett was well on his way to his goal, living in the affluent Virginia suburbs of DC, making enough money to keep him in fine style, all the while conducting affairs on the side while his wife and one-year-old daughter were waiting at home. Meanwhile, his work kept him in contact with the uppermost echelons of the nascent military-industrial complex. His work still involved multiple worlds—but now they were the many worlds of the Cold War operations researcher, gaming out different hypothetical scenarios of nuclear apocalypse. As always, Everett proved adept at his work, coauthoring an influential early study on the disastrous effects of nuclear fallout that made it all the way to President Eisenhower himself. The universal wave function was, to all appearances, far behind him.

But Everett did eventually make it to Copenhagen in March 1959, three years after Wheeler first urged him to go. With his wife Nancy and infant daughter Liz in tow, Everett took a vacation to Europe—and Denmark was their first stop. Everett spent two weeks in Copenhagen and spent a couple of days talking with Bohr, Petersen, Rosenfeld, and several others of Bohr's circle. He also visited Misner, who was working at Bohr's institute at the time (and who had just become engaged to a young Danish woman, Suzanne Kemp, the daughter of one of Bohr's friends). In Misner's recollection, there were no fireworks or showdowns between Bohr and Everett. Bohr was remarkably difficult to talk with, speaking very quietly and constantly interrupting himself and others to relight his pipe. "You didn't get the chance to say something, and then he would relight his pipe seventeen times," Misner recalled later. "He was hard to hear. You had to lean close." And Everett didn't like public speaking, so there was no opportunity for a dramatic public exchange of views. Nor would a lecture by Everett have changed much anyhow. As Misner pointed out, "Bohr's view of quantum mechanics was essentially totally accepted throughout the world by thousands of physicists doing it every day. And to expect that on the basis of a one hour talk by a kid he was going to totally change his viewpoint would be unrealistic." Everett agreed, albeit more colorfully. There is only one existing recording of Everett's voice, a tape of an informal interview by Misner in 1977. When Misner asks him about the Copenhagen visit, Everett and Misner's laughter swallow the recording temporarily, and only a few words of Everett's reply come through: "That was a hell of a—doomed from the beginning."

Bohr's inner circle simply dismissed Everett as a misguided young man. "With regard to Everett, neither I nor even Niels Bohr could have any patience with him, when he visited us in Copenhagen . . . in order to sell the hopelessly wrong ideas he had been encouraged, most unwisely, by Wheeler to develop," wrote Rosenfeld years later. "He was undescribably stupid and could not understand the simplest things in quantum mechanics." Indeed, Bohr himself had elevated his own principle of complementarity to such lofty heights it was a wonder he could see Everett at all from his perch. "On one of those unforgettable strolls during which Bohr would so openly disclose his innermost thoughts,"

Rosenfeld wrote several years later, "Bohr declared, with intense conviction, that he saw the day when complementarity would be taught in the schools and become part of general education; and better than any religion, he added, a sense of complementarity would afford people the guidance they needed." And Bohr was still unwilling to entertain the idea of a fully quantum world. "The brilliant demonstration given by Bohr of the limitation of classical concepts is not accompanied by even the slightest indication of new concepts by which to replace them," complained Vladimir Fock, one of Bohr's disciples. Ultimately, the differences in aims and assumptions between Everett and the Copenhagen camp virtually guaranteed mutual incomprehension and frustration.

After a long day of fruitless discussion with Bohr, Everett walked back to his Copenhagen hotel under the steel-gray afternoon twilight of the Danish sky, leaving quantum physics behind him. Drinking and smoking incessantly at the hotel bar—"He was sloppy and had a cigarette *all the time*," recalled Suzanne Misner—Everett had another brilliant alcohol-fueled idea, totally unrelated to the universal wave function. Jotting notes on hotel stationery while downing several pints of beer, Everett developed a new optimization algorithm for allocating military resources. It was easy to apply and fast to run on the bulky and slow computers of the time. When he returned home, Everett secured a patent for his algorithm, and it ultimately made him and his circle of military-industrial colleagues rich. Everett finally had what he wanted: a never-ending supply of booze, food, and cigarettes. Life was good.

Meanwhile, Everett's quantum ideas languished. Wheeler's prediction did not come to pass: there was even less discussion of Everett's theory than there had been of Bohm's. One of the few times Everett's theory was remembered was at a conference on the foundations of quantum physics in 1962 at Xavier University, organized by Boris Podolsky, the "P" in EPR. This was one of the first conferences to discuss the philosophical underpinnings of quantum theory since the Einstein-Bohr debates thirty years earlier. But unlike those conferences, this conference was decidedly low profile—as Podolsky mentioned in his opening remarks, "We want the participants to feel free to express themselves spontaneously . . . without things getting out in the newspapers." After all, the foundations of quantum physics were settled, and investigating them was at best a waste of time—and, at worst, a sign you were a Communist. Yet there were also a surprising number of big names there: the foundations of quantum physics were still troubling to some. In addition to Podolsky, Rosen (the "R" of EPR) was there, as was Paul Dirac, the father of relativistic quantum field theory; so was Wigner. And, while Bohm was still in exile and couldn't attend, his former student Aharonov was there. The conference attendees spent three days discussing the measurement problem, the inconsistencies of the Copenhagen interpretation, and alternatives like Bohm's pilot-wave theory. At one point on the first day, discussing the trickiness of wave function collapse, someone pointed out that Everett had a theory in which there was no collapse at all. The organizers decided to extend a belated invitation to Everett, and he flew up to Xavier from DC for the second day of the conference. The assembled luminaries interrogated Everett. "It looks like we would have a non-denumerable infinity of worlds," commented Podolsky. "Yes," Everett replied. At this point, one of the attendees, Wendell Furry, voiced his disbelief at the number of worlds involved. "I can think of various alternative Furrys doing different things, but I cannot think of a non-denumerable [infinity] of alternative Furrys." The conference continued, and Everett's ideas were discussed with genuine interest for the remainder. But other than the small group of attendees, nobody knew what had happened: the proceedings remained publicly unavailable for the next forty years (by which point everyone who had been present, aside from Aharonov and one or two others, was dead).

Everett's theory slipped into a deep obscurity for the next decade, provoking almost no immediate reaction—certainly not the kind of fierce backlash that Bohm's papers had prompted. The universal wave function was simply ignored for years, while Everett himself hid in plain sight as a cold warrior. Occasionally the idea would come up in conversation with one of his physicist–turned–war-gamer colleagues; when it did, Everett was reluctant to talk about it, and he never took the debate to a larger arena. He was a dark jester, amused by paradox, perverse arguments, and private jokes. The wider stage of academia held no allure for him—and, anyhow, he didn't like public speaking. He felt no particular need to remedy the physics community's misguided manner of thinking about quantum physics. That task would take a different sort of person: not just an academic but someone with a stronger sense of moral obligation and integrity, someone who didn't mind voicing unpopular opinions on a large stage, someone who could speak and write compellingly and who understood exactly how to approach the problems at hand in a way that other physicists would pay attention to. It took someone who had always known Copenhagen was rotten, who had seen David Bohm do the impossible. It took a person like John Stewart Bell.

7

The Most Profound Discovery of Science

John and Mary Bell arrived to a nation in mourning. The day before they landed in California, President Kennedy had been shot and killed in Dallas. "It was the worst possible moment to have come," John said later. John and Mary, both specialists in the physics of particle accelerators, had been invited to spend a year as academic guests at the Stanford Linear Accelerator Center (SLAC), half a world away from their usual professional home in Switzerland. Despite the tragedy, they set to work. "Mary was quickly integrated into the accelerator division," John recalled, "and I into the [particle] theory group."

John used the change of scenery as an opportunity to explore scientific ideas that had been weighing on his mind for over a decade. Ever since reading David Bohm's papers in 1952, Bell had known there was something wrong with von Neumann's famous proof that purportedly showed theories like Bohm's pilot-wave interpretation couldn't work. Yet other physicists still regularly cited von Neumann as justification for ignoring Bohm's ideas. Shortly before Bell left Switzerland, he had spoken with Josef Jauch, a physicist at the University of Geneva who had recently published a "strengthened" version of von Neumann's proof. Jauch defended his own ideas and pointed Bell toward another proof that also supposedly ruled out Bohm's version of quantum physics. "For me, that was like a red light to a bull," Bell said. "I wanted to show that Jauch was wrong. We had gotten into some quite intense discussions." Now, surrounded by the unfamiliar and stark California landscape, Bell set about doing just that. In the process, he ended up discovering a remarkable truth about the quantum world—and ultimately loosened the grip of the Copenhagen interpretation on the collective psyche of physics.

John Stewart Bell was born on June 28, 1928, the second of four children in a working-class Protestant family in Belfast, Northern Ireland. By his own account, he came from a long line of "carpenters, blacksmiths, laborers, farm workers, and horse dealers." Bell was the first in his family to attend high school—his father had left school at the age of eight, and his siblings all found jobs by fourteen. By sixteen, Bell had graduated from the least expensive high school in the area, but the local university, Queens, wouldn't admit anyone younger than seventeen. So he went looking for work. "I applied to be office boy in a small factory, some starting job at the BBC—things like that. But I didn't get any of the jobs I applied for," Bell recalled years later. Ultimately, he found work as a lab assistant in the physics department at the university. "It was a tremendous thing for me, because there I met, already, my future professors. They were very kind to me. They gave me books to read, and in fact, I did the first year of my college physics when I was cleaning out the lab and setting out the wires for the students."

Toward the end of his formal studies at Queens, Bell had his first encounter with the mathematics of quantum physics, and the Copenhagen interpretation that invariably accompanied it. He was not happy with what he found. "You learn about the periodic table of the elements—all the practical aspects of the theory," Bell recalled. "Then the puzzles start." Bell's instructors and textbooks were vague about the nature of the wave function itself. They were never clear "whether it [the wave function] was something real or some kind of bookkeeping operation." And if the wave function was just a bookkeeping device, just information, then whose information was it? And if there really was no quantum world, as Bohr had insisted, then what was that information about? Bell even got into an argument with one of his instructors. "I was getting very heated and accusing him, more or less, of dishonesty. He was getting heated too and said 'You're going too far.' But I was very engaged and angry that we couldn't get all that clear."

Frustrated, Bell started reading works by the founders of quantum physics, in hopes of clearing up his confusion. What he found there was not particularly helpful. Bohr was unclear about where the division between the quantum and the classical world lay. "[Bohr] seems to have been extraordinarily insensitive to the fact that we have this beautiful mathematics, and we don't know which part of the world it should be applied to," said Bell. "Bohr seemed to think he had solved this question. I could not find the solution in his writings. But there was no doubt that he was convinced that he had solved the problem and, in so doing, had not only contributed to atomic physics, but to epistemology, to philosophy, to humanity in general." And Heisenberg's writing was "perfectly obscure" to Bell. The measurement problem was clearly a serious issue, but the Copenhagen interpretation treated it like it was trivial. Bell wanted rigor and honesty; instead, his deep questions were waved away with insubstantial answers.

Then Bell encountered von Neumann's proof—really Max Born's account of von Neumann's proof, since Bell couldn't read German. "I was very impressed that somebody—von Neumann—had actually *proved* that you couldn't interpret quantum mechanics" in some other way, Bell said. So he moved on. "For me, it was a big risk that I would get hung up on these questions once I learned about them. . . . I rather deliberately walked away from them," Bell recalled. "I had the feeling then that getting involved in these questions so early might be a hole I wouldn't get out of."

After graduating from Queens, Bell found a job with the Atomic Energy Research Establishment in Harwell, England, working on nuclear reactors with Klaus Fuchs, a veteran of the Manhattan Project. But several months after Bell arrived, Fuchs confessed to passing atomic secrets to the Soviets, and Bell was reassigned to the accelerator physics division. While there, he met Mary Ross, a fellow physicist and his future wife. And it was while John and Mary were working at Harwell that John encountered Bohm's pilot-wave papers in 1952, shortly after they were first published.

Figure 7.1. John Bell in Harwell, c. 1952.

Bell was shocked by the chilly reception that Bohm's ideas received. "For twenty-five years people were saying that [alternatives to Copenhagen] were impossible. After Bohm did it, some of the same people said that now it was trivial. They did a fantastic somersault." After reading Bohm's papers, Bell immediately recognized that von Neumann's proof must be wrong, but it still wasn't available in English. So he found a colleague at Harwell who spoke German, Franz Mandl. "Franz . . . told me something of what von Neumann was saying," Bell recalled later. "I already felt that I saw what von Neumann's unreasonable axiom was."

But von Neumann's proof wasn't published in English for another three years, and by the time it was, Bell had started entirely different work for his PhD. When Bell arrived at graduate school, his PhD advisor, Rudolf Peierls, asked him to give a talk on what he'd been working on lately. Bell said he could talk about either accelerator physics or interpretations of quantum physics. Peierls told him he would much prefer Bell give his talk about accelerators. Bell complied, and for the next few years he stayed away from questions about the meaning of quantum physics.

Several years later, Bell met Bohr himself at CERN (the European Laboratory for Particle Physics, best known today as the home of the

Large Hadron Collider), in Geneva, Switzerland. The Bells had just started working there, and Bohr was among the many luminaries who arrived for the inauguration of the then-new research center. Bell ran into Bohr in an elevator and wasn't sure how to talk to the living legend. "I didn't have the nerve to say, 'I think your Copenhagen interpretation is lousy,'" he recalled later. "Besides, the lift ride wasn't very long. Now, if the lift had gotten stuck between floors, that would have made my day! In which way, I don't know."

When the Bells arrived for their sabbatical in California three years later, Bell took the opportunity away from his usual work at CERN to finally figure out where von Neumann had gone wrong, and to show up Jauch. He found that von Neumann's revered proof, consistently invoked to defend against any heresy, was hardly a proof of anything at all. "The von Neumann proof, if you actually come to grips with it, falls apart in your hands!" said Bell. "There is *nothing* to it. It's not just flawed, it's *silly*!" The great John von Neumann, as it turned out, had simply made a mistake—he made assumptions in his proof that were entirely unwarranted. "When you translate [von Neumann's assumptions] into terms of physical disposition, they're nonsense. . . . The proof of von Neumann is not merely false but *foolish*!"

Bell didn't merely show von Neumann and Jauch were wrong—he left a new proof in place of the old ones. Von Neumann's proof and its ilk (including Jauch's "strengthened" proof and the proof Jauch had mentioned to Bell, by Andrew Gleason) purported to rule out any interpretation of quantum physics that used so-called hidden variables. A hidden-variables interpretation assigns definite locations or other properties to quantum objects before they are observed, even if those properties can't be calculated from the theory itself. These properties go unseen in the mathematics of quantum physics, hence "hidden" variables. Bohm's pilot-wave interpretation is a prime example of such a theory: in Bohm's world, particles always have positions, even though those positions are largely hidden from view and can't be calculated from Schrödinger's equation. The proofs of von Neumann, Jauch, and Gleason all suggested that this kind of scheme must be impossible—yet Bohm's pilot-wave interpretation clearly worked, as Bell knew quite well. Something must be wrong, and Bell thought he knew what it was. He meticulously disassembled the no-hidden-variables proofs, delicately prodding at their component pieces until he found one that easily snapped in half—an unjustified assumption at the foundation. Turning this assumption on its head, Bell showed that the purported "no-hidden-variables" proofs collectively suggested something else entirely, something the original creators of these proofs did not intend or fully understand. Specifically, Bell found that a hidden-variables theory could avoid the traps laid by these proofs if it had a rather peculiar property, later dubbed *contextuality*.

Contextuality means that the outcome of a measurement on a quantum system depends on the other things you measure about that system at the same time. In other words, if you measure a property of a thing, the outcome of your measurement can depend on what other stuff you measure about that thing at the same time. In a contextual world, if you measure the energy of a neutron along with its momentum, you'll get an answer about the neutron's energy—but if you had measured the energy along with the location, you could have gotten a completely different answer about the neutron's energy, simply by virtue of the context in which you made the energy measurement.

To get a better handle on contextuality, let's forget about neutrons and instead talk about something larger and more familiar: a roulette wheel. Imagine your friend Flo is at a roulette wheel in a casino, and you're talking with her over the phone. You can't see the wheel, but you can ask her questions about how the ball landed in the wheel after any particular spin. You can ask whether the ball landed on an even or odd number, a high or low number, or a red or black number. (Roulette wheels are constructed in such a way that half the numbers are red and half are black, but they're not split along evens-odds or highs-lows—half the highs and half the lows are red, and the same is true for evens and odds. See Figure 7.2.) But Flo is being strangely reticent to tell you what's happening at the casino: for any spin of the wheel, she'll only tell you the answers to two of your questions, not all three. Normally, you wouldn't think this mattered: no matter what Flo tells you, the ball landed in a particular slot on each spin. Therefore, even though the ball's actual

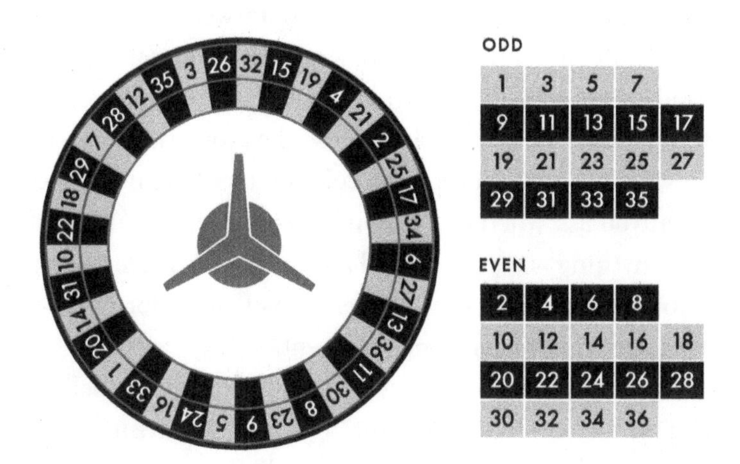

Figure 7.2. A fair roulette wheel. The numbers are split evenly between high-low, black-red, and even-odd, with no 0 or 00 slots.

state is hidden from you, the answers to all three of your questions are already set once the ball stops moving. If the ball lands on 34, then the ball landed on a high, even, red number, even if Flo will tell you only two of those three things.

But if the roulette wheel is contextual, all that goes out the window. For a contextual roulette wheel, the answer to the question, "Is the ball on a red number?" depends on what else you ask at the same time. Say you ask whether the ball is on a red number and whether it's on an even number after a particular spin of the wheel. As it turns out, the answer to both your questions is yes. But if you had asked different questions after the same spin of the wheel—if you had instead asked whether the ball is on a red number and whether it's on a high number—both answers would have been no. Somehow, the answer to the question, "Is the ball on a red number?" is actually affected by which other question you ask! This is contextuality: the answer to a question depends on its context of surrounding questions asked at the same time. In demolishing the no-hidden-variables proofs, Bell also demonstrated that quantum physics describes a contextual world.

At first blush, the fact that quantum physics is contextual seems to support the Copenhagen interpretation, or something like it. If the answers to questions depend on the other questions asked along with them, doesn't that suggest that there are no answers to questions until they're asked? After all, if the quantum world is contextual, it can't really be like a roulette wheel—there can't be a ball that's on a particular number, passively waiting for us to look, because the properties of the number depend on what we ask about it. The color of a roulette number doesn't change when you ask whether the number is odd; 34 is red, whether or not you ask anything at all about it. Thus, in the quantum world, there can't be a roulette ball until you look for it. As Pascual Jordan said, "We ourselves produce the results of measurement."

Despite the appeal of this Copenhagen-ish argument, Bell handily dismissed it in a "judo-like maneuver" by citing Bohr himself. In the same paper where Bell established contextuality as a key feature of the quantum world, he also pointed out that contextuality shouldn't be a surprise, because, as Bohr said, it's impossible to draw "any sharp distinction between the behaviour of atomic objects and [their] interaction with the measuring instruments." You can't look at the quantum world without altering it—but that doesn't mean the quantum world isn't there before you look. Quite the opposite: if it weren't there, you wouldn't be able to alter it by looking! A contextual roulette wheel can exist—it's just that the ball's location will change when you look at it in different ways, because you can't separate the behavior of the ball from its interaction with you when you look at it. That doesn't mean the ball doesn't exist or that it doesn't have a location before you look; it just means the ball is kind of jumpy and sensitive, moving around dramatically at even the slightest disturbance. The hidden variables in Bohm's pilot-wave interpretation behave in exactly this way. Particles, according to Bohm, always have positions—but those positions can be dramatically altered by small disturbances and changes in experimental setups. Ask a slightly different set of questions to an electron, in Bohm's world, and you can get an enormously different set of answers—but the electron has a definite position all the while. And because Bohm's theory is contextual, it evades all of the proofs that supposedly rule it out. "What is proved by impossibility proofs," concluded Bell, "is lack of imagination."

Despite his definitive demonstration that Bohm's theory was not impossible, Bell was still concerned about the strangest feature of pilot-wave theory: it was "hideously nonlocal." "Terrible things happened in the Bohm theory," said Bell. "For example, the [paths of] particles were instantaneously changed when anyone moved a magnet anywhere in the universe." Was the nonlocality in Bohm's theory an essential feature of quantum physics? Bell asked this question at the conclusion of his paper demolishing von Neumann's proof, leaving it unanswered as a possible avenue for future work.

Nobody saw Bell's question about nonlocality for a long while; his paper demolishing von Neumann's proof spent two years sitting in an editor's desk due to a series of clerical errors. But Bell couldn't leave the question alone—he wanted to know the answer now. For his next project, he set about finding it. "I knew, of course, that the Einstein-Podolsky-Rosen setup was the critical one [for nonlocality], because it led to distant correlations," Bell recalled later. "So I explicitly set out to see if in some simple Einstein-Podolsky-Rosen situation I could devise a little model that would complete the quantum-mechanical picture and leave everything local."

In his work, Bell used a simplified version of the EPR setup, devised by Bohm in the textbook that he'd written just before developing his pilot-wave interpretation. Bohm's version of the EPR experiment made the whole thing easier for Bell to play around with in his head. Instead of two particles colliding and flying away from each other with entangled momentum, Bohm's version of EPR involved photons with entangled polarization.

Polarization is a property of light—light is an electromagnetic wave, and the polarization is the direction that the wave is doing its waving in. But, for our purposes, all that really matters is that it's directional: polarization is sort of like a little arrow that each photon carries with it that can point in different directions. But it's not quite that simple. For one thing, we can't actually tell what direction a photon's polarization arrow is pointing. All we can do is measure a photon's polarization along one particular axis at a time in a somewhat indirect manner, by shooting it at a polarizer (like a lens in a pair of polarized sunglasses). When a photon hits a polarizer, it either passes through or gets blocked;

the closer the photon's polarization is to the polarizer's axis, the more likely it is to pass through.

In Bohm's version of EPR, two photons with entangled polarization go flying off in opposite directions from a common source, toward two polarizers. The two polarizers are set to measure polarization along the same axis. Because the photons have entangled polarization, when they arrive at the polarizers, they will always do the same thing—they'll both either be blocked by the polarizers or pass through them. It doesn't matter what axis the polarizers are set to: as long as the polarizers' axes match, a pair of entangled photons will always do the same thing at both of them. And, crucially, it doesn't matter how far apart the polarizers are either: the two photons will always pass through together or be blocked together, regardless of distance.

And this is just as quantum physics says it should be. The single wave function shared by the two entangled photons guarantees that they will always behave in the same manner when encountering two polarizers with matching axes. But that wave function does not specify what they will do—merely that they will do the same thing.

Now, Einstein's forced choice, the one he had feared was obscured in the EPR paper itself, comes out in vivid relief. Assuming that nature is local, then the only explanation for the perfectly synchronized long-distance choreography of the entangled photons is that they have a prearranged dance routine, one that they agreed upon before flying off from their common source. But the wave function shared by the entangled photons says nothing about any kind of prearrangement. It just guarantees that the photons will always do the same thing at polarizers with the same settings, that they will be perfectly correlated. Therefore, if nature is local, the wave function is not everything—there must be hidden variables. So either quantum physics is incomplete, or nature is nonlocal. We cannot have both locality and completeness in quantum physics. This is Einstein's forced choice, the heart of the EPR argument.

Bell toyed with this EPR-Bohm thought experiment, trying to construct a model that would maintain all the predicted results of quantum physics while remaining purely local. "Everything I tried didn't work," Bell said. "I began to feel that it very likely couldn't be done. Then I constructed an impossibility proof."

Einstein had proven that quantum physics must choose between locality and completeness, but Bell's impossibility proof showed that the choice is actually between locality and *correctness*. Starting from the assumption that nature is local, Bell derived an inequality, a mathematical condition that any local theory of nature has to meet. Then Bell shrewdly altered Bohm's version of the EPR thought experiment to create a situation where the predictions of quantum physics violate that inequality.

Bell's stroke of brilliance was to consider imperfection, rather than perfection. After all, the perfect correlations in the EPR-Bohm setup are easily compatible with locality—the photons could be sharing hidden instruction sets at their common origin. But if you rotate the axis of one of the polarizers, quantum physics predicts that pairs of entangled photons arriving at the polarizers will no longer behave in exactly the same way every time. And Bell showed that the imperfect correlations predicted by quantum physics were too strong for any local theory of nature to be able to account for them. So either the predictions of quantum physics are wrong and nature can be local, or quantum physics is right and "spooky action at a distance" is real. Bell had discovered a remarkably profound and counterintuitive truth about the world.

Bell had also shown that there was an experimental test that could decide between the two options. All someone had to do was actually construct and perform Bell's modified version of the EPR thought experiment, or another experiment along those lines involving entangled particles. If the results showed that Bell's inequality was violated, quantum physics was safe but nature was nonlocal; if his inequality held, then quantum physics was wrong but nature could be local. Bell's impossibility proof had taken the question of nonlocality out of the realm of debate and turned it into an experimental challenge. This proof, now known as Bell's theorem, has rightly been called "the most profound discovery of science."

Bell's result is both unexpected and problematic. Locality is a basic assumption of physics, and indeed all of science. Without locality, it would be difficult to perform any controlled experiments at all—no matter how well you control the surroundings of your experiment, there would always be the possibility of a far-distant, instantaneous influence affecting your results. Einstein, in particular, emphasized that locality must be a core principle of science, not given up unless absolutely necessary, for exactly this reason. "Without such an assumption of the mutually independent existence (the 'being-thus') of spatially-distant things, an assumption which originates in everyday thought, physical thought in the sense familiar to us would not be possible," he wrote. "Nor does one see how physical laws could be formulated and tested without such a clean separation. . . . The complete suspension of this basic principle would make impossible the idea of the existence of (quasi-) closed systems and, thereby, the establishment of empirically testable laws in the sense familiar to us."

Even putting Einstein's philosophical concerns aside, Einstein's scientific work made it clear that locality was a key feature of the world. According to Einstein's special relativity, physical objects can't be pushed up to or past the speed of light, on pain of a whole host of paradoxes involving infinite energy. You might try to get around this by finding something that is already moving faster than light—but no such object has ever been found. Indeed, relativistic particle physics states that such objects would be spectacularly unstable, precluded from existence by their own special infinite-energy paradoxes. And even if you somehow get around these problems and manage to send a signal faster than light, you still run the risk of paradox. Relativity dictates that merely sending a faster-than-light signal would immediately make it possible to construct a "tachyonic antitelephone" that would let you send messages back in time.

But Bell's theorem doesn't mean we call ourselves yesterday or send a DeLorean to 1955. Bell and others later proved that it was impossible to use quantum entanglement for faster-than-light signaling. And the specific sort of nonlocality displayed by entangled particles is so delicate and subtle, appearing only under such specific conditions, that it doesn't pose the sort of existential threat to science itself that Einstein had feared. But the fact remains that in a world where special relativity has easily passed every single test we've thrown at it—a world that appears local—the specter of nonlocality raised by Bell's theorem is profoundly disturbing. If the quantum prediction for Bell's experiment is correct, and Bell's inequality is violated, then something is nonlocal, and locality is merely an illusion. That suggests a need for a radical revision of our conception of space and time, far beyond Einstein's relativity. Any story of the world that could incorporate a violation of Bell's inequality would have to be truly strange.

How could Bell possibly prove something so unexpected and so vast in its implications? To fully understand his proof, we'll need more than a roulette wheel—we'll need a whole casino. (If you aren't interested in following the details of the proof, feel free to skip the next section entirely—it won't affect your ability to understand any of the rest of this book. But following through the argument in the next section will give you a greater understanding of how Bell proved what he did.)

A new casino has opened up in the small town of Bellville, California, in the sparsely populated northeastern corner of the state—and it's owned by Ronnie the Bear, who is suspected of having mob connections. Fatima and Gillian, two inspectors from the California Gaming Bureau, head up to Bellville to check out the casino before it opens, because they know Ronnie is probably up to something.

Ronnie's casino floor has an overcomplicated roulette setup, possibly to impress the inspectors. In the center of the room is a large machine, with a chute extending from each side to the roulette tables at either end of the floor. At each of the two roulette tables, there are three roulette wheels, with a smaller spinning dial in the center. In accordance with state law, the roulette wheels only have alternating squares of red and black on them, not numbers—roulette wheels with numbers on them are illegal in the state of California (Figure 7.3). Once Fatima and Gillian are each seated at one of the tables, Ronnie presses a button on the machine, and a roulette ball appears in each of the chutes, rolling toward the tables. The inspectors spin the center dial while the balls are

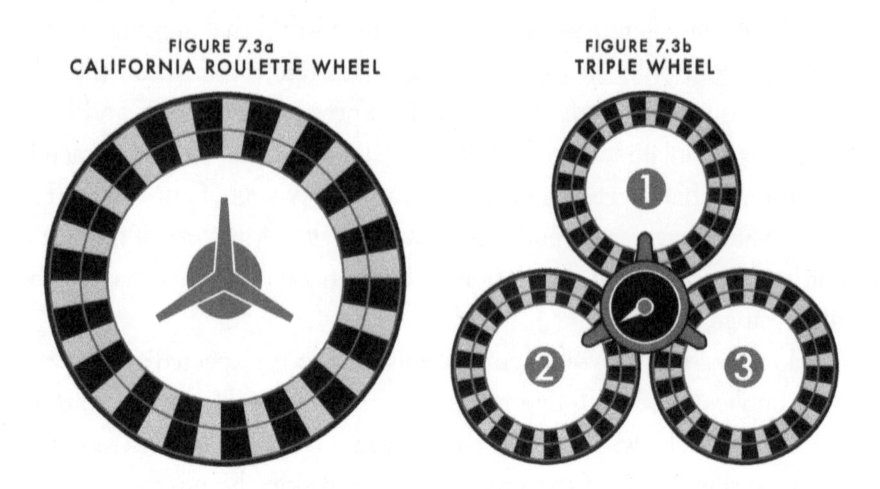

Figure 7.3. (a) A California roulette wheel. (b) The "triple wheel" at Ronnie's casino, with the selector dial in the center.

en route, and each roulette ball lands automatically in whichever wheel is selected, eventually settling on a red or black square (Figure 7.4).

Gillian and Fatima do this many times over, to inspect the properties of these wheels thoroughly, and they take detailed notes on the outcomes: which wheel was used and what color came up on each run. Black and red show up in roughly equal proportions, and after several dozen runs, the inspectors go back to their office to compare their notes.

The inspectors find that each table's roulette wheels really do seem completely random—red and black each came up almost exactly half the time. But there are strange correlations between Fatima's notes and Gillian's notes. Each time the small dials at the two tables selected the same wheel number, the two roulette balls landed on the same color. For example, on run 87, both dials pointed to wheel 2—and both roulette balls

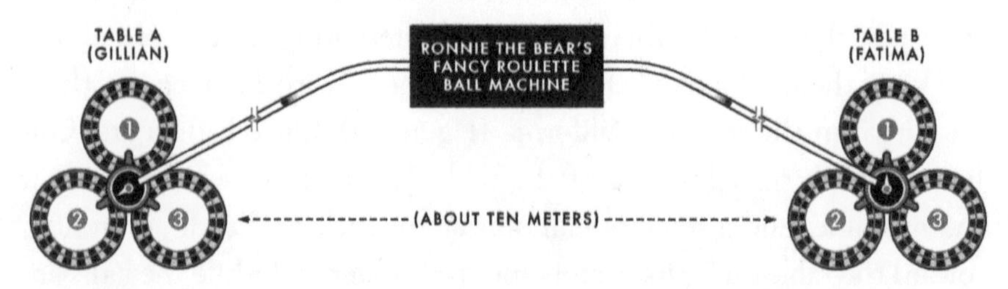

Figure 7.4. The roulette tables at Ronnie the Bear's casino in Bellville.

	GILLIAN		FATIMA	
RUN	WHEEL	COLOR	WHEEL	COLOR
83	3	RED	3	RED
84	3	RED	1	BLACK
85	1	BLACK	1	BLACK
86	3	BLACK	2	RED
87	2	RED	2	RED
88	1	BLACK	2	RED
89	1	BLACK	3	BLACK

Figure 7.5. A sample of Gillian's notes and Fatima's notes, compared.

landed on red (Figure 7.5). The inspectors conclude that the balls are preprogrammed in the giant roulette-ball machine to ensure they always land on the same color when they go to matching wheels.

But then Fatima notices a second pattern in the results. When Fatima and Gillian didn't use corresponding wheels, they only got the same outcome 25 percent of the time. Fatima doesn't think that this can be right. She writes out the eight different possible instruction sets that the roulette balls can be carrying (Figure 7.6).

A ball with the first of these instruction sets, "Red Red Red," would always land in a red slot no matter what wheel it was placed in. A ball with the second set, "Red Red Black," would always land in a red slot if it were placed into wheel 1 or 2, but would always come up black in wheel

WHEEL 1	WHEEL 2	WHEEL 3
RED	RED	RED
RED	RED	BLACK
RED	BLACK	RED
RED	BLACK	BLACK
BLACK	RED	RED
BLACK	RED	BLACK
BLACK	BLACK	RED
BLACK	BLACK	BLACK

Figure 7.6. The only possible instruction sets for the roulette balls.

3, and so on. Fatima points out that no matter which of these instruction sets the roulette balls are sharing, her results and Gillian's results should match more than 25 percent of the time when they don't use corresponding roulette wheels:

- If the two balls are sharing instruction sets Red Red Red or Black Black Black, then they'll match 100 percent of the time, even when they end up in differently numbered wheels.
- If the two balls are sharing one of the other instruction sets, then when Fatima and Gillian are using differently numbered wheels, the balls should land on the same color one-third (33 percent) of the time. For example, say the instruction set is Black Red Red. Then Fatima and Gillian will get different colors if they're using wheel combinations 1&2, 2&1, 1&3, or 3&1. But they'll get the same color if they're using wheel combinations 2&3 or 3&2—two out of the six total possibilities, or one-third. The other instruction sets (other than Black Black Black and Red Red Red) work the same way.

Therefore, when Fatima and Gillian aren't using corresponding roulette wheels, they should be getting the same color at least 33 percent of the time, because there are no instruction sets that would make such matches less common than that. And yet, they only match 25 percent of the time under those circumstances. The inspectors are forced to conclude that the roulette balls are not sharing instruction sets. Yet the roulette balls always land on matching colors when Gillian and Fatima are using corresponding roulette wheels, so they clearly do have some kind of coordination going on—that's what led the inspectors to suspect the balls were sharing instruction sets to begin with. Therefore, to account for these results, the roulette balls must be sending signals to each other after they know which wheel they're arriving at.

The preceding section is a proof of Bell's theorem, thinly disguised. The pairs of roulette balls are pairs of photons with entangled polarizations. The roulette wheels are polarizers that measure polarization along three different directions, selected randomly while the photons are in flight toward the polarizers. And Bell's theorem is the proof embedded in the story, the one that Fatima figured out. If your roulette balls really behave that way, there's something strange going on, and it can't be accounted for by assuming the roulette balls have hidden instructions—hidden variables—that they carry with them from the moment they separate. And entangled photons really do behave this way, so something very strange must be going on in quantum physics. But what, exactly, did Bell prove? To understand this, let's take a closer look at what happened at Ronnie's casino.

We started out with the assumption that roulette balls can't magically communicate with each other instantaneously across long distances (though we never explicitly stated this until the end). In other words, we started out with the assumption of locality. This led us to the idea that there must be hidden instruction sets in the roulette balls themselves, because that was the only way to account for the perfectly matched outcomes when Gillian and Fatima used the same roulette wheels. But the strange correlations in outcomes when Gillian didn't use the same wheel as Fatima ruled out the possibility of hidden variables. Therefore, something must be wrong with our assumption: locality must be violated. In the case of Ronnie's casino, the roulette balls could still be communicating by radio, of course. But, in real experiments, the "roulette balls" are photons, traveling at the speed of light, and the "roulette wheels" are polarizers that can be very far apart—in some experiments, as far as hundreds of kilometers. No light-speed signal sent after one photon arrives at a polarizer could possibly reach the other one before it too reaches a polarizer and makes its decision about what to do. In short, the results of real experiments with entangled photons mean that something, some influence, is going faster than light. Entanglement is not just an artifact of the mathematics of quantum physics: it's a real phenomenon, an actual instantaneous connection between far-distant objects.

This is an astonishing result. How can it be right? What story of the world could account for this? The most obvious answer is: a nonlocal one. Bohm's pilot-wave interpretation of quantum physics has no trouble at all with Bell's theorem, because Bohm's theory is explicitly nonlocal. This turns one of the apparent weaknesses of pilot-wave theory—its instantaneous connections between particles separated by huge distances—into a strength. Bell's theorem strongly suggests that quantum physics must be nonlocal; pilot-wave theory merely makes this strange quantum behavior so obvious that we can't ignore it.

But the cost of nonlocality is high. Relativity is one of the best-tested and most solid foundations of modern physics; nonlocality would put it at risk. Is there any other way out of Bell's theorem? Is locality really the only assumption? Well, Gillian and Fatima did assume that their notes on what happened at the casino were a complete record of everything that happened there. Specifically, they assumed that there was only one outcome for each spin of a roulette wheel, the outcome they wrote in their notes. If there is somehow more than one outcome for each spin of a roulette wheel, each time a photon hits a polarizer, Bell's proof falls apart. And this is exactly what happens in Everett's many-worlds interpretation. According to Everett, each spin of a roulette wheel leads to both possible outcomes, red and black, branching in the many worlds of the universal wave function. So Bell's theorem does suggest that the strangest piece of Everett's scheme could be a necessary feature of the world if we don't want to abandon locality.

Now we have two assumptions: locality and living in a single universe. One of them must be wrong, since Bell's inequality is violated in real experiments. Is that the choice forced on us by Bell's theorem? Or is there some weird third thing? It's possible that the roulette wheel selector wasn't truly random, that the roulette balls knew what roulette wheels they would enter in advance. This kind of conspiracy between the roulette wheels and the roulette balls could account for the results that Fatima and Gillian saw. But translate that into real physics, and it starts to look problematic: a conspiracy between photons and polarizers sounds far-fetched, to say the least. What if a human experimenter is deliberately selecting which polarizer to use each time? How could the photons know in advance? We like to think that we are freely choosing the conditions of our experiments—and, even if that's an illusion, the idea that the photons somehow have all our actions pre-encoded within them is difficult to imagine. But technically, this kind of "superdeterminism" may be a logically possible option for escaping Bell's theorem, and a small handful of physicists do work on fleshing out this kind of theory (though there are concerns about whether such a vast natural "conspiracy" would make it impossible to do science in the first place).

Is there anything else? Are there any other ways out of Bell's remarkable theorem? Many books and papers claim that there is another assumption in Bell's proof, that of hidden variables. Don't assume that there are any hidden instructions in the roulette balls at all, the argument goes, and Bell's theorem has no force. But this isn't correct. We didn't assume that there were any instructions in the roulette balls, at least not at the start of the proof. Instead, we merely assumed locality, and this inevitably led us to the conclusion that there must be instructions in the roulette balls, in order to account for the perfect match between Gillian's results and Fatima's when they were both using the same roulette wheels. If this sounds familiar, that's because it's simply the EPR argument. If pairs of roulette balls always land on the same color, then either they're sharing instruction sets from the start, or they're somehow communicating faster than light once they reach their destinations. There was no assumption of hidden variables—there was merely an assumption of locality, and the behavior of the roulette balls forced us to consider hidden variables. "It is remarkably difficult to get this point across: that [hidden variables are] not a *presupposition* of the analysis," Bell complained fifteen years after his theorem was first published. "My own first paper on this subject [Bell's theorem] starts with a summary of the EPR argument *from locality* to deterministic hidden variables. But the commentators have almost universally reported that it begins with deterministic hidden variables."

Another, related claim is that Bell's theorem assumes *realism* of some kind. This is an especially popular claim among supporters of the Copenhagen interpretation. If you don't assume that the quantum world has real properties—or if you don't assume that there's a quantum world at all—then, they claim, Bell's theorem doesn't work. This is also incorrect. The problem here is that phrase "realism of some kind." What, exactly, is meant by "realism"? Some physicists claim that Bell's theorem assumes the idea that quantum objects have well-defined properties before they're measured, and that this is what is meant by "realism." But this is simply not true, as previously stated—there's no assumption of pre-existing properties (i.e., hidden variables) in Bell's theorem at all. That idea comes out of the assumption of locality, just as it does in the EPR argument. Others claim that the form of realism assumed by Bell's theorem is the very idea that anything at all exists independently of observation. Denying this, they claim, is the true insight of the Copenhagen interpretation, and this is what allows the Copenhagen interpretation to remain local despite Bell's ingenious proof. Ignoring the problem of solipsism that this introduces into physics—whose observations make things real?—another problem arises. Without the assumption that reality exists independently of observation in some form, the idea of locality itself is meaningless. How can it mean anything to talk about effects moving faster than light from one place to another when neither the objects nor their locations exist at all? Denying realism to break Bell's proof invariably breaks the concept of locality as well—a Pyrrhic victory for the antirealists determined to keep physics local at all costs. As Bell himself said, "I don't know any conception of locality which works with quantum mechanics. So I think we're stuck with nonlocality."

Even quantum physics itself isn't an assumption in Bell's proof. After all, Fatima didn't have to appeal to quantum physics when explaining the impossibility of Ronnie's roulette balls. Bell's theorem is merely a claim about the world, independent of quantum physics. If the world works in a particular way—if Ronnie's roulette balls, or entangled photons, obey the statistics observed at the casino—then either locality is broken or nature works something like the many-worlds interpretation (or, maybe, nature is conspiratorial and superdeterministic). The only place quantum physics enters into the argument is that according to the mathematics of quantum theory, entangled photons will behave in the way Ronnie's roulette balls did. Thus, if quantum physics is right, or at least right about this particular kind of situation, we have to give up on locality or on living in a singular universe (or perhaps both).

In short, Bell's theorem really leaves only three unequivocal possibilities: either nature is nonlocal in some way, or we live in branching multiple worlds despite appearances to the contrary, or quantum physics gives incorrect predictions about certain experimental setups. No matter the outcome, Bell's work presents a threat to the Copenhagen interpretation. Perhaps because it contradicts the widely received wisdom, physicists have long had particular difficulty understanding the true implications of Bell's theorem—in fact, the misunderstandings began before it was even published.

Once Bell had written up his revolutionary theorem, he wasn't sure which research journal he should send it to. The obvious choice was *Physical Review,* the premier physics journal, where the EPR paper, Bohr's reply, and Bohm's pilot-wave papers had all appeared over the previous thirty years. Nearly every physicist in the world read *Physical Review;* it was the journal of record. But the journal charged a fee for publishing papers, one usually paid for by the author's institution. As a visitor to SLAC, Bell didn't want to impose by asking his hosts to pay such a fee, especially for such an unorthodox paper. "I was embarrassed to ask them to pay for my article," said Bell. Instead, Bell published his paper in *Physics,* a brand-new and still-obscure journal.

Physics—more properly, *Physics Physique Fizika: An International Journal for Selected Articles Which Deserve the Special Attention of Physicists in All Fields*—was an unusual journal. It was founded by two prominent solid-state physicists, Philip Anderson (who would go on to win a Nobel Prize in 1977) and Bernd Matthias. Anderson and Matthias wanted their journal to be a physicist's version of "a journal of literature and general information, such as *Harper's,*" covering material from all subfields of physics, as the subtitle of the journal suggested. And in the style of *Harper's,* Anderson and Matthias paid their authors a (very small) fee, rather than having them pay to be published. This was perfect for Bell. "I thought that I would submit my paper to *Physics,* and that would be a good way to avoid embarrassment."

When Anderson received Bell's paper, he was impressed—but not for the reasons Bell would have hoped. "I was pleased by the possible refutation of Bohmism [*sic*]," recalled Anderson, "and believed that it was basically right." Acting as editor and reviewer, it seems that Anderson approved Bell's paper for publication precisely because he had profoundly misunderstood its content.

Making matters worse, *Physics Physique Fizika* did not survive long—within a few issues, Anderson and Matthias were forced to rebrand it as a traditional solid-state physics journal, and, by 1968, plagued by distribution problems and saddled with a publisher unwilling to publicize the journal, it had folded altogether. Languishing in the forgotten back issues of a little-circulated and out-of-print academic journal, Bell's work was almost totally ignored for several years. Bell received absolutely no correspondence about it for nearly five years after it first appeared. But the few people who did read it took it and ran—and, by the middle of the 1970s, Bell's work had inspired a full-blown quantum rebellion, the first truly widespread and serious challenge to the Copenhagen interpretation from within the physics community since the Bohr-Einstein debates.

But before that happened—in fact, before Bell had even thought of his theorem—another rebellion had started. This academic fight quickly grew into a revolution, toppling the previous order, with huge implications for the foundations of quantum physics. Nonetheless, it escaped the notice of John Bell and most other physicists. In fact, it hardly involved physicists at all. Yet the overthrow of logical positivism and the rise of scientific realism radically changed philosophy of science—and ultimately struck a major blow at the root of the Copenhagen interpretation itself.

8

More Things in Heaven and Earth

The air smelled of stale hops, as usual, and the sky over the city was a dull, low gray. The cobblestone street below had a slight rise to it as it curved around the hill. The hill itself was notable simply for existing, since the entire city was built on a low-lying island. Yet there it was, a small green hill surrounded by a low stone wall, impossibly situated on the outskirts of Copenhagen. Around the corner, a man appeared. He was on the early side of middle-aged, wearing a suit and a pair of thick, black-rimmed glasses. His hair was dark and receding noticeably. He walked along the wall, then crossed the street and went up to the gates of the Carlsberg Brewery itself. It was November 17, 1962, a Saturday, and Thomas Kuhn was here to see the man who had resided in the Carlsberg House of Honor for the past thirty years: Niels Bohr.

Kuhn was the director of the brand-new Archives for the History of Quantum Physics at UC Berkeley. A physicist by training, Kuhn had taken an interest in the history of his field while studying for his PhD at Harvard, and now, fifteen years later, he was a professor of history at Berkeley. For the past few months, and for the next two years, Kuhn and his team of assistants were traveling the world, interviewing the surviving members of the heroic generation who had first uncovered the laws of quantum physics: Heisenberg, de Broglie, Born, Dirac, and many more. Einstein and Schrödinger were already dead by the time the project started, as was Pauli, but Kuhn and his team worked to compile their papers as well and assemble a sketch of the work the men had done, all with an eye toward aiding contemporary and future historians. Bohr, of course, was the single most important living subject of their research. Even putting aside his seminal work in quantum physics and the enormous influence he exerted among his colleagues, Bohr's institute in Copenhagen was home to many important papers from the hundreds of scientists who had been guests there over the past four decades. No wonder, then, that Kuhn and his team set up temporary headquarters in Copenhagen while they wandered around Europe, collecting interviews and papers.

Today, Kuhn was interviewing the great man himself, again. Bohr had already given four recorded interviews across three weeks, and Kuhn was planning to speak with Bohr several more times. Once inside the Carlsberg house, Kuhn sat down with Bohr and his two assistants, Aage Petersen and Erik Rüdinger; after a few minutes of idle conversation among the four men, Kuhn turned on the tape recorder. The subject quickly turned to Bohr's debates with Einstein over quantum physics.

"When I met Einstein for the first time," Bohr recalled, "I said to him, what is he really after, what is it that he is trying to do? Does he think that, if he could prove [quantum objects] were particles, he could induce the German police to enforce a law to make it illegal to use diffraction gratings or, opposite, if he could maintain the wave picture, would he simply make it illegal to use photo-cells?" Einstein had never denied the importance of both particles and waves to quantum physics—in fact, he had been an early champion of both ideas. His criticisms of quantum physics had more to do with locality and completeness, criticisms that Bohr had never adequately responded to. Yet to Bohr, his debate with Einstein had long been settled, and Einstein had lost. "The whole thing with Einstein is so difficult to me because really Einstein had a lot of criticism, and he was shown at every single point, to my mind, that he was entirely wrong. But he did not like it." Bohr lamented the years Einstein had wasted in fighting against quantum physics with his endless series of thought experiments, culminating with the EPR paper. "It was terrible that [Einstein] fell in that trap to work with Podolsky," Bohr said. "Rosen is worse, from my point of view. Rosen even today believes [the

EPR thought experiment]; Podolsky has given it up, as far as I know. . . . The whole idea is absolutely nothing when one really gets into it. You may think that I say it too strongly but it is true; there is absolutely no problem in it."

Bohr also talked about complementarity, and his hope that it would become "common knowledge," a necessary part of the workings of all fields of human inquiry. In physics, he saw complementarity as a simple consequence of the supposed fact that quantum physics couldn't describe large objects, like measurement devices. "I really think that by these few arguments—that the measuring apparatus are [*sic*] heavy bodies and thereby outside the description—one gets at once into the complementary description. And I do not—but perhaps I am wrong, perhaps I am unjust—I do not know why the people don't like it." In particular, he was unhappy that philosophers did not seem to understand his ideas, complaining that "no man who is called a philosopher really understands what one means by the complementary description." (Later in the interview, when Petersen asked Bohr for a clear statement of the principle of complementarity, Bohr ducked the question: he said that he'd given a simple explanation of complementarity to Einstein, who "did not like it." Then Bohr changed the subject, and the question was dropped.)

Despite Bohr's complaint, many prominent philosophers of the day were quite friendly to the Copenhagen interpretation. But that was changing. Part of the reason was a new book, *The Structure of Scientific Revolutions*, published earlier that year. The book argued for a radically new picture of how science works, railing against the philosophical conventional wisdom of the time. Though the position the book argued for was not widely accepted among philosophers, the standard it argued against—known as logical positivism—was already ailing when *Structure* came out, and the book hastened its demise. Logical positivism, like the Copenhagen interpretation, held that talking about unobservable things was meaningless; positivism-inspired arguments were frequently used to defend the Copenhagen interpretation by physicists and philosophers alike. Though *Structure* didn't have the Copenhagen interpretation in its crosshairs—in fact, the book was largely favorable toward Copenhagen—its incisive critique of positivism was potentially ominous news for the quantum orthodoxy.

Kuhn's interview with Bohr would have been a fantastic opportunity to find out what Bohr thought of *Structure*'s arguments against positivism, because the author of *Structure* was none other than Kuhn himself. Unfortunately, Kuhn didn't talk with Bohr about positivism that day, and he never had another chance to ask him about it—nor about anything else. Bohr took a nap after lunch the next day and never woke up. He didn't live to see the toppling of logical positivism—and the subsequent erosion of support for the Copenhagen interpretation among philosophers of physics.

When Moritz Schlick returned to Vienna in October 1929, his colleagues rejoiced. Their leader had returned. Schlick, the chair of *Naturphilosophie* at the University of Vienna, had been visiting Stanford for the past term. While there, he considered a generous job offer from the University of Bonn, in Germany. Schlick waffled for several months but ultimately decided to stay at his post in Vienna. Whatever the charms of the position at Bonn, it couldn't match Schlick's unique informal position in Vienna, as the head of a group of scientists and philosophers known as the Vienna Circle, champions of the new philosophy of logical positivism. Schlick's gentle demeanor, elegant charm, and formidable intellect made him an ideal leader for a group of feisty academics. As a "token of gratitude and joy" for their leader's decision to return, several of the circle's most senior members—Otto Neurath, Rudolf Carnap, and Hans Hahn—wrote a manifesto, articulating the shared philosophical, scientific, and political vision of the group, to present to Schlick upon his return. Like any good manifesto, "The Scientific Conception of the World: The Vienna Circle" declared not only what the group was for but what it was resolutely against, and painted both itself and its opposition as part of vast emerging global movements:

> Many assert that metaphysical and theologising thought is again on the increase today, not only in life but also in science. . . . The assertion itself is easily confirmed if one looks at the topics of university courses and at the titles of philosophic publications. But likewise the

> opposite spirit of enlightenment and *anti-metaphysical factual research* is growing stronger today. . . . In some circles, the mode of thought grounded in experience and averse to speculation is stronger than ever, being strengthened precisely by the new opposition that has arisen. In the research work of all the branches of empirical science this *spirit of a scientific conception of the world* is alive.

The rising tide of "metaphysical and theologising thought" that the members of the circle pit themselves against in their manifesto was not merely religious. German idealism was among the most influential styles of philosophy in Central Europe at the time—and it was entirely incompatible with the Vienna Circle's down-to-earth empiricism. German idealists believed in the primacy of ideas over the material world; they were the intellectual descendants of G. W. F. Hegel, the famous German philosopher of the early nineteenth century. Hegel believed in a world-spirit that arises out of the course of history and steers it to some ultimate end. Prone to grand pronouncements about the nature of reality, the positivists found him needlessly vague and difficult to understand. For example, in one of his best-known works, *Lectures on the Philosophy of History,* Hegel proclaimed that "reason . . . is substance, as well as infinite power, its own infinite material underlying all the natural and spiritual life; as also the infinite form, that which sets the material in motion." To the positivists, this seemed like nonsense.

In addition to Hegel and his followers, there was a contemporary German philosopher, Martin Heidegger, whose philosophy ran counter to the Vienna Circle's ideals. Though Heidegger disagreed with Hegel on many subjects, they both emphasized abstract ideas and intuition over empirical data and material substance, exactly the reverse of the Vienna Circle's ideals.

The Vienna Circle manifesto was a call to arms against what it perceived as regressive, involuted, and deliberately obscure philosophy. "Neatness and clarity are striven for, and dark distances and unfathomable depths rejected," it proclaimed. The work of Hegel, Heidegger, and their ilk, removed from the everyday world of sight and sound, was dismissed as "metaphysics." "The view which attributes to intuition a superior and more penetrating power of knowing, capable of leading beyond the contents of sense experience and not to be confined by the shackles of conceptual thought—this view is rejected. . . . [T]here is no way to genuine knowledge other than the way of experience; there is no realm of ideas that stands over or beyond experience." In place of idealism or theology, the Vienna Circle promoted a "scientific world-conception," with two important features. "*First* [the scientific world-conception] is *empiricist* and *positivist:* there is knowledge only from experience. . . . This sets the limits for the content of legitimate science. *Second*, the scientific world-conception is marked by application of a certain method, namely *logical analysis*." Hence, "logical positivism."

The logical positivists were understandably opposed to philosophical castles in the sky and the tortuous prose they were often defended with. But the logical positivists weren't merely against metaphysics—they believed they could actually dismiss metaphysical claims as meaningless. Meaning, they held, was a matter of *verification:* knowing what a statement means is equivalent to knowing how to verify it using your senses. According to the positivists, when you say "it's hotter outside than it is in here," you really mean "if you go outside, you will feel hotter than you do in here." The statement's meaning is the method of verifying it empirically—and if there's no way of verifying a statement against your senses, then that statement has no meaning. So abstruse statements like Hegel's pronouncements about substance and form, and other metaphysical claims like "there is a God," are meaningless, since they make no contact with the observable world.

But idealist and theological claims aren't the only kinds of statements that make no contact with the senses. There are also more straightforward claims, like "the couch is in the living room even when nobody's there," that can't be confirmed directly. Statements like these, about the existence and persistence of material objects independent of perception, are *realist* claims—they're statements about a real world that exists whether or not there are any humans around. These statements are fundamental to science. Yet some of the positivists, throwing the baby out with the bathwater, dismissed realist claims as meaningless too, because they can't be verified by experience. All that is meaningful, on the positivists' account, are statements about perceptions, along with the purely logical statements of mathematics.

This left positivists in a pickle. They thought it was meaningless to talk about a world that existed independently of perception, but they also wanted to be able to say that science worked. They got around this problem by developing a view of scientific practice that meshed well with their verification theory of meaning. Science, on their account, was about organizing perceptions. Scientific theories were merely methods of predicting future perceptions by churning past perceptions through mathematical machinery. Science wasn't about an objectively real world that existed independently of our perceptions, because anything that existed beyond perception—even a putatively "real" world—was simply metaphysics. Any statements that scientists made about nonobservable yet "real" things on the basis of their scientific theories were dismissed as unnecessary hypotheses, extraneous metaphysical baggage irrelevant to the true task of science. Electrons, for example, weren't real—they could not be seen. Only visible tracks in particle detectors like cloud chambers could be considered real, since that was all that could be directly perceived. Physicists certainly talked about electrons as if they were real, but this was merely a shorthand for their perceptions and not to be taken literally. Science was an instrument for predicting perceptions, nothing more. This view of science came to be known as *instrumentalism*.

The positivists also held that scientists and philosophers should strive for "unity of science"—a single, consistent worldview, based on science and observation, with the different sciences forming a continuous and consistent whole. Biology should be grounded in chemistry, which should be grounded in physics, and so on. This seems relatively innocent and uncontroversial now, but, at the time, there were strong movements within the sciences that pushed against this. Physics and chemistry were at odds for much of the nineteenth century—chemists mostly believed in atoms, whereas physicists were often skeptical that atoms existed—and only in the first decades of the twentieth century did the two fields start to build a consistent picture of chemical interactions. And biology was still not totally on board. Some biologists of the time believed in *vitalism*—the idea that living organisms were not subject to the same laws of physics as inanimate matter, that there was something nonphysical in cell division and inheritance that defied thermodynamics. The positivists rejected this claim, and others like it, as meaninglessly vague metaphysics. Even philosophy itself was to be subsumed by the unity of science, according to the Vienna Circle manifesto: "*There is no such thing as philosophy as a basic or universal science alongside or above the various fields of the one empirical science.*" Philosophy, like the natural sciences, should be grounded in statements about observation and sensations.

Despite its emphasis on empiricism and logic, the Vienna Circle didn't limit its concerns to science and philosophy—the unity of science extended to all human activity. "We witness the spirit of the scientific world-conception penetrating in growing measure the forms of personal and public life," the manifesto boldly claimed, "in education, upbringing, architecture, and the shaping of economic and social life according to rational principles." The Circle's members forged connections with artistic and social movements that shared a similar ethos, like the Bauhaus school of architecture and design. And the Circle also had politics to match its revolutionary rhetoric. Its philosophical opponents, like the German idealists, often had regressive right-wing politics. Heidegger, for example, was a staunch nationalist and agrarian traditionalist, who saw industrialization as a dehumanizing force. He urged a return to traditional cultural values, set himself against modern trends like representative democracy, and ultimately joined the Nazi Party in 1933. The Vienna Circle thought the horrors of the political far right went along with the unscientific and outmoded philosophies that the circle was fighting against. The members saw themselves in the tradition of the great Enlightenment empiricist philosophers such as Hume and Locke, and promoted Enlightenment values: international cooperation over nationalism, reason over faith, humanism over fascism, and democracy over authoritarianism. They saw industrialization not as an oppressive force, but as a modernizing one. The Vienna Circle believed these political causes were intimately connected to its philosophical work. Neurath, for example, had been the economist for the short-lived revolutionary socialist Bavarian state in 1919, and he was nearly sent to prison for his troubles. "Endeavours toward a new organization of economic and social relations, toward the unification of mankind, toward a reform of school and education, all show an inner link with the scientific world-conception," he wrote in the Vienna Circle's manifesto. "The representatives of the scientific world-conception resolutely stand on the ground of simple human experience. They confidently approach the task of removing the metaphysical and theological debris of millennia."

True to their humanist and internationalist politics, Schlick and his colleagues reached out to the world. "The Vienna Circle does not confine itself to collective work as a closed group," their manifesto declared. "It is also trying to make contact with the living movements of the present, so far as they are well disposed toward the scientific world-conception and turn away from metaphysics and theology." In this, they succeeded for a time. Philosophers like Hans Reichenbach in Germany (who had his own Berlin Circle of philosophers) and A. J. Ayer in England visited Vienna, then returned home and promoted logical positivism across national borders and language barriers. Rudolf Carnap became the leading exponent of the Vienna Circle's views; his landmark 1929 book *The Logical Structure of the World* established him as a towering figure in the positivist movement, and many of his students went on to become important philosophers in their own right. Carnap and Reichenbach managed to take over an existing philosophy journal, *Annalen der Philosophie,* and turned it to their own ends, renaming it *Erkenntnis* ("knowledge" or "realization") and publishing articles on positivism from their own circles and elsewhere. Meanwhile, Otto Neurath, ebullient and enormous by nature—"as untidy and rumbustious as Schlick was elegant and urbane," Ayer recalled, "a giant of a man who used to sign his letters with a drawing of an elephant"—worked on several ambitious schemes to change the world in the name of the unity of science. He started a grand encyclopedia project, the *International Encyclopedia of Unified Science*, meant to explain the ideas of positivism and the sciences in one authoritative multivolume reference. He worked to develop an international symbolic language, ISOTYPE, that would precisely specify sense data in unambiguous ways, to aid international collaboration in science and philosophy. And Neurath also organized a series of conferences—International Congresses for Unified Science—where positivists from around the world met and discussed the progress of their philosophical and social program. For a brief moment in the late 1920s and early 1930s, the promise of the Vienna Circle's manifesto burned bright.

Many of the positivists' ideas—their emphasis on observation, their dismissal of "reality" and unseen entities as metaphysics, and their idea of science as a mere instrument for organizing perceptions—sound similar to some of the ideas associated with the Copenhagen interpretation. Logical positivism and quantum physics came out of the same time and place: the Vienna Circle and Berlin Circle both formed in the 1920s, the same decade Heisenberg and Schrödinger (who were German and Austrian, respectively) first developed full-blown theories of quantum physics. This is not a coincidence, but it's not a conspiracy either. There were hazy ideas floating around the intellectual culture in their shared time and place that may have contributed to the ideas of both the early positivists and the first quantum physicists. But there were definitely specific common inspirations for both groups—most importantly, the work of Ernst Mach.

Mach, who worked at the University of Vienna a generation before the Vienna Circle, demanded that all scientific theories refer only to observable entities. (We first encountered him in Chapter 2; he denied the existence of atoms because they couldn't be seen, much to Ludwig Boltzmann's dismay.) Mach's observables-only philosophy of science was a direct inspiration for the development of logical positivism. The Vienna Circle even mentions him by name in its manifesto as one of its direct forerunners and most important influences. But Schlick, Neurath, and the rest weren't the only ones inspired by Mach. Mach was also godfather to a young Viennese mathematical prodigy by the name of Wolfgang Pauli. Mach's views permeated Pauli's philosophy of science. "There is no point in discussing . . . quantities [that] cannot, in principle, be observed experimentally," a young Pauli wrote, fresh out of college, in 1921. Such quantities, he maintained, would be "fictitious and without physical meaning." Over thirty years later, Pauli dismissed Einstein's concerns about the ability of quantum physics to describe what happens between measurements, saying that "one should no more rack one's brain about the problem of whether something one cannot know anything about exists all the same, than about the ancient question of how many angels are able to sit on the point of a needle."

Mach was not the only shared source of ideas for both the Vienna Circle and the physicists of Copenhagen. There was someone else both groups held in similar regard: Einstein, who had himself taken some inspiration from Mach in formulating special relativity. By holding fast to what could be observed—clocks and meter sticks—and dismissing the luminiferous aether as an unobservable phantom, Einstein revolutionized science, and the success of special relativity was seen as a triumph for the Machian view of physics. Certainly Heisenberg thought of relativity this way, as he told Einstein when they met in Berlin in 1926 (as we saw in Chapter 2). And he was not alone—Pauli also counted relativity as a vindication of his godfather's views. The positivists viewed Einstein's work this way too. Moritz Schlick had first made his reputation as a philosopher with his book *Space and Time in Contemporary Physics*, a celebrated exposition of relativity and its philosophical implications. And the rest of the Vienna Circle was confident enough in Einstein's support of their ideas that it took the liberty of naming him one of the "leading representatives of the scientific world-conception" at the end of its manifesto.

Yet, despite the fact that he had borrowed from Mach, Einstein was not enamored of Mach's philosophy of science—at least, not later in life. "You know what I think about Mach's little horse," he wrote to a friend in 1919. "It cannot give birth to anything living. It can only exterminate harmful vermin." When Philipp Frank, a founding member of the Vienna Circle, asked Einstein about his philosophy of science, he was astonished to find that Einstein was not a positivist. Frank protested that Einstein had invented the positivist approach to physics in his theories of relativity. "A good joke should not be repeated too often," Einstein replied, much as he had to Heisenberg several years before.

Einstein certainly thought that science was about more than organizing perceptions. "What we call science," he said, "has the sole purpose of determining what *is*." In an essay replying to Bohr and other critics in 1949, Einstein wrote that what he found unsatisfactory about quantum physics was that it denied the possibility of "the programmatic aim of all physics: the complete description of any (individual) real situation (as it supposedly exists irrespective of any act of observation or substantiation)." Einstein knew this was wildly out of step with the philosophical trends of the day—immediately after stating his beliefs about the aim of physics, he gave a sardonic aside about what he had just said, as seen from the point of view of an imagined positivist who takes both relativity and quantum theory as validation of his philosophical position:

> Whenever the positivistically inclined modern physicist hears such a formulation, his reaction is that of a pitying smile. He says to himself: "there we have the naked formulation of a metaphysical prejudice, empty of content, a prejudice, moreover, the conquest of which constitutes the major epistemological achievement of physicists within the last quarter-century. Has any man ever perceived a 'real physical situation'? How is it possible that a reasonable person could today still believe that he can refute our essential knowledge and understanding by drawing up such a bloodless ghost?

Then he implored, "Patience!" and went on to defend his views skillfully, carefully explaining the unanswered challenge of the EPR experiment yet again. But, despite his own views, Einstein's work remained a source of positivist inspiration for Heisenberg, Pauli, Frank, the Vienna Circle, and a whole generation of German physicists.

Einstein's work also inspired other positivist-leaning philosophies to spring up of their own accord outside of Vienna and Copenhagen. In 1927, Percy Bridgman, an experimental physicist working at Harvard, articulated a philosophy of science that he called *operationalism*. In his book, *The Logic of Modern Physics*, he opened the discussion by explicitly ascribing his inspiration to Einstein's theories of special and general relativity. "There can be no doubt that through these theories physics is permanently changed," Bridgman wrote. "Although [Einstein] himself does not explicitly state or emphasize it, I believe that a study of what he has done will show that he has essentially modified our view of what the concepts useful in physics are and should be." Bridgman went on to claim that Einstein had shown that all scientific concepts must have operational definitions, definitions in terms of some kind of concrete experimental procedure. So "temperature," for example, must be defined as "what a mercury thermometer measures." To Bridgman, the deep insight of relativity was that operational definitions were the most fundamental definitions available for scientific concepts. "In general, we mean by any concept nothing more than a set of operations; *the concept is synonymous*

Figure 8.1. The Second International Congress for the Unity of Science in June 1936, at Niels Bohr's house in Copenhagen. Jørgen Jørgensen is standing, Niels Bohr is on the far right of the front row, and Philipp Frank is second from right in the front. Karl Popper is immediately to the left of Jørgensen. Otto Neurath is third from the left in the fourth row and Carl Hempel is seated immediately behind Neurath. The empty chairs in front are likely for Schlick, Carnap, and Reichenbach, all of whom wanted to attend and none of whom were able to do so.

with the corresponding set of operations." Bridgman was a leading American physicist, who went on to win a Nobel Prize in 1946; the Vienna Circle, naturally, was thrilled to find such a prominent physicist espousing a philosophy of science so similar to their own and invited Bridgman to its International Congress for the Unity of Science in 1939.

The positivists and the founders of quantum physics didn't just share common sources of inspiration—they also had direct contact with one another, discussing common interests in science and philosophy. Neurath visited Copenhagen a few times, meeting Bohr in 1934, and afterward corresponded with Bohr for several years. Writing to Carnap after first meeting Bohr, Neurath said that Bohr "possesses certain basic attitudes which agree with mine." Writing to Neurath later, Bohr expressed his pleasure that their views were not too far apart. In the summer of

1936, Neurath and Bohr teamed up with Jørgen Jørgensen, a Danish positivist, to organize the Second International Congress for the Unity of Science. The conference, naturally, was in Copenhagen—in fact, it was held in Niels Bohr's house, the Carlsberg House of Honor (Figure 8.1). At that conference, Frank presented a paper on behalf of Schlick, titled "Quantum Theory and the Knowability of Nature." Schlick's paper argued that "in physics it is meaningless to speak about factors which are unknowable in principle," and that speaking about undetermined quantities in quantum physics was "neither true nor false, but *meaningless*"—claims that sound awfully similar to the Copenhagen interpretation.

None of this is to say that logical positivism was the philosophical motivation behind the Copenhagen interpretation. Bohr, in particular, may not have been much of a positivist. While it's hard to say what Bohr's position actually was—there's a small mountain of papers attempting to decipher his views on any given subject, and few of them agree—Bohr did seem to flirt with certain ideas that the positivists found distasteful, such as vitalism. (In fact, Bohr spoke favorably about vitalism with complementarity-based reasoning at the 1936 conference at his own house, while Schlick's aforementioned paper at the same conference argued against vitalism.) And Neurath thought that Bohr's "printed remarks are full of crass metaphysics," and that he "express[es] himself somewhat unclearly." But Bohr also seemed sympathetic to the positivists, and at times came close to declaring himself one of them. When Frank asked whether Bohr's reply to the EPR paper was based on positivist reasoning, Bohr told him that "you have caught the sense of my efforts very well."

Regardless of Bohr's true philosophical convictions, the Copenhagen interpretation was certainly defended with arguments and slogans derived from logical positivism. The verification theory of meaning—especially the idea that unverifiable statements are meaningless—was introduced to physics students as a fundamentally new insight about how the world worked, an inextricable part of the success of quantum physics. According to a very popular midcentury quantum physics textbook, the old physics before the quantum revolution assumed that particles like photons always had a definite position at every moment of time, but "quantum mechanics . . . asserts instead that the position of a photon has meaning only when the experiment includes a position determination." Heisenberg himself often used operationalist language when talking about the quantum world. "There is no way of observing the orbit of the electron around the nucleus" of an atom, he declared, and "therefore, there is no orbit in the ordinary sense." According to him, presuming the electron has an orbit or any kind of path at all between observations "would be a misuse of language which . . . cannot be justified."

Yet, in truth, the physics community hadn't really embraced positivism—it had embraced a knockoff version, convenient for its own purposes. The verification theory of meaning couldn't actually justify most formulations of the Copenhagen interpretation. And few physicists truly believed, as the Vienna Circle did, that electrons didn't exist. Physicists had simply adopted a caricature of the positivist attitude. If something can't be seen, why worry about it? Things that can't be seen are meaningless anyhow. And if anyone still wasn't convinced, there was a large pile of borrowed and bastardized arguments from the positivists about why this kind of reasoning worked, enough to keep most people from worrying—especially with the wide variety of interesting work to be done using the mathematical machinery of quantum physics.

This cartoonish parody of positivism, despite its flaws, worked well for the practical-minded physics that was encouraged during and after World War II. And some of the Vienna Circle's members, like Schlick and Frank, did argue that the Copenhagen interpretation had a solid philosophical grounding in accepted tenets of logical positivism. But, just as the war brightened the prospects for the Copenhagen interpretation, it dimmed the fortunes of the positivists themselves.

The Vienna Circle started running into serious trouble in the mid-1930s, as fascism crept over Europe. The deteriorating political situation convinced some of the leaders of the circle and their allies that it was time to leave Europe altogether. Reichenbach was forced out of his job in Berlin when Hitler came to power in 1933 and fled to the University of Istanbul for several years. Fascists seized power in Austria at roughly the same time, and by 1934 Czechoslovakia was in a precarious position as the only functioning democracy anywhere in Eastern Europe. Carnap, who had moved to the University of Prague several years earlier, saw the writing on the wall. With the help of the American positivist philosopher Charles Morris, Carnap moved to the United States in 1935 and took a job at the University of Chicago shortly thereafter. Schlick stayed on in Vienna but faced increasing political problems: both the fascist government and the Austrian Nazi Party saw him (rightly) as a political and ideological opponent, and the Nazis falsely claimed he was Jewish. In 1936, the Austrian government denied Schlick a travel visa to attend the conference in Copenhagen at Bohr's house. On the morning of the first full day of the conference, while Bohr and Frank were presenting papers in Copenhagen, Schlick was approached by a former student, Johann Nelböck, on the steps of the University of Vienna. Nelböck shot Schlick four times at point-blank range; he died on the spot. Nelböck was apprehended, confessed, and found to be of sound mind, but the Austrian Nazis took up his cause and distorted the facts of the case in the Viennese press. Nelböck was sentenced to a mere ten years in prison for murder. When Austria became part of Nazi Germany in the Anschluss of 1938, Nelböck applied for a pardon, stating in his application (where he referred to himself in the third person) "that by his act and the resulting elimination of a Jewish teacher who propagated doctrines alien and detrimental to the nation he rendered National Socialism a service and also suffered for National Socialism as a consequence of his act." The Nazis pardoned him after he had served only two years of his sentence.

By the time war broke out in 1939, the only core member of the Vienna Circle still living in continental Europe was Otto Neurath. After the fascists took over Austria, he had fled to the Netherlands, hoping to continue his internationally minded work from The Hague. In 1940, he and his assistant managed to escape to England in a boat while Rotterdam burned, hours before the Nazis arrived in The Hague. After the war, there were attempts to reorganize the circle, but Neurath's sudden death in December 1945 largely put an end to those efforts. Positivism continued on as a philosophy under the new name "logical empiricism," but the Vienna Circle's grand dream of an organized political, philosophical, and scientific movement was dead.

Any remaining hope of reviving a unified movement around positivism was dashed by the postwar political environment in the United States. Anticommunist hysteria in the United States rose sharply after World War II, and the nascent Cold War had chilling effects on all arenas of intellectual discourse, including philosophy. To some, the Unity of Science movement, with its left-wing politics, its antireligious philosophy, and its internationalist aspirations, sounded suspiciously like a Communist Party front. During the "red scare" that effectively exiled David Bohm, J. Edgar Hoover's FBI compiled dossiers on Carnap, Frank, and other leading positivist lights. Under enormous pressure to refrain from all political activity, the positivists were forced to focus solely on issues in logic and the philosophy of science—pushed to what their now-distant manifesto had called "the icy slopes of logic."

But the final blow against positivism came from within philosophy, not from the external forces of geopolitics and chance that had left it battered. New arguments against some of the central tenets of positivism came from a new generation of philosophers. These arguments laid bare the inadequacy of the verification theory of meaning and the instrumentalist account of science—and turned philosophers of science against the Copenhagen interpretation.

One of the young philosophers who had visited the Vienna Circle in its heyday was a brilliant American student with the improbable name of Willard Van Orman Quine. Quine had written a PhD thesis on mathematical logic at Harvard in 1932; for the year afterward, he traveled Europe on a fellowship, meeting with Schlick, Frank, Ayer, and other leading positivists. He spent six weeks studying with Carnap in Prague—"my first really considerable experience of being intellectually fired by a living teacher rather than by a dead book," Quine said later. Returning from Europe "an ardent disciple of Carnap" (and with only seven dollars in his pocket), Quine went back to Harvard and taught classes on positivist philosophy. He also did important work in mathematical logic. But as Quine continued to work and teach—interrupted only by World War II, when he cracked coded messages from Nazi submarines—doubts about positivist dogma seeped in and accumulated. Finally, in 1951, the dam burst: Quine wrote a paper that brought positivism to its knees.

Quine's paper, "Two Dogmas of Empiricism," took aim at the core of the positivist program, the verification theory of meaning. Quine pointed out that there was no way to verify single statements—all attempts to verify a statement inevitably involve the assumed truth of other statements, which are themselves subject to the same problem. For example, say the remote control for your TV isn't working, and you can't turn the TV on. You suspect that the batteries in the remote are dead. You can verify this by replacing the batteries and trying to turn on the TV with the remote again. You do this, and the TV turns on. Does that mean you were right? No. It's entirely possible that the batteries in the remote weren't dead at all. Maybe the remote has a short and will only work intermittently, no matter what batteries you use. Or maybe the old batteries were in backward, and you didn't notice; you just popped them out and put in new ones pointing the right way. Or maybe something more exotic is going on: maybe the remote always worked, but when you tried it before, the TV mysteriously shifted its picture to the infrared and its sound to the ultrasonic, so you couldn't see or hear it; the TV happened to go back to normal after you changed the batteries, but not *because* you changed them. This last idea is clearly preposterous—how could that happen?—but the point remains that in testing out the new batteries in the remote, you've assumed a wide variety of basic facts about the world, all based on previous experience, and any of which could, in principle, be wrong. This isn't merely true of suppositions about batteries in remote controls; verification of any statement behaves this way. Looking out the window and saying, "It's raining outside," assumes that your view through the window's glass gives you an accurate picture of the outside world, and that your eyes are functioning properly, and that the dimmed light and falling droplets are in fact caused by a rain cloud and not an alien spaceship blotting out the Sun and dropping some exotic substance onto your front lawn. So you can never verify a single statement: you're always stuck verifying the entirety of your knowledge about the world, or at least a very large fraction of it. As Quine put it, "Our statements about the external world face the tribunal of sense experience not individually but only as a corporate body."

With the verification theory of meaning left gasping for air, Quine dismissed the idea that it's meaningless to talk about unobservable things. Unverifiable statements must have meaning, because no individual statement is verifiable. Thus, the "metaphysics" so dreaded by the positivists came roaring back: rather than speaking simply of sensations, Quine contended that it was perfectly intelligible to speak of physical objects with existence independent of the speaker.

Quine's paper emboldened other thinkers who harbored doubts about logical positivism. One of them was a younger colleague of Quine's at Harvard: Thomas Kuhn. Kuhn had spoken with Quine at length while he was writing "Two Dogmas," and he was very impressed with Quine's arguments. Quine's paper "had a considerable impact on me because I was wrestling already with the problem of meaning," Kuhn said later. Kuhn had first become interested in the history and philosophy of science while he was a graduate student studying solid-state physics. He had been roped into being a teaching assistant for a new course in the history of science, and he ended up reading Aristotle's *Physics.* There, Kuhn encountered a strange world, one where heavy things fell because they were supposedly trying to return to their "natural place" at the center of the universe—the Earth. "What Aristotle could be saying baffled me at first, until—and I remember the point vividly—I suddenly broke in and found a way to understand it, a way which made Aristotle's philosophy make sense," Kuhn recalled. He abruptly found himself looking at the product of a first-rate mind at work, struggling to understand the physical world around him, just like a modern scientist. The key difference, Kuhn realized, was that Aristotle was starting from an entirely different worldview. Within that worldview, Aristotle's ideas made a great deal of sense. Kuhn came to believe that the whole picture he had of the progress of science, a cartoonish picture he'd picked up from his coursework as a physicist-in-training, was simply wrong. Science didn't progress by simply piling one victorious theory upon another. It was far more complex and subtle than that.

Once Kuhn had finished his PhD in 1949, he switched fields entirely, becoming a historian and philosopher of science. After spending several years researching the history of physics, especially the period around the Copernican revolution, he set out to expound his new view of science, which ran counter to the positivist conception of scientific progress. Ironically, he was given the perfect opportunity to do so by the positivists themselves. Charles Morris, the American positivist who had helped Carnap come to America, approached Kuhn and asked him to contribute a monograph on the history of science to the *International Encyclopedia of Unified Science,* still limping along after Neurath had started it over twenty years earlier. Morris had tried to find someone to write this monograph for several years, to no avail; his working title for it was *The Structure of Scientific Revolutions.*

Despite its appearance in Neurath's encyclopedia, Kuhn's book expounded a view that was totally at odds with positivist ideas about science. Kuhn argued that both the observable and unobservable content of scientific worldviews—what he called "paradigms"—play vital roles in the actual practice of science. These scientific paradigms influence what experiments are done, how they're performed, and how the results are interpreted. Going back to the example of the malfunctioning remote control, replacing the batteries is a reasonable thing to do because your knowledge of remote controls, televisions, and batteries suggests that dead batteries are the most likely reason for the remote to stop working. That same set of knowledge—your "paradigm of home entertainment systems"—also tells you that it's impossible for your TV to suddenly display all pictures in infrared and emit all sound in ultrasonic. Kuhn argued that paradigms guide scientific practice in a similar way. For example, chemists in the nineteenth century believed in atomic theory, which dictated that there was a limited number of elements, each composed of identical atoms, and those atoms combined to form compounds that had fixed ratios of each element. These ideas were central to the practice of chemistry at the time and, according to Kuhn, had the power of "setting the problem of atomic weights, bounding the admissible results of chemical analyses, and informing chemists what atoms and molecules, compounds and mixtures, were." At every step of the way—forming hypotheses, designing and conducting experiments, even simply observing the results of those experiments—the paradigm of atomic theory informed the actions of the nineteenth-century chemists. And they were wildly successful, discovering the periodic table of the elements decades before physicists discovered electrons or learned anything about atomic structure. Yet, according to the best science of the time, atoms were unobservable. So, Kuhn concluded, it's not just the observable part of a theory that matters—the full content of scientific paradigms influences how science is done. The interpretation of a physical theory, like quantum physics, is important for the day-to-day practice of science itself. Logical positivism couldn't account for this.

What, precisely, Kuhn advocated in place of positivism is not entirely clear. And some of his bolder claims, about the inability to rationally compare competing scientific theories, were dismissed as mistakes and didn't catch on with professional philosophers of science. But Kuhn's criticism of positivism and his observations on scientific practice were widely seen as accurate—and he wasn't the only one pointing these things out. Other philosophers, including J. J. C. Smart, Hilary Putnam, Karl Popper, Grover Maxwell, Norwood Russell Hanson, and Paul Feyerabend all piled onto the positivists' philosophy of science in the late 1950s and 1960s, building on each other's works and pointing out irreparable flaws in the positivist account of scientific work and progress. Hanson had made many of the same points as Kuhn several years prior to *Structure* in his book *Patterns of Discovery.* (Hanson and Kuhn knew each other and each acknowledged the other's work in their books.) Hanson dubbed the role that nonobservable entities play in scientific work the "theory-laden" practice of science, a name that stuck. This new crop of philosophers largely agreed that the practice of science was theory laden and that the history and practice of actual science was an important guide in developing the philosophy of science. And, while there was much they didn't agree on, a new consensus began to form among professional philosophers of science, a position in opposition to logical positivism, which they called *scientific realism.*

Scientific realism is what it sounds like: the view that there is a real world out there, independent of our observations of it, and that science gives us an approximate description of that world. When a new scientific theory is accepted in place of an old one, this is generally because it gives us a better approximation of the true nature of the world in some important way. This is not to say that the world is entirely insensitive to our probings of it—quantum contextuality assures us that our measurements have some impact on the world—but, by and large, the world proceeds on whether or not we interfere with it. And the contents of that world are approximately described by the content of our best scientific theories, both the observable and unobservable parts.

The realists also argued that the distinction between what is observable and what is unobservable was neither meaningful nor relevant to science. This, of course, was anathema to the positivists. Some positivists had even gone so far as to say that objects seen in microscopes were not truly real, because they were not "directly" perceived. The scientific realists thought this was preposterous. "If this analysis is strictly adhered to, we cannot observe physical things through opera glasses, or even through ordinary spectacles, and one begins to wonder about the status of what we see through an ordinary windowpane," wrote Grover Maxwell, one of the most vocal defenders of scientific realism. Maxwell also pointed out that the very idea of something that is "unobservable in principle" is subject to revision with new theories and new technology. Before the development of optics and the microscope, he pointed out, something "too small to be seen" would have been unobservable in principle. "It is theory, and thus science itself, which tells us what is or is not . . . observable," Maxwell wrote, echoing Einstein's words to Heisenberg. "There are no *a priori* or philosophical criteria for separating the observable from the unobservable."

With a better understanding of and appreciation for the history of science and the theory-laden practice of science, the realists also made quick work of positivist ideas about the functioning of science like instrumentalism and operationalism. The realists pointed out that if operational definitions are the ultimate definitions of scientific concepts, then there's no way to improve measuring processes or to design them in the first place, since that would require going beyond the operational definition. If, for example, length is defined as the thing that existing measuring sticks measure, then there's no way to design a better measuring stick, because we have a perfect one by definition. Yet scientists develop new and improved measurement devices all the time. The ideas of length, time, mass, and so on aren't merely defined by experimental operations: they are inherent in the theories used to design and test new measurement devices.

The realists' dismissal of operationalism was not fundamentally new—in fact, many positivists, including Carnap, had dismissed operationalism years earlier as too simplistic to account for the success and practice of science. But many positivists still clung to instrumentalism, the view that science is simply a tool for organizing and predicting perceptions, and that the metaphysical content of theories was unnecessary. The realists pointed out that this was untenable as well. If the unobservable "metaphysical" content of our best scientific theories—stuff like electrons—really bears no relation at all to the actual stuff in the world, then why do our scientific theories work at all? The theories themselves suggest explanations for the phenomena that we see based on the unobserved stuff. But if the unobserved stuff is just a convenient picture that just happens to come along with the "real" content of the theory (i.e., the predictions about the observable world) and that picture really doesn't line up with the stuff in the world in any way, then we're astonishingly lucky that our theories work so well!

Here's an example. Lighting a magnesium sparkler and sticking it into a mixture of powdered rust and aluminum triggers a runaway chemical reaction that quickly reaches about 2,500°C—nearly half as hot as the surface of the Sun—giving off a blinding light and bringing the iron and aluminum close to their boiling points. This is called the thermite reaction; it is weird and dangerous (seriously, never do this!) but it gets even weirder. Not only is the thermite reaction amazingly intense, but it continues to run until the rust and aluminum are used up, no matter what you do to it. You can put it underwater, you can cover it with sand, you can even put it in the vacuum of space—it will keep burning. (In fact, one of the main industrial applications of thermite is underwater welding.) This is because the reaction doesn't need anything other than the rust and the aluminum, along with a little heat to get it started (hence the sparkler).

The thermite reaction works because aluminum desperately wants to react with oxygen. Rust is nothing but iron and oxygen, so the aluminum rips the oxygen off of the rust, leaving you with aluminum oxide, iron, and a tremendous amount of heat. Or, at least, that's what quantum chemistry has to say on the matter. But if you want to be an instrumentalist, that explanation isn't the real answer. There is no "real" answer to be had. All you care about is the fact that quantum chemistry correctly predicts that you'll get a violent reaction from sticking a magnesium sparkler into a pile of rust and aluminum. The deeper explanation that quantum chemistry gives—why aluminum wants to bond with oxygen so desperately, which has to do with electron orbitals—not only isn't of interest, but isn't real.

But if quantum chemistry's account of thermite isn't real, then the instrumentalist has a serious problem. The theory doesn't just predict that the thermite reaction will happen—it also predicts, in great detail, how it happens and what happens. It tells you how hot the magnesium sparkler has to be to set off the reaction. It tells you exactly how hot the reaction will become, and how long it will last. It even tells you other kinds of rust you can use (different metal oxides) along with the aluminum, and exactly how those will alter the reaction. And all of these excruciatingly detailed answers given by quantum chemistry, down to the fifth decimal place, are explained by the behavior of electron orbitals in the constituent atoms of the powder you started with. Now, you can be an instrumentalist—you can deny that electron orbitals are real things that are involved in this reaction—but then how can you explain the fabulous match between the theoretical prediction and the experimental outcome? Why does quantum chemistry work so well to explain the thermite reaction if atoms and electron orbitals aren't real? "If [instrumentalism] is correct we must believe in a *cosmic coincidence*," said J. J. C. Smart. "Is it not odd that the phenomena of the world should be such as to make a purely instrumental theory true? On the other hand, if we interpret a theory in a realist way, then we have no need for such a cosmic coincidence . . . a lot of surprising facts no longer seem surprising." Smart went on, clearly impatient with instrumentalism:

> Suppose that a detective finds a lot of footprints, bloodstains, and so on. If the criminal were a theoretical fiction for relating footprints and bloodstains hitherto found to one another, then it would seem too good to be true that it should actually issue in true predictions

> of further footprints and bloodstains and even of missing five-pound notes. But if there really were a criminal, then these predictions would no longer be surprising.

Hilary Putnam put it more succinctly. "Realism," he claimed, "is the only philosophy that doesn't make the success of science a miracle."

Smart didn't just see positivism as problematic philosophically—inspired by Feyerabend, he saw it as a practical problem as well. "A positivist attitude has frequently been inimical to progress," wrote Smart in 1963. "Positivism would once have supported the Ptolemaic [Earth-centered] theory against the Copernican one, by showing that at the time it was the better [at making predictions] of the two. It supported phenomenological thermodynamics and resisted the [atomic] theory of gases. And today it opposes, *a priori*, any attempts to construct alternatives to the prevailing Copenhagen interpretation of quantum mechanics." This, to Smart, and to Putnam, Feyerabend, and other leading philosophers of the day, was a serious problem—because with the collapse of positivism, the Copenhagen interpretation was indefensible. How could our everyday world of things that exist be composed of a quantum world where nothing was real? "The great and compelling reason for refusing to regard the elementary particles as theoretical fictions," wrote Smart, "is that unless something like what quantum mechanics tells us is true of some underlying reality, then [the fact] that the macroscopic laws are what they are . . . [is] too much of a coincidence to be believed."

And there was another problem: the most straightforward way to make "measurement" fundamental to a theory was to be an operationalist—but operationalism was manifestly wrong. "Measurements are a subclass of physical interactions—no more or less than that," wrote Putnam in 1965. "'Measurement' can never be an *undefined* term in a satisfactory physical theory, and measurements can never obey any 'ultimate' laws other than the laws 'ultimately' obeyed by *all* physical interactions." Fixing the problem through a split between the micro-world of the quantum and the macro-world of everyday classical physics, as Bohr had insisted, didn't help either. "Besides pushing the problem back to exactly the same problem in classical physics . . . the suggestion is completely unacceptable," Putnam said. "We can hardly refer to one theory (classical physics) in the [foundations] of another [quantum physics] if the first theory is supposed to be incorrect and the second is designed to supersede it. . . . Quantum mechanics must, if correct, apply to systems of arbitrary size. . . . In particular, it must apply to macro-systems." But if that's the case, Putnam continued, then "what of macro-observables that are isolated for a long time, say, a system consisting of a rocket ship together with its contents out in interstellar space? We cannot seriously suppose that the rocket ship begins to exist only when it becomes once again observable from the earth or some other outside system." For the Copenhagen interpretation, measurements were a serious problem. Smart agreed, with withering criticism of the idea that measurements must be described classically:

> Proponents of the Copenhagen interpretation of microphysics have been wedded to *classical* physics. They argue that since this is the physics of macroscopic instruments whereby we interpret our observations, this must remain stable whatever the advances in microphysics. That this is not so can be shown (as by Feyerabend) by posing one simple question: why *classical* physics? Why not, for example, Aristotelian physics, or even witchcraft, which was once just "scientific common sense"? Similarly, we must reject the view that there are laws at the [instrumental] or macro-level which are sacrosanct, and which the micro-theories explain. We must insist . . . that the micro-theories can directly explain observations, such as the outcome of the two-slit experiment.

Smart and Putnam were clear-eyed about the challenges faced by any alternative to the Copenhagen interpretation. "Any realistic philosophy of the theoretical entities must not be too naïve. It must reckon with the very real difficulties in giving a non-[instrumentalist] interpretation of physics," wrote Smart. "One way out of the dilemma may lie in the development of deterministic theories of microphysics on the lines foreshadowed by such writers as D. Bohm and J.-P. Vigier." Putnam agreed that "something is wrong with the [quantum] theory." But he thought von Neumann's proof ruled out Bohm's pilot-wave interpretation—at the time, Bell's refutation of the proof was still languishing in an editor's desk—and he was wholly unaware of Everett's many-worlds interpretation (as were Smart and almost everyone else). Putnam concluded that "*no* satisfactory interpretation of quantum mechanics exists today." But he was hopeful that the problem would be solved. "Human curiosity will not rest until those questions [of quantum interpretation] are answered. . . . The first step toward answering them has been attempted here. It is the modest but essential step of becoming clear on the nature and magnitude of the difficulties."

Yet this is precisely what physicists, as a whole, were still unclear about. The philosophers had successfully overthrown positivism and had a good understanding of the mathematical intricacies of quantum physics—but the physicists were still blinkered, walled off from philosophy and the developments there. They had no idea any of this had happened. While Einstein and Bohr's generation was widely schooled in philosophy, the push toward specialization after World War II had taken its toll on the liberal arts education of the new crop of physicists. Academic departments had become Balkanized as they had grown in the postwar boom, and physicists, busy with enormous grants and hard-nosed calculations, were generally dismissive of philosophy. So physics trudged along, not knowing that a major revolution had happened in an adjacent field. And philosophers were, mostly, not surprised by this. "Unless the real difficulties in quantum mechanics can be dealt with," wrote Smart, "the philosophical objections to the Copenhagen interpretation, which consist only in exposing the positivistic preconceptions thereof, will be found unsatisfactory by physicists." If the physicists were to pay attention to the problems at the heart of their field, there would have to be more than philosophy at stake. There would have to be the chance of overturning accepted physics, of finding something fundamentally new, something shiny and exciting, something preferably involving a laboratory experiment—something like finally putting John Bell's ideas to the test.

Part III

The Great Enterprise

The aim remains: to understand the world. To restrict quantum mechanics to be exclusively about piddling laboratory operations is to betray the great enterprise. A serious formulation will not exclude the big world outside the laboratory.

—John Bell, 1989

9

Reality Underground

It was the Summer of Love in New York City, and John Clauser was cooped up in a room at the Goddard Institute for Space Studies on 112th Street, teasing secrets from the oldest light in the universe. Clauser, a physics graduate student at Columbia, was attempting to measure the recently discovered cosmic microwave background (CMB) radiation, an echo of the Big Bang itself. It was difficult and painstaking work at the cutting edge of science—the CMB, a faint static hiss of radio across the sky, had been discovered only three years previously, by a pair of physicists working at Bell Labs. Only one other group of physicists had managed to detect it since. Clauser and his graduate advisor, Patrick Thaddeus, were intent on being the next to hear the beginning of the universe and to do it more accurately than any before them. But, one day in 1967, Clauser made a different kind of discovery altogether. Looking for the latest research in the Goddard Institute library, Clauser stumbled upon a journal with the unusual name *Physics Physique Fizika*. Intrigued, he flipped it open, and a paper caught his eye: "On the Einstein-Podolsky-Rosen Paradox," by one J. S. Bell.

Clauser was young, brash, voluble—and had already been doubting the Copenhagen interpretation for years. Clauser's father, Francis, had a PhD in aeronautics from Caltech (as did Francis's twin brother Milton) and had trained John to be skeptical. "Son, look at the data," Francis told him. "People will have lots of fancy theories, but always go back to the original data and see if you come to the same conclusions. . . . Common wisdom is frequently a poor interpretation of what is actually observed." Francis specialized in the physics of fluids, and was suspicious of the mathematically simple yet difficult-to-visualize quantum theory. "There were very strong similarities between the mathematics of fluid flow and the mathematics of quantum mechanics, and [my dad] didn't understand quantum mechanics," the younger Clauser recalled. "And he kind of pre-programmed me as the guy who might help try to solve the problem that he couldn't solve." When it came time for John to go to Caltech, he learned quantum physics at the feet of Richard Feynman himself—but he could never shake the suspicion that something was wrong with the theory. His doubts followed him to his PhD program at Columbia, where he learned more about the debates that had hounded quantum physics since its inception. "I . . . was struggling, trying to understand quantum mechanics. I had read EPR's paper, and also had read Bohm's and de Broglie's work. While I had difficulty understanding the Copenhagen interpretation, the arguments by its critics seemed far more reasonable to me at that time," Clauser recalled. "I also found EPR's arguments much more persuasive than Bohr's. . . . Hidden variables thus seemed (to me then) to be a perfectly logical solution to the problem. By holding that opinion . . . I was certainly branded as a heretic by many, and undoubtedly as a quack by others."

Given his background, the title of Bell's paper immediately captured Clauser's attention—and the elegant proof contained within the short paper came as a massive shock. "I said, 'this can't possibly be true,'" Clauser recalled later. "I [thought] it should be easy to find a counter-example. I worked and worked and worked. . . . I can't find a counter-example. Well, Bell's got to be wrong on the proof. But nope, can't find anything wrong with it. Bounced back and forth between those two [ideas] and finally it just hit me: Jesus Christ, this is a very important result." Clauser, an experimental physicist down to his bones, immediately wondered: Could Bell's idea be put to the test?

Clauser knew there was a chance that Bell's theorem had already been tested inadvertently, as part of some earlier experiment. And, even if it hadn't been, Clauser needed to search the relevant experimental literature, to find out how such an experiment could best be done. He already knew that Chien-Shiung Wu, a professor at Columbia renowned for her work in nuclear physics, had done an experiment similar to the EPR thought experiment fifteen years earlier. Clauser asked Wu whether she had any unpublished data from her experiment that could be used to test Bell's theorem. She didn't, and the experiment couldn't easily be adapted to perform such a test. Then Clauser went a few blocks north to Yeshiva University, where a friend introduced him to a young professor working there: Yakir Aharonov, Bohm's former student. When Clauser told Aharonov that he was hoping to test Bell's theorem, "he thought it was really quite interesting and would be well worth doing," Clauser recalled. But Aharonov was a theoretical physicist, and was working on different problems anyhow; he couldn't help Clauser much. Finally, an old college buddy of Clauser's told him about a group of physicists doing work up at MIT that seemed like it could be adapted to test Bell's theorem. Clauser went up to Cambridge and gave a talk on Bell's work; afterward, he was introduced to Carl Kocher, a newly arrived postdoc. "Carl Kocher . . . had just finished his doctorate at Berkeley under Gene Commins. And they had done a polarization correlation experiment with photons," Clauser recalled. "So they told me about Carl's experiment, and said 'might that work as an alternative?' And I said, 'you betcha it would! That's exactly what I'm looking for.'" Reading the paper Kocher and Commins had written about their experiment, Clauser realized their experiment could have tested Bell's theorem—but they hadn't. "I looked at Kocher-Commins's result and they hadn't really been aware of what Bell's theorem said." With a few tweaks, Clauser could adapt their setup and test Bell's theorem.

Finally satisfied that the experiment hadn't been done, but could be done, Clauser went to his advisor at Columbia, Pat Thaddeus, and asked him for advice. Thaddeus had already caught wind of Clauser's strange extracurricular activities. "He was pissed," Clauser recalled. "The first thing he said was, 'Well, this is all nonsense. Tell you what you do: write up a letter to Bell and de Broglie and these guys, and they'll set you straight. This is a waste of time.'" So on Valentine's Day 1969, Clauser wrote a love letter of sorts to Bell, asking him whether he thought it was worthwhile to perform a test of his inequality, whether he knew of any existing experimental results on the subject, and proposing an extension of the Kocher-Commins experiment that would provide such a test. It was the first correspondence of any kind that Bell had received about his paper in the four years since it had been published. Several weeks later, Clauser found a letter waiting for him at the Institute for Space Studies, on CERN letterhead, from Bell himself.

"I think that the experiment you propose is of very high interest. I do not know of other relevant experiments that have been done," Bell wrote. "In view of the general success of quantum mechanics, it is very hard for me to doubt the outcome of such experiments. However, I would prefer these experiments, in which the crucial concepts are very directly tested, to have been done and the results on record." Bell, intimately familiar with the workings of quantum physics, knew that it was unlikely that the theory would be proven wrong. But he also knew better than to dash the obvious hopes of the young man who had written to him out of the blue. "Moreover," his letter concluded, "there is always the slim chance of an unexpected result, which would shake the world!"

"The Vietnam War dominated the political thoughts of my generation," Clauser wrote later. "Being a young student living in this era of revolutionary thinking, I naturally wanted to 'shake the world!'" John Clauser's mind was made up. He was going to perform the experiment—and, he hoped, prove quantum physics wrong.

On the other side of the Atlantic, a young German physicist named Dieter Zeh was having similar doubts about the Copenhagen interpretation. "It was a slow process, and not just a sudden thing," he said later. "I always had these doubts, but of course I didn't dare to draw the conclusion that these people are all nuts." Thoughtful, humble, and unfailingly polite, Zeh had little in common with the loud and blustery Clauser, aside from a shared skepticism of Copenhagen. Clauser's day-to-day work as an observational astrophysicist involved building and testing sensitive experimental equipment. Zeh, by contrast, was a theoretical nuclear physicist. His work involved detailed quantum calculations; he was very much at home with the abstract mathematics behind quantum physics. And these differences between the two men were also reflected in their ultimate aims. While Clauser was uncomfortable with quantum physics and wanted to prove it wrong in the laboratory, Zeh understood the theory intimately—and found something truly surprising lurking within it.

Zeh had been puzzling over a problem in nuclear physics, one in which an atomic nucleus was in a Schrödinger's cat–like state of superposition, pointing in a multitude of directions at once. Meanwhile, the protons and neutrons within the nucleus were highly entangled with one another, so finding the position of just one of them would determine the positions of all the rest. "This made me think," Zeh recalled. "I said, let's assume that the universe is a closed system, like a nucleus. It was for me a very important step." Zeh didn't think that the universe was literally a single atomic nucleus. But he realized that the general idea—a system in a superposition, with its components strongly entangled—could explain how measurement works in quantum physics, without resorting to any of the tricks the Copenhagen interpretation used, like wave function collapse or a split between the physics of the small and the large. Treat a measurement device as a quantum system, and the act of measurement as a normal physical interaction, and quantum physics says that the measurement device will become strongly entangled with the thing it's measuring—and the overall system of measuring-device-and-thing-being-measured will be in a Schrödinger's cat state. But, Zeh realized, it went beyond that: the measuring device interacts with the experimenter, and everything else in the room, and eventually the entire universe—so when a small quantum system interacts strongly with a large object, ultimately, the entire universe ends up like Schrödinger's cat, splitting into dead-cat and alive-cat "branches." And the inhabitants of each branch of the universe only see one outcome: the dead cat or the living cat, depending on which branch they're in. But the wave function never collapses, and the different branches of the universe are extraordinarily unlikely to interact. "If you make a measurement, you get an entanglement between the system and the apparatus and the observer," Zeh said. "The observer sees only one component [of the Schrödinger's cat state] and not the superposition of all the others. So, that solves the measurement problem." Zeh had unknowingly reinvented Everett's many-worlds interpretation from scratch—and, along the way, he had also developed a mathematically sophisticated account of the interactions between small quantum systems, like atoms, and the relatively large quantum objects around them, like rocks and trees and measurement devices. This, in turn, explained why the different branches of the universal wave function would not interact, and did so in a much more detailed way than Everett ever had. Zeh's approach to these interactions was later dubbed "decoherence."

Zeh excitedly wrote up his account of decoherence and the universal wave function, but he wasn't sure where to turn for feedback on his work. "Of course, I could not say such ideas to colleagues," Zeh said. "They would [say] 'hey, you're completely crazy,' they would not accept even to think about that." Zeh instead took his work to his mentor, J. Hans Jensen, a Nobel Prize–winning physicist who had supervised Zeh's PhD at Heidelberg several years earlier. But Jensen was no expert on the quantum theory of measurement, so he sent the paper to a friend of his who knew more about the subject: Léon Rosenfeld, Bohr's former right-hand man and rabid defender of the Copenhagen interpretation. Rosenfeld, who had been insulting to Bohm and dismissive of Everett, was no kinder to Zeh. "I established a rule in my life never to step on anybody's toe," he wrote to Jensen, "but a preprint written by a certain 'Toe' [Zeh, in German] from your institute that I have received makes me digress from that rule. I have all the reasons in the world to assume that such a concentrate of wildest nonsense is not being distributed around the world with your blessing, and I think to be of service to you by directing your attention to this misfortune." Zeh knew Jensen had written to Rosenfeld, but not what Rosenfeld had said in reply. "I knew there was an answer, but Jensen never showed it to me," Zeh said. "But he showed [Rosenfeld's letter] to some other colleagues, and I noticed them chuckle about that. But for me it was very strange. I knew that there must have been some very negative comments, but I didn't even know what they [were]." Shortly thereafter, Jensen told Zeh that further work on the subject would extinguish his academic career. After that, Zeh said, "our relationship deteriorated."

Zeh was a polite man, but a stubborn one. After Rosenfeld's disastrous letter to Jensen, Zeh decided to submit his paper to several research journals anyhow. It didn't go well. One journal rejected it with a short note, stating that "the paper is completely senseless. It is clear that the author has not fully understood the problem and the previous contributions in this field." Another claimed that "quantum theory does not apply to macroscopic objects." And some journals just politely declined to publish it, with no reason given. In desperation, Zeh sent his paper off to one of the few prominent physicists who was interested in the quantum measurement problem: Eugene Wigner.

Wigner was still at Princeton, where he had, thirty years earlier, first heard about nuclear fission while in the university infirmary. In the intervening decades, Wigner's star had risen considerably: he was one of the foremost experts in mathematical physics alive and had won the Nobel Prize in 1963 for his contributions to the mathematical underpinnings of quantum physics. All the while, though, Wigner had advocated the view of quantum physics that he attributed to his friend and compatriot von Neumann (who had died in 1957). He thought that wave function collapse was a real phenomenon, and the fact that this was not incorporated into the quantum theory, he believed, merely pointed to its incompleteness. In fact, Wigner had been among the first to use the term "measurement problem," in a paper in 1963 discussing exactly this point.

The solution to the measurement problem, Wigner was convinced, lay somewhere in the special qualities of human consciousness—a view he also attributed to von Neumann. Furthermore, he didn't think there was anything controversial about this—he called this the "orthodox" view. By claiming that this was all perfectly orthodox—and by the respect that his own name commanded—Wigner managed to keep his work from being dismissed out of hand by the wider physics community, although he was not terribly successful at convincing others that consciousness had anything to do with wave function collapse. But Wigner was not dogmatic. He was willing to entertain different ideas about how quantum physics worked and how to interpret it. And he spent more time pointing out the real problem around quantum measurement than he did on promoting his own preferred solution. Over the course of the late 1950s and into the 1960s, Wigner published several papers detailing the nature of the quantum measurement problem, and pointed out flaws in various proposed solutions that claimed to resolve the problem without altering the Copenhagen interpretation or adding to the mathematical formalism of the theory. This had not earned him any friends in Copenhagen, nor had his disparaging comments on complementarity decades earlier. Zeh's mentor, Jensen, had shared the Nobel Prize in Physics with Wigner in 1963 and had sat next to Wigner at the award banquet in Stockholm afterward. The subject turned to Bohr's institute, and Jensen was surprised to hear Wigner say, "I have never been *invited* to Copenhagen."

Rosenfeld, unsurprisingly, couldn't let Wigner's heresies lie. In a series of papers in the mid-1960s, Rosenfeld and Wigner exchanged barbs, with Rosenfeld claiming that there was no such thing as the measurement problem, and that recent work by a trio of Italian physicists had explained in detail what Rosenfeld claimed Bohr had originally been driving at: that "measurement" occurred when any quantum system came into contact with a large classical object. The proof Rosenfeld and the Italians gave relied heavily on nonquantum statistical physics; Wigner and others (including Bell's old sparring partner Jauch) pointed out that this was simply incorrect—the math didn't work. For Wigner, dismissing Rosenfeld's claims was not just a matter of pointing out bad physics, nor about maintaining his own reputation. He was also concerned about the reputations of his own students, some of whom had published work on the measurement problem that Rosenfeld and the Italians were directly attacking. "It is not good taste to say about a set of articles that they do not make substantial contributions to a subject," he said in a letter to Jauch complaining about the Italians. "Needless to say, I am less concerned about myself than about other people who are much younger than I am and whose future careers such statements may hurt." Despite this back-and-forth in the physics journals of the time, the perception among the wider physics world was not that there was anything wrong with the Copenhagen interpretation. Thanks to Wigner's positioning of his view as the "orthodox" view, the perception was that there was a dispute within the orthodoxy, that there were different versions of the Copenhagen interpretation, the "Copenhagen" and "Princeton" camps, that disagreed on certain details in the quantum theory of measurement, nothing more. To be sure, a lot of unorthodox work on the foundations of quantum physics had come out of Princeton in the 1950s—Bohm and Everett chief among them—but Wigner wasn't generally associated with any of it. Indeed, Wigner's conservative Republican Party politics were diametrically opposed to Bohm's—Wigner had received a letter from President Nixon himself thanking him for his support of the Vietnam War—and the two men had little contact at Princeton. Wigner had discussed quantum physics with Everett, but their proposed solutions weren't terribly similar, and few people had heard of Everett's ideas anyhow. To the rest of the world, Wigner looked like an orthodox quantum physicist, even as he supported the work of students and colleagues who questioned Copenhagen.

"The only man who responded positively about [my paper] was Eugene Wigner. I sent him a copy," Zeh said. "I knew already that he was against the Copenhagen [interpretation]. . . . Then he encouraged me to get this published." Wigner suggested Zeh submit the paper to a new journal, *Foundations of Physics,* where Wigner was on the editorial board. Zeh translated his paper into English and added in a reference to Everett's work (which he had discovered in the meantime while researching general relativity). Zeh's paper appeared in the first issue of *Foundations of Physics,* in 1970. He hoped his ideas would now get a better hearing than Rosenfeld and Jensen had given them. He didn't have to wait long.

One of the "much younger" people Wigner was concerned about protecting was Abner Shimony. Shimony had earned his PhD in physics working with Wigner at Princeton—but he had already earned a PhD in philosophy before that. Shimony had studied with Rudolf Carnap himself in Chicago, then written a doctoral thesis on the philosophy of probability at Yale. During the course of that work, he read Max Born's book *Natural Philosophy of Cause and Chance,* which rekindled his long-standing interest in physics. "I was in the process of typing up my [philosophy] thesis (I typed the technical part and my wife, Annemarie, typed the prose part) and I told her after I read Born's book, 'When I finish this thesis and get my doctorate, I'm going back to school to get a doctorate in physics,'" Shimony recalled. "Any normal wife would have said, 'It's about time for you to get a job.' She didn't say that. She said, 'If that's what you want to do, that's what you should do.' I thought it was wonderful. I told her, 'That was your finest hour,' in Churchill's phrase. . . . [It was] an incredible act of indulgence and understanding!"

Once he arrived at Princeton's Physics Department in 1955, Shimony quickly discovered that his perspective on quantum physics was slightly different from that of most of the physicists there. "I wanted to do a thesis with Wightman," Shimony said. "His first assignment to me was an exercise: read the Einstein-Podolsky-Rosen paper and find the flaw in the argument. . . . So that was my first reading of the EPR paper, and I didn't think anything was wrong with the argument. It seemed to be a very good argument. I never saw anything wrong with it."

Shimony quickly found himself overwhelmed by the density of the mathematics that Wightman was working with and decided to switch to a different area of physics. "I . . . turned to Wigner for a problem in statistical mechanics," Shimony recalled. "One of the great side-effects of studying with Wigner was that I learned about his ideas on the foundations of quantum mechanics, especially on the measurement problem. . . . He took a stand contrary to the orthodoxy of the time that the measurement problem had not been solved by the Copenhagen Interpretation." Despite the fact that Shimony's thesis wasn't related to the interpretation of quantum physics, Shimony became Wigner's informal philosophy consultant for Wigner's papers on the measurement problem. The two men shared similar beliefs on the subject, to a point. "I already was inclined to doubt the Copenhagen solution," Shimony wrote, "because it was cognate to some of the positivist epistemological arguments of Mach, Russell, Carnap, Ayer, etc., which I had earlier studied and rejected. . . . I had long adhered to a version of realism."

But Shimony split with Wigner on the solution to the measurement problem. Shortly after completing his physics PhD in 1962, Shimony wrote a paper on the measurement problem, affirming that the problem was real—and rejecting the idea that consciousness could provide a solution. "There is no empirical evidence that the mind is endowed with the power . . . of reducing superpositions," Shimony wrote, "and furthermore, there is no obvious way of explaining the agreement among different observers who independently observe physical systems." (Shimony had never been one to shy away from disagreeing with his teachers or from voicing unpopular opinions: when he was in high school in Memphis in the 1940s, he caused trouble with his spirited defense of evolutionary theory in the classroom.) But Wigner, true to form, didn't mind Shimony's dissent—indeed, he had encouraged Shimony to write the paper. Shimony, in turn, needed the encouragement in the face of the vast indifference of most physicists to such work at the time: "It certainly was important for morale to have Wigner's endorsement of the importance of research in foundations of quantum mechanics."

Shimony took a position in MIT's philosophy department while still working on his physics PhD with Wigner and started teaching classes there on the foundations of quantum physics, aimed at upper-level undergraduate students. He also made friends in the physics and philosophy departments at several of the other universities in the Boston area, so he was not terribly surprised when, sometime in the 1964–1965 academic year, he received an envelope containing a preprint from Brandeis University written by a physicist visiting there from CERN named John Bell. "I thought, 'Here's another kooky paper that's come out of the blue.' I'd never heard of Bell," Shimony recalled. "And it was badly typed, and it was on the old multigraph paper, with the blue ink that smeared. There were some arithmetical errors. I said, 'What's going on here?' But I re-read it, and the more I read it, the more brilliant it seemed. And I realized, 'This is no kooky paper. This is something very great.'"

Shimony started thinking about how to test Bell's theorem in the laboratory "almost immediately," by his own recollection. "As soon as I'd understood what he had done, I thought 'Now, that's really interesting. . . . Have the predictions of quantum mechanics been examined carefully in such situations?' Then I thought I knew one other relevant piece of literature." Shimony asked his friend Aharonov whether Wu's old experiment could be modified to test Bell's theorem; Aharonov told him (incorrectly) that the experiment already provided such a test. "Aharonov is a very fast thinker and a very fast talker, and I was in awe of him," Shimony recalled. "I . . . thought, 'He's right. Maybe he's right. But maybe he isn't right.' The more I thought of it, the less convinced I was."

Shimony worked on understanding the issue on and off for several years, not getting anywhere for a while, until 1968. That year, he moved to Boston University for his dream job: a joint appointment in the Physics and Philosophy Departments. Shortly thereafter, he took on a physics graduate student named Michael Horne and assigned him the task of looking into how to test Bell's theorem. "The more he read, the less optimistic he was that Wu's experiment could be used to test Bell's Inequality," Shimony recalled. Shimony and Horne hit the library and soon found the Kocher-Commins experiment, which Shimony immediately recognized was what they needed. "By March 1969 the main lines of the work of Horne and myself were complete," Shimony said. "I told Mike Horne . . . nobody else is working on such far out things, and we can simply prepare a good paper at our leisure. I was wrong." Looking through the program for the upcoming meeting of the American Physical Society that April, Shimony saw an abstract titled "Proposed Experiment to Test Local Hidden-Variable Theories," describing precisely the experiment that he and Horne were preparing to do. The author was another physicist Shimony had never heard of: one John Clauser.

"As soon as the abstract appeared, I got a phone call from Abner Shimony," Clauser said. Shimony had gone to Wigner when he saw Clauser's abstract, fearing Clauser had scooped him; Wigner suggested joining forces with him instead. Shimony invited Clauser to meet with him, Horne, and Richard Holt, a graduate student at Harvard that Shimony had also recruited to the cause. Clauser agreed, and the four men started work on a paper together. "I was very pleased that Clauser agreed," Shimony wrote to Wigner after the meeting. "This is certainly the civilized way to handle the matter of independent discovery." After finishing his PhD thesis at Columbia, Clauser spent several weeks in Boston working with Shimony and the others, hammering out a draft of their paper. But Clauser had taken a position as a postdoc at Berkeley and couldn't stay long enough to polish the draft of the paper. An avid sailor, Clauser had made plans to sail his boat (which he had lived in, docked at the East River, during his time at Columbia) to his new job in California. "Originally, we were just going to sail the boat all the way to Galveston and put it on a truck there, and truck it across to LA and sail it up the coast to Berkeley. It turns out we ran into Hurricane Camille, so we got kind of stopped at Fort Lauderdale," Clauser said. "Abner [Shimony] knew my schedule. And . . . he would send off his re-drafts to all of the various marinas in the next city where we put in, some of which I picked up, and some of which are probably still sitting there for all I know. While I was sailing, I would be writing furiously away and editing various things. And we'd get on the phone and chatter about various versions, and we'd keep swapping drafts." By the time Clauser arrived in Berkeley, the paper was complete, and Shimony sent it off for publication.

The Clauser-Horne-Shimony-Holt (CHSH) paper recast Bell's mathematics in a form more amenable to a laboratory test and laid out a detailed proposal for an experiment that would determine whether Bell's inequality was violated. The experiment CHSH proposed was similar in spirit to the setup at Ronnie the Bear's casino from Chapter 7. Rather than pairs of roulette balls, the CHSH experiment uses pairs of photons with entangled polarizations. The CHSH experiment proposed sending each photon through a polarizer pointing in one of two different directions (see Figures 9.1 and 9.2), and repeating the experiment with many pairs of entangled photons. Just as each roulette ball at the casino landed on red or black, each photon would either pass through a polarizer or be blocked by it. Comparing the behavior of many pairs of these photons would test Bell's theorem. If each pair of entangled photons had a prearranged plan for how to behave at each of the two polarizers, then the results would satisfy Bell's inequality. But quantum physics predicted that the photons would violate Bell's inequality, just as the roulette balls did at Ronnie's casino.

Whatever the outcome, Clauser, Shimony, and the others knew that the experiment would be immensely important: either it would show that quantum physics was wrong, blowing up a cornerstone of modern physics and garnering a nearly instant Nobel Prize, or the quantum predictions

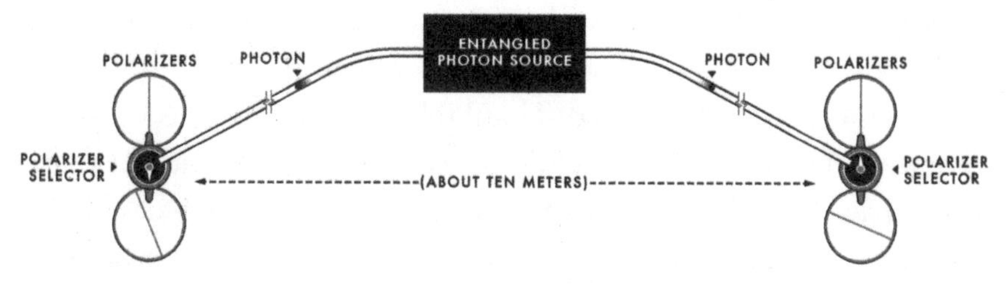

Figure 9.1. Testing Bell's theorem in the laboratory.

would be correct and Bell's inequality would be violated, meaning nature must be nonlocal (or something even stranger is going on). Clauser was still optimistic that the experiment would show that Bell's inequality was not violated—he estimated a 50 percent chance that quantum physics was wrong. But, like Bell, Shimony suspected that the experimental results would match the quantum-mechanical prediction—as did nearly everyone else. "Aharonov bet [Clauser] $100 against $1 that the outcome would favor quantum mechanics," Shimony wrote to Wigner. "I am much more conservative than Clauser in estimating the outcome. However, in view of the difficulty of the measurement problem in quantum mechanics and of the fact that hidden variable theories do offer a solution, I would not entirely discount the possibility of an outcome in favor of [local] hidden variables."

Actually performing the experiment fell to Clauser. He had been hired as a postdoc at Berkeley to do radio astronomy with Charles Townes, an astrophysicist who had won the Nobel Prize several years earlier for inventing the laser. When Clauser arrived, he told Townes about his hopes to adapt the Kocher-Commins experiment—which had been done at Berkeley—to perform a test of Bell's theorem. "I said, 'Hey look, I've got this great experiment I want to do,'" Clauser recalled, "and [Townes] goes, 'Well, why don't you give my group a seminar and tell me how it's all gonna work. And in the meantime, we'll haul up Gene Commins.'" So Clauser gave a talk explaining Bell's inequality and how the Kocher-Commins experiment could be modified to test it, hoping to get Townes interested and persuade Commins to loan him his experimental equipment. But Commins was thoroughly unimpressed by Clauser's talk. He had originally envisioned his experiment with Kocher as a mere classroom demonstration, not as a real test of EPR. That experiment had ended up being far more difficult and time-consuming than he had anticipated, and the last thing he wanted to do was to sink more time and money into a project that he saw as pointless. "Commins thought it was a total crock," Clauser said. Thankfully, Townes disagreed. "Townes thought 'Y'know, this is an interesting experiment.' And without that, I would have been dead. . . . At the end of [my seminar], Townes kind of puts his arm around Gene Commins and says, 'Well, what do you think of this, Gene? It looks like a very interesting experiment to me.'" Townes

Figure 9.2. John Clauser and one of his Bell experiments in Berkeley, 1975.

managed to convince the reluctant Commins to loan Clauser the equipment, to split the cost of the experiment—and to lend Clauser a graduate student from Commins's group, Stuart Freedman. Clauser and Freedman spent the next two years scrounging up the rest of the equipment they'd need for the experiment—"I've gotten pretty good at dumpster diving," Clauser boasted later—including an old telephone relay that they repurposed to control the motions of the polarizers. Once the equipment was assembled and tested, Clauser and Freedman assiduously collected two hundred hours' worth of data. Finally, in 1972, Clauser and Freedman published their results: quantum mechanics had survived. Bell's inequality was violated—and something awfully strange was going on in nature.

Shortly before Dieter Zeh's paper on decoherence was published in 1970, he received an invitation to speak at a summer school on the foundations of quantum physics, sponsored by the Italian Physical Society. Strangely, this summer school had its origins in the political and cultural tumult that had swept the world in 1968. Left-wing Italian physicists, generally younger, had been agitating for a reevaluation of the relationship between physics and the wider world, the social responsibilities of physicists, and the philosophical foundations of physics itself. The older, more conservative Italian physicists were uninterested in upsetting the status quo. With the society on the brink of splitting entirely, the board accepted the proposal to hold a summer school in Varenna on the foundations of quantum physics. They invited the French physicist Bernard d'Espagnat, a former student of de Broglie's and a colleague of Bell's at CERN, to organize the school, and, in turn, Wigner had suggested inviting Zeh to d'Espagnat.

The 1970 Varenna summer school was later dubbed "the Woodstock of quantum dissidents," and with good reason. The speakers, aside from Zeh, included David Bohm, Louis de Broglie, Eugene Wigner, Abner Shimony, Josef Jauch, Bryce DeWitt, and John Bell himself. "When I arrived at Varenna, I found the participants (John Bell included) in hot debates about the first experimental results regarding the Bell inequalities," Zeh recalled later. "I had never heard of them." Nonetheless, Zeh was relieved and gratified to find that Bell and the others valued his work, even if some of them disagreed with his conclusions. Wigner, in his keynote address to the conference, laid out six possible solutions to the measurement problem; he included Zeh's combination of decoherence and many-worlds among them.

But when Zeh returned to Heidelberg, he found his colleagues there as disdainful as ever of his work in quantum foundations—so much so that his career had stalled entirely. "There was a lot of naiveté on my side," Zeh recalled. "I thought when you have a good idea and you publish it, then everybody should read that and accept that, which is of course quite wrong." Zeh soldiered on, trying to look on the bright side. "I concentrated on these issues because I had decided that my career was destroyed," Zeh said. "I would never get a [full] professorship because of these things already, and so I said, 'Now I can just do what I like.'" Zeh's own job was safe as long as he stayed at Heidelberg; he had tenure, though he was denied promotion. "I did not have to suffer," he recalled. "[But] my students never had a chance. I had not expected that." When Zeh's students went looking for academic work, they were denied job after job, since they had not done "real" physics. "This," Zeh said, "was something I will never be ready to forgive." Zeh dubbed these the "dark ages of decoherence." They would not lift for over a decade.

Despite his groundbreaking experiment, John Clauser's career had also stalled out—and unlike Zeh, he didn't have a permanent position. When his postdoctoral position at Berkeley ended, Clauser struggled to find another job. "I was sort of young, naive, and oblivious to all of this," Clauser recalled. "I thought it was interesting physics. I had yet to recognize just how much of a stigma there was, and I just chose to ignore it. I was just having fun." Clauser's PhD advisor, Pat Thaddeus, wrote a "recommendation" letter for Clauser's new job search, in which he warned prospective employers that Clauser's Bell experiments were "junk science." Thankfully, Clauser was alerted to the problem in the letter and didn't use it for his job applications. Instead, Shimony, d'Espagnat, and others wrote glowing letters in support of Clauser. But Thaddeus wasn't the only one who thought Clauser's work wasn't truly scientific. "When I saw d'Espagnat last week he had a letter from the Dep't Chairman at San Jose, inquiring whether what you have been doing is real physics," Shimony wrote to Clauser. "Needless to say, he'll write a strong letter answering the question in your favor." But their efforts led nowhere: Clauser couldn't get a permanent academic job.

Clauser, at least, was not suffering from the same isolation that Zeh faced in Heidelberg. After arriving in Berkeley, Clauser had fallen in with a group of eccentric physics students and junior faculty who shared his interests in the foundations of quantum physics. Inspired by the counterculture of the time and place—Haight Ashbury, the center of the hippie movement, was just a short hop across the San Francisco Bay—these physicists hoped that their investigations would lead to a new way of doing physics, in line with their interests in Eastern philosophy, extrasensory perception, and the mind-expanding powers of psychedelic drugs. They called themselves the "Fundamental Fysiks Group," and their discussions centered around turning on, tuning in, and dropping out of the Copenhagen interpretation.

Although this group certainly provided a moral support network for Clauser, it couldn't help him find a job—indeed, most of the other members of the group had trouble securing permanent positions for themselves. The prejudice against work in quantum foundations wasn't the only reason: if anything, the lack of work was part of what drove their interest in the subject. The postwar boom in physics funding that had fueled the rise of "shut up and calculate!" was coming to a sudden and precipitous end. As the Cold War cooled off into détente, deep cuts in defense spending at the end of the 1960s meant less funding for physicists from the US government. And protests against classified research on university campuses across the country weakened the link between academia and the military-industrial complex. The upshot was a dramatic shortage of jobs for physicists. Immediately after World War II, physics had grown faster than any other academic field—now, it shrank faster than the rest. From the end of the war up through the mid-1960s, there were never enough freshly minted PhDs in physics to fill all the available jobs for physicists. But by 1971, the American Institute of Physics job placement service had 1,053 applicants for a grand total of fifty-three jobs. No wonder, then, that the pragmatic allure of good work in other fields was no longer enough to keep the Fundamental Fysiks Group from puzzling over the foundations of quantum physics—and no wonder that John Clauser had trouble finding work when even "respectable" physicists were out of a job.

Finding a job wasn't Clauser's only worry. His experimental results were also being called into question. A second test of Bell's inequality, conducted by Holt and Francis Pipkin at Harvard, directly contradicted Clauser's results—they found that Bell's inequality held, suggesting that nature was local and quantum physics was wrong. Another experiment was needed to break the tie. At Berkeley, Clauser set up a modified version of Holt and Pipkin's experiment, hoping again to find that quantum physics was wrong. Meanwhile, Ed Fry and Randall Thompson at Texas A&M University set up a similar experiment but used cutting-edge "tunable" lasers to dramatically cut down on the time needed to collect the data. In 1976, both Clauser and the Texas team announced their results: quantum mechanics was vindicated, and Clauser and Freedman's original result stood. Quantum nonlocality was real.

But Clauser's continuing work in quantum foundations still stood in the way of finding a permanent job. Few physicists valued his work. One of the exceptions, unsurprisingly, was John Bell. In the spring of 1975, Bell and d'Espagnat started organizing a conference on experimental tests of the foundations of quantum physics, to be held the next spring in Erice, a small town on the coast of Sicily, with Clauser planned as one of the guests of honor. Bell sent a letter inviting him, but Clauser, still hunting for a job, didn't write back immediately, because he wasn't sure where he'd be the next year. After waiting for a month, Bell, worried by Clauser's silence, sent him an urgent telex. "WITHOUT YOU MEETING WOULD BE HAMLET WITHOUT PRINCE," Bell wrote. "MAY WE PUT YOUR NAME ON THE POSTER?" Clauser happily accepted, and traveled to Erice in April 1976 to bask in the professional recognition that he had thus far been denied.

Zeh, Clauser, and the Fundamental Fysiks Group weren't the only ones facing professional consequences for investigating the foundations of quantum physics. Nearly every physicist of the time learned to avoid such questions as part of their training. But it was rarely an explicit part of their training—there was no concerted, deliberate effort to keep young physicists away from research on the foundations of quantum physics. There were other factors at work that merely had the side effect of keeping such research outside of the mainstream of professional physics. These historical factors are the same ones we've seen throughout this book. There was the lack of interest in foundational questions that came out of the postwar scientific funding model, which rewarded research with clear and concrete results in particular areas of physics. There was the ascendancy of American physics, which always had a more pragmatic bent than its European counterpart had. Philosophy played a role: positivism provided a variety of convenient ways to dismiss concerns about the Copenhagen interpretation. And the association of hidden-variables theories with communism (especially after Bohm), the overwhelming quantity of military funding in physics, and the still-fresh memories of the McCarthy era created a toxic brew: anyone flirting with hidden-variables theories opened themselves up to suspicion about their political convictions, a suspicion that could threaten the sources of funding that kept the lights on at nearly every physics department in the United States.

Young physicists were also discouraged from looking into the foundations of quantum physics precisely because the theory was so successful. With so many other fruitful avenues of research available, why bother with something as stubborn and abstract as quantum foundations, especially when Einstein himself hadn't been able to make sense of it? "As part of the 'common wisdom' taught in typical undergraduate and graduate curricula, students were told simply that Bohr was right and Einstein was wrong. That was the end of the story, and the end of the discussion," Clauser recalled. "Any student who questioned the theory's foundations, or, God forbid, considered studying the associated problems as a legitimate pursuit in physics was sternly advised that he would ruin his career by doing so." And the astonishing success of quantum physics in the laboratory, and the sheer power of its theoretical apparatus in explaining an enormous variety of phenomena, made questioning its foundations an even more uncomfortable task. As J. J. C. Smart had noted (see the end of Chapter 8), it was unreasonable to expect purely philosophical arguments against the Copenhagen interpretation to sway the vast majority of physicists in reevaluating the philosophical foundations of such a successful theory. Arguments for an alternative interpretation would be needed as well. But most physicists were still convinced that alternatives to Copenhagen were impossible—Bell's detailed takedown of von Neumann's proof wasn't well-known yet. There was also the suspicion that quantum foundations wasn't "real" physics, because it was entirely removed from experimental work. Bell had shown this wasn't true either, but recognition of this fact was also slow. And until that recognition was wider, careers suffered—especially those of young physicists. Zeh and Clauser pursued work in quantum foundations despite constant discouragement, but they had done that work only after safely earning a PhD. Many physicists who were interested in such questions were discouraged from pursuing them even earlier in their careers—and those who didn't listen paid a price.

David Albert was a physics PhD student at Rockefeller University, in New York City, in the late 1970s. Albert had always been interested in philosophy, and one night early in graduate school, he was up at four in the morning reading a book by the philosopher David Hume when the true gravity of the quantum measurement problem hit him with full force. Thinking about Hume "somehow made it vivid that what happens to the wave function during a measurement ought to be a straightforward mechanical consequence of the Schrödinger equation, and not something that requires a separate postulate," he recalled. "It became very vivid to me that this isn't going to work, and this was the moment when I understood the measurement problem. . . . This night changed my life. I said, OK, that's what I want to work on. I want to work on the measurement problem."

None of the physicists at Rockefeller worked on the foundations of quantum physics, so Albert wasn't sure how to proceed. "There was nobody to talk to at Rockefeller. [A friend] said why don't you write to Aharonov? He was the only person people could think of at the time in physics, and I had no idea anybody was interested in these things in philosophy." Albert sent a letter off to Aharonov, who was in Israel at the time, without having met him—and Aharonov replied. "He was very generous to me," Albert said. The two started a long-distance research collaboration, working on locality and the measurement problem. "We actually published a couple of papers in *Physical Review* together, by snail mail in those days, before we had ever met each other."

But when Albert suggested that his work with Aharonov would make a good basis for a PhD thesis, the Rockefeller Physics Department balked. "I said to them, I've been working on this measurement problem stuff with Aharanov, I'd like to do my thesis about that," Albert recalled. "A few days after that I was asked to come to the Dean's office at Rockefeller, the Dean of Graduate Students, and told that under no circumstances was anybody going to do a thesis about that in the Physics Department at Rockefeller University, and if I further insisted on that, I was going to be kicked out of the program." They handed Albert a different subject for his thesis. "It was a very calculation-heavy problem about Borel resummation in φ^4 field theory . . . which was clearly being assigned because it was thought it would be good for my character," Albert said. "There was an explicitly punitive element there. And they said, here are your choices: you can do *this* problem and no other, or you can leave the program."

After discussing it with Aharonov, Albert decided to stick it out at Rockefeller. "[Aharonov said] why don't you just put your head down, do this problem that they've assigned to you, I can give you a postdoc at Tel Aviv as soon as you get your PhD, and you'll be on your way," Albert recalled. "And that's what I did. But it was made very clear to me what the rules of the game were, and that there was to be no more talk about the measurement problem in the Physics Department at Rockefeller."

Ultimately, Albert used his postdoc with Aharonov as a launchpad to switch careers to philosophy of physics. But other physics students interested in foundations were not so lucky. And the means used to suppress inquiry into the foundations of quantum physics were not merely limited to career stagnation and withholding degrees. As Zeh discovered when he tried to publish his first paper on decoherence, physics journals were generally reluctant at best and hostile at worst when faced with submissions on the foundations of quantum physics. *Physical Review* actually had an explicit editorial policy barring papers on quantum foundations unless they could be related to existing experimental data or made new predictions that could be tested in the laboratory. "It should not be overlooked that physics is an experimental science," wrote *Physical Review*'s editor in chief in 1973, Samuel Goudsmit, the Dutch physicist who had led the Alsos mission in World War II. "No physical theory is significant unless it can be related to experimental data." (Clauser pointed out that these restrictions would have barred *Physical Review* from publishing Bohr's response to the EPR paper, as they had done forty years earlier.) There were only a handful of journals that would accept papers on quantum foundations, among them *Foundations of Physics*, where Zeh's paper had finally ended up.

To solve this problem, the quantum underground founded a new ersatz "journal" called *Epistemological Letters*. Billing itself as a permanent written symposium on "hidden variables and quantum uncertainty," this samizdat was hand-typed, published by mimeograph, and overseen by an informal collection of editors, including Shimony. "*Epistemological Letters* are not a scientific journal in the ordinary sense," it boldly declared on the back cover of every issue (and referring to itself in the third person plural). "They want to create a basis for an open and informal discussion allowing confrontation and ripening of ideas before publishing in some adequate journal." In its pages, the verboten was discussed: the measurement problem, the true meaning of Bell's theorem, and more. Papers by Bell, Shimony, Clauser, Zeh, d'Espagnat, and Karl Popper appeared in its pages over its eleven-year run. "The variety of the contributions and the vigor of the debates showed that the purpose [of the journal] was very well accomplished," Shimony said later. "The reputation of the written symposium spread rapidly, and many people throughout the world wrote to be added to the list of recipients."

For the first time since 1935, there was a cohesive community of physicists working on the foundations of quantum physics. They had a shared theoretical and experimental research program, they had a journal of their own (such as it was), and they even had occasional conferences. But it was still not safe to publicly identify as a member of that group, especially for young researchers—at least not yet.

In 1974, a young French physicist named Alain Aspect arrived at the Institut d'Optique just outside of Paris. He was freshly returned from three years teaching in Cameroon, and he was looking for a research topic to earn his PhD while he worked as a lecturer at the institute. A professor mentioned an interesting seminar he had just heard given by an American physicist named Shimony, and from there Aspect found Bell's paper. "When I read the paper of Bell, I was absolutely fascinated. I thought it was the most exciting subject I had ever read," Aspect recalled. "It's like love at first sight. . . . So, at that point, I say, OK, I want to do my PhD on that." Aspect read Clauser and Freedman's paper and the conflicting results of Holt and Pipkin's experiment, and decided not to compete with them. "I was sure that somebody would settle the conflict much before I would start," Aspect said. "If I want to enter the game, I have to do something different. And I look carefully [at] the paper of Bell, and clearly in the conclusion, Bell said what was the important experiment to do. It was changing the orientation of the polarizers while the photons were in flight."

Bell's idea was simple in theory but enormously difficult to carry out in practice. When Clauser and the others had performed their tests of

Bell's inequality, they had selected the polarizer angles randomly—but that random selection had happened *before* the pair of entangled photons was emitted from the source. In theory, the photons could have somehow detected those randomly selected settings before leaving the source, through some as yet unknown physics. If that had happened, there was no need to invoke nonlocality to explain the results of Clauser's experiment—some new purely local physics could explain it. The only way to rule this out would be to set the polarizers randomly while the entangled photons were already flying apart from each other. That way, no signal traveling at the speed of light could reach both photons after the polarizers had been set. "John Bell, I think, believed that . . . possible discrepancies with quantum predictions would show up in experiments where you rotated the [polarizers] rapidly," Clauser said later. The problem was that this required switching the polarizers enormously fast, faster than the time that it took light to travel from the photon source to the polarizers themselves. Typically, that distance was about ten meters, meaning that the polarizers needed to be switched in less than forty nanoseconds. The technical challenge was huge. "I began to think, carefully, how could I do it," Aspect recalled. "Finally I came to a conclusion that it might be possible." Aspect went back to the professor who had pointed him in this direction, Christian Imbert, and asked whether he could try to do this experiment in his lab. Imbert "said look, I don't understand what you tell me, but it looks interesting, so go to Geneva, talk to John Bell," Aspect said. "If John Bell tells you that it is interesting, then I will offer you the possibility to do it in my lab."

So Aspect went down to Geneva in the spring of 1975 to meet with Bell, just as Bell was starting to organize the conference in Erice. "I explained to him my idea, and he said nothing, he was very quiet," Aspect recalled. "And then the first question [he asked] was, 'do you have a permanent position?'" Aspect was confused. "I said, why are you asking me [this] question? He said, 'answer first.'" So Aspect explained that his position was, in fact, permanent—despite the fact that he was still working on his PhD, his lecturer position at the Institut d'Optique had the French equivalent of tenure. Satisfied, Bell explained his question to Aspect. "This kind of physics is not popular at all," he said, "and so you are going to have difficulties. So I would not recommend that you go into that if you are not tenured." Keenly aware of the professional dangers that came with work in foundations of quantum physics, Bell had made a habit of discouraging young physicists from pursuing the subject until they were more established in their careers. But Aspect, thankfully, was already safe. "Then he strongly encouraged me," Aspect recalled. "He told me that it was really *the* experiment to do. He told me, if you can do an experiment where you change the orientation of the polarizers while the photons are in flight, yes, this is [the] real experiment to do."

Aspect returned to Paris and got to work assembling his experiment in Imbert's lab. "Basically I borrowed everything, except for one thing, at one point I needed to buy a laser," Aspect said. "So I got the money to buy one laser. That's the only grant I got. All the rest is equipment borrowed here and there. Or built in the workshops of the Institut. I had no competition, so I was not under pressure. Nobody was interested." Over the course of the next six years, Aspect assembled and tested the delicate experimental equipment, eventually pulling in an undergraduate, Phillip Grangier, an intern, Jean Dalibard, and a research engineer, Gérard Roger, to help. Meanwhile, unbeknownst to Aspect, Imbert was shielding him from the criticisms and concerns of the rest of the institute. "Imbert acted as an umbrella," Aspect said. "He protected me against all these people who were telling him, you are guilty to let this young guy waste his time, he should do some real physics rather than doing that. And I did not realize that so much." Finally, in 1982, Aspect and his collaborators published their results: Bell's inequality was still violated, even when the polarizers were switched while the photons were in flight.

Aspect followed up on his experimental tour de force with an even more astonishing and difficult act. "If you speak to an 'ordinary' physicist about hidden variables and testing hidden variable theories against quantum mechanics, basically, they are not interested," Aspect said. "But, if you tell them, there is a nice experiment looking for correlations, and these correlations are extraordinary, then they are likely to listen to you, because physicists like nice experiments, and [testing Bell] is a nice experiment, there is no doubt." Aspect, a teacher at heart—"I was myself fascinated, so when you are fascinated, you should be able to transmit your fascination, right?"—found a way to talk with other physicists about Bell's theorem. "I like to explain. And I think I found the right way to explain . . . why [this experiment] is interesting, in less than 30 minutes," Aspect said. "I got a way of explaining to an ordinary physicist why it is really interesting. And so, after a while, you are invited to give a seminar, and if the seminar is well received, there are other people in the room who will invite you to give another seminar here and there, and [ultimately] I really gave many many many seminars to explain Bell's inequalities, and the interest of doing these experiments, the way I was understanding it." Aspect's series of talks turned out to be one of the final cracks in the edifice of silence that Copenhagen had constructed. In the 1980s, for the first time in half a century, large numbers of physicists began to openly question the Copenhagen interpretation. Copenhagen still held a strong majority, and not all of those who questioned it thought it was wrong. But the avalanche of dissent, held back for so long, finally came barreling down the mountain. The new field of quantum foundations had arrived.

10

Quantum Spring

Reinhold Bertlmann starts each day with a tiny act of rebellion. He doesn't look like a rebel at first glance—his impeccably trimmed facial hair and his professorial taste in clothes match the formal style of his hometown, Vienna, which has never really shed its imperial facade. But Bertlmann's sartorial conformity stops just short of his shoes: his socks are always mismatched. "I've worn socks of different colors since my early student days. And I am a student of the so-called '68 generation," Bertlmann says. "And this was my little protest. My hidden protest. To wear socks of different colors, because I realized that whenever somebody sees this, they were either shocked—they said, 'how stupid, how can you do it?'—or they laughed about it and thought I am crazy."

Forty years ago, Bertlmann's rebellion was more obvious. With shoulder-length hair and an unruly beard, he stuck out when he first arrived at CERN in 1978. "An American would say [I was] a hippie or something," he recalled. Nonetheless, Bertlmann's open, friendly smile attracted many friends at CERN, and most eventually noticed his socks. But John Bell never mentioned them. Bertlmann and Bell worked together for two years on a thorny calculation in particle physics, totally unrelated to Bell's theorem. "He did not say one word [about my socks], not one word," Bertlmann recalled. And Bertlmann, in turn, did not ask Bell about the rumor he had heard in the CERN canteen: that Bell had done some kind of important work in the foundations of quantum physics. "People said, 'Oh, you're collaborating with Bell? He is somehow

Figure 10.1. John Bell's cartoon of Bertlmann's socks, 1980.

famous in quantum physics.' And I always asked, 'what did he do?' 'Oh, he did something, you don't have to worry about it, because quantum mechanics works anyhow.' Nobody at CERN could explain what the Bell inequalities are." But one day in the fall of 1980, while Bertlmann was visiting Vienna for several weeks, he was suddenly confronted by Bell's theorem in an unexpectedly personal way. A colleague of Bertlmann's came running down to his office brandishing a new paper by Bell. "He just came in waving this [paper]," Bertlmann recalled. "And he said, "Reinhold, look what I have! Now you are famous!"

Bertlmann, astonished, read and reread the title of the paper: "Bertlmann's Socks and the Nature of Reality." The paper even came with a small cartoon, drawn by Bell himself (Figure 10.1).

"The philosopher on the street, who has not suffered a course in quantum mechanics, is quite unimpressed by Einstein-Podolsky-Rosen correlations," Bell wrote. "He can point to many examples of similar correlations in everyday life. The case of Bertlmann's socks is often cited. Dr. Bertlmann likes to wear two socks of different colors. Which color he will have on a given foot on a given day is quite unpredictable. But when you see that the first sock is pink you can be already sure that the second sock will not be pink. . . . There is no accounting for tastes, but apart from that, there is no mystery here. And is not the EPR business just the same?" Bell briefly outlined the Copenhagen interpretation and its history, explaining that "influenced by positivistic and instrumentalist philosophies, many came to hold not only that it is difficult to find a coherent picture [of the quantum world] but that it is wrong to look for one—if not actually immoral then certainly unprofessional. Going further still, some asserted that atomic and subatomic particles do not *have* any definite properties in advance of observation." Then Bell brought it back to Bertlmann's socks:

> It is in the context of ideas like these that one must envisage the discussion of the Einstein-Podolsky-Rosen correlations. Then it is a little less unintelligible that the EPR paper caused such a fuss, and that the dust has not settled even now. It is as if we had come to deny the reality of Bertlmann's socks, or at least of their colors, when not looked at. And as if a child has asked: How come they always choose different colors when they *are* looked at? How does the second sock know what the first has done?

Bell himself had answered the question of why entangled particles can't be like Bertlmann's socks—his theorem, and the experiments of Clauser and Aspect, showed that something much stranger must be going on. "Certain particular correlations, realizable according to quantum mechanics, are *locally inexplicable.* They cannot be explained, that is to say, without action at a distance," Bell wrote. "You might shrug your shoulders and say 'coincidences happen all the time' or 'that's life.' Such an attitude is indeed sometimes advocated by otherwise serious people in the context of quantum philosophy. But outside that peculiar context, such an attitude would be dismissed as unscientific. The scientific attitude is that correlations cry out for explanation."

Aspect's charm offensive had done wonders for quantum foundations, but indifference to the subject was still widespread among physicists. And, as Clauser knew well, there was little hope of finding a full-time job doing work on quantum foundations. Bell himself spent nearly all of his time at work doing particle physics with relativistic

Figure 10.2. John Bell in his office at CERN discussing tests of his theorem, 1982.

quantum field theory, which he knew worked very well—"for all practical purposes" as he said—just as he had done with Bertlmann at CERN. But Bell's pressing concern about the foundations of his field was never far from his mind. "I am a quantum engineer," he once announced at the start of a talk, "but on Sundays I have principles." Bell, normally soft-spoken, could turn on a dime if a visiting speaker said something silly about quantum foundations. "In conferences . . . he would usually say nothing," recalled another one of his younger colleagues, Nicholas Gisin. "But if someone would say wrong things, especially on [quantum] interpretations . . . he was erupting, and then making with his Irish accent very sharp comments and very down to the point and when that started, the speaker could just dissolve and liquify."

But this kind of fire didn't come from anger. It came from Bell's deep moral convictions about the integrity of science, the same kind of moral convictions that had led him to become a vegetarian decades earlier. While the Copenhagen interpretation was unwilling to grapple with the measurement problem, Bell was unwilling *not* to grapple with it. He had no patience for the vagueness of the Copenhagen interpretation and its willingness to kick the can down the road. Though he was wary of encouraging young physicists to devote their careers to foundations, he was patient and kind with anyone who wanted to talk with him on the subject. "When I was asking my questions about foundations, he would be extremely nice and take the time to answer," Gisin recalled. "And when he was coming to my lab to talk . . . he had this red hair, and this hat, and this little pom pom on top. He was not at all looking like, you know, The Great John Bell."

Bell "was always smiling . . . and he had a weakness for nonconforming people," Bertlmann said. "We had discussions, not only about physics, but also about politics, about art, and so on." Yet until Bertlmann saw Bell's paper, they had not discussed Bell's work in foundations. "When I saw [that paper], it kicked me out of my socks," he recalled. "Totally knocked me down, you can imagine. I was so excited, my heart trembled, and then I remember I went to the telephone and phoned with him. I was excited, he was very calm." Once Bertlmann had recovered, he resolved to learn more about quantum foundations. "I was shocked, and then I had to dig into this field."

Young physicists like Gisin and Bertlmann weren't the only ones attracted to quantum foundations. Older, established physicists were turning their attention to the field too—even physicists who had previously dismissed the field as irrelevant or impractical. Back when John Clauser was working on his first experimental test of Bell's inequality, in the very early 1970s, he went down to see his family in Pasadena one Christmas. Clauser's father, Francis, was a professor at Caltech at the time. "I got there, and [my dad] said 'Oh, I set up an appointment for you with Feynman!'" Clauser recalled. "I said, 'Oh, Jesus. . . .'" Richard Feynman was a legend, among the most prominent and brilliant physicists alive. He had been one of the architects of quantum electrodynamics, the theory of how light and matter interact—an achievement that had garnered him a Nobel Prize in 1965. Feynman had started his career as a student of John Wheeler, and, like his mentor, he had few qualms about the Copenhagen interpretation. Clauser was worried that his work on Bell's little-known theorem would be dismissed out of hand—and he wasn't wrong. "I walk into Feynman's office, and he was instantly hostile," Clauser said. "He said 'What are you doing? You don't trust quantum mechanics? Once you show something's wrong with it, come back and we'll talk about it. Get out of here, I'm not interested.'"

But by the time Alain Aspect came to speak at Caltech in 1984, Feynman had changed his tune. "He was extremely friendly," Aspect recalled. "He made interesting comments." After the talk, Feynman invited Aspect back to his office, where they discussed his work further. Once Aspect returned home, he received a letter from Feynman, following up with further praise: "once again let me say, your talk was excellent."

Though it's unlikely that he learned much from Clauser's ill-fated visit to his office, Feynman was certainly well aware of Bell's theorem by the time Aspect arrived at Caltech. In the aftermath of the first Bell experiments, there had been a flurry of articles on the subject, explaining Bell's theorem to both physicists and the public at large. D'Espagnat had written the first popular account of Bell's work, in *Scientific American* in 1979. Shortly thereafter, popular books on quantum physics published by physicists and writers associated with the Fundamental Fysiks Group in Berkeley, such as *The Tao of Physics* and *Quantum Reality*, covered the subject as well. And a celebrated series of articles on Bell's theorem by the distinguished Cornell physicist N. David Mermin elucidated the subject for his fellow physicists through a set of particularly simple thought experiments, which rapidly became the standard way of teaching the subject. Feynman, respected among physicists for the clarity of his teaching as well as the depth of his physical insight, was an instant fan of Mermin's work. "One of the most beautiful papers in physics I know of is yours," Feynman wrote to Mermin in 1984. "All my mature life I have been trying to distill the strangeness of quantum physics into simpler and simpler circumstances. . . . I was recently very close to your description when your ideally pristine presentation appeared."

Feynman himself had explained Bell's theorem during his keynote speech at a Caltech conference in 1981 (though strangely he did not actually mention Bell himself in the course of doing so). The conference was on a seemingly unrelated subject—the physics of computation—yet Feynman showed that Bell's theorem held the answer to a crucial question in this field. "Can physics be simulated by a universal computer?" Feynman asked the conference. "[The] physical world is quantum mechanical, and therefore the proper problem is the simulation of quantum physics—which is what I really want to talk about," he continued. The answer, for a normal computer operating under ordinary conditions, was no: using simple ones and zeros in the usual fashion, without strange long-range connections within the computer or some other kind of trick, limited the computer to simulating local physics, rendering it impossible to fully simulate quantum effects. But, Feynman speculated, there might be another way to get the job done. "Can you do it with a new kind of computer—a quantum computer?" Feynman wondered. "I'm not sure. . . . So I leave that open."

Several years later, a young physicist named David Deutsch took up where Feynman had left off. In 1985, Deutsch proved that a quantum computer—a computer taking full advantage of the difference between quantum physics and classical physics—could perform tasks more efficiently than an ordinary classical computer. Deutsch's proof opened up the possibility of practical technological applications for Bell's ideas, a feat that Bell had never foreseen. But Deutsch didn't actually provide an example of how a quantum computer could outperform a classical one—he had merely proven that it could be done in theory. Finding an algorithm for a computer that hadn't been built, to outperform all existing ones, was a tall order.

Nearly a decade later, a brilliant mathematician named Peter Shor filled that order in spectacular fashion. In 1994, he devised a quantum algorithm that could rapidly factor extremely large numbers—an extraordinarily important result. Not only was this a true demonstration of what Deutsch had proven possible, but Shor's algorithm had massive practical consequences. It's difficult for ordinary computers to factor large numbers—and, as Shor well knew, this difficulty was the basis for nearly all forms of practical cryptography, especially for secure communications over the newly burgeoning Internet. Shor had demonstrated that any kind of secure financial transaction over a computer network—from buying books to trading stocks—would be impossible to accomplish by conventional means in a world with working quantum computers.

But, by that time, quantum information theory had also yielded a solution to this problem: quantum cryptography. In fact, two forms of absolutely secure communication had been devised based on work first done in quantum foundations. One method, developed by Charles Bennett and Giles Brassard in 1984, was based on a result known as the "no-cloning theorem," which had been proven in response to work done by the Fundamental Fysiks Group. Another method, developed by Artur Ekert in 1991, was based directly on Bell's theorem. Both held the promise of perfectly secure communication, with the possibility of undetected eavesdropping forbidden by the fundamental laws of physics themselves.

Suddenly, entanglement and Bell's theorem weren't just concerns for a handful of physicists and philosophers in an abstruse and neglected corner of science. Practical questions of computing technology and cryptography were at stake, and naturally, governments and militaries took a fierce interest in the subject. Mastering control of entanglement, decoherence, and other phenomena first described by researchers in quantum foundations was potentially big business—and the race to build a quantum computer was on. The funding floodgates opened. Within a decade of Shor's breakthrough, the Department of Defense was funding a $20 million initiative in quantum information. By 2016, multiple US government agencies, both military and civilian, were funding quantum information technology; the EU was funding €1 billion of research and development in the subject; and China was testing a quantum communication satellite. Private corporations, like Google and Microsoft, were also doing work in this field. In short, quantum information processing was no longer a part of quantum foundations—it had spun off and become its own billion-dollar industry.

But little of this money was going to quantum foundations. The flood of new grants were almost entirely for developing practical things like building quantum computers themselves, not for new approaches to the measurement problem. Quantum foundations had borne this new fruit, it had proven it wasn't useless, but the advances in quantum information processing didn't have any immediate bearing on the mysteries at the heart of quantum theory. And many physicists, even those working in the new fields spun off from Bell's work, still took the Copenhagen interpretation's approach to physics, as summed up by Mermin: "Shut up and calculate!"

Quantum foundations were affecting computers—but computers were also affecting quantum foundations. In 1978, three of David Bohm's colleagues at Birkbeck College in London—Chris Dewdney, Chris Philippidis, and Basil Hiley—started looking over Bohm's old pilot-wave papers from the 1950s. Hiley had been working closely with Bohm at Birkbeck for over a decade; he knew about Bohm's pilot-wave work, but he was under the impression that the theory didn't work, because Bohm had abandoned that approach years before they'd ever met. The Chrises, younger and more foolish, looked at Bohm's old papers anyhow. "[Dewdney and Philippidis] came to me one day with Bohm's '52 paper in their hand," Hiley recalled. "And, they said, 'Why don't you and David Bohm talk about this stuff?' And I then started saying, 'Oh, because it's all wrong.' And then they started asking me some questions about it and I had to admit that I had not read the paper properly. Actually I had not read the paper at all apart from the introduction! . . . And so I went back home and I spent the weekend working through it. As I read it, I thought, 'What on earth is wrong with this? It seems perfectly all right.'" Come Monday, Hiley said, "I went back again to see the two Chrises again, I said, 'Okay, let's now work out what the trajectories are.'" Dewdney used a computer to generate the trajectories of particles guided by pilot-waves in various scenarios, including the double-slit experiment (see Figure 5.4). "Of course once you've got [those] images, they are worth more than a thousand words," Hiley said. Hiley and the Chrises took the images to Bohm, who was astonished. "His eyes suddenly popped open," Hiley said, "and then he and I started talking about this in earnest." After letting it languish for twenty years, Bohm took up his pilot-wave interpretation again, dusting it off and working with Hiley to find a way forward.

Bohm's renewed interest in his old ideas was followed soon after by new work on pilot-wave theory from a small handful of other physicists as well. But where Bohm and Hiley tried to forge a connection between pilot-wave theory and Bohm's ideas about "implicate order" that he had developed over the course of the 1960s and 1970s, these new Bohmians reworked Bohm's original 1952 theory, altering the language and mathematics, and developed powerful defenses against the various arguments that had been leveled against Bohm's interpretation over the years. Some found ways of deriving pilot-wave theory from more basic assumptions, putting the lie to accusations that the theory was inelegant and ad hoc. Others worked to succeed where Bohm had failed in the 1950s, attempting to extend the theory to the realm of relativistic quantum field theory, which had by then continued to be astonishingly successful in predicting the diverse phenomena seen in particle accelerators.

But sadly, Bohm didn't live to see much of this work. He died of a heart attack in the back of a London taxicab in 1992, at age seventy-four. He had survived the blacklist, he had suffered four decades of exile with dignity and integrity—and he had unequivocally proven that alternatives to Copenhagen were possible. His work put the lie to von Neumann's proof and directly prompted Bell's marvelous theorem. If John Bell was the father of the quantum revival, then surely David Bohm had been its grandfather.

Bohm's pilot-wave interpretation wasn't the only old idea dusted off as quantum foundations gained ground in the wake of the Bell experiments. Dieter Zeh also found new recognition for his work on decoherence, thanks to an unlikely source: John Wheeler. After his failure to reconcile his student Everett's work with the ideas of his mentor Bohr, Wheeler had set aside his interest in the foundations of quantum theory. But the Bell experiments, as well as long talks with Eugene Wigner, his colleague at Princeton, had brought Wheeler back to his former interest in the subject. Shortly after moving to the University of Texas in 1976, Wheeler started teaching a class on quantum measurements, and, as he had done at Princeton, he attracted a group of brilliant students—some of whom were profoundly affected by Wheeler's course. "Until I met John Wheeler in Austin, Texas, I had assumed that all of the deep questions were understood—or in any case, not an appropriate subject for a student," said Wojciech Zurek, one of Wheeler's students. "Wheeler changed that. . . . [In his class] we read Bohr and Einstein, but we also discussed connections between quantum theory and information, and played with ideas. . . . I became gradually convinced that questions about quantum mechanics, the role of the observer, and the nature of information in physics are important and largely open."

Zurek's work in Wheeler's class, along with a David Deutsch lecture he attended in Texas, got him thinking about the relationship between entanglement and measurement in quantum physics, and specifically the effects of entanglement between a quantum system and its wider environment—in other words, decoherence. Talking extensively with Wheeler about his ideas—"Wheeler was absolutely essential in defining the problem, or rather, the whole set of problems," he recalled—Zurek drafted a paper on decoherence in early 1981. Though Zurek hadn't been directly aware of Zeh's earlier work on the subject, Wheeler certainly was. After hearing about Zeh's ideas from Wigner, Wheeler had gone to visit Zeh in Heidelberg the previous May. Soon after completing his draft, Zurek heard about Zeh's work from Wheeler and Wigner. When Zurek's paper on decoherence was published later that year, Zurek cited Zeh's still-obscure work as a forerunner of his own.

Although the content of their work had strong similarities, Zurek's approach to decoherence was very different from Zeh's. Zeh had promoted his idea that the many-worlds interpretation was an unavoidable consequence of decoherence in his first papers on the subject. But Zurek was fairly agnostic about the interpretation of quantum physics. According to Zurek, "The whole point of [my] paper (and more broadly of my approach to decoherence) was that I could say things that were relevant to foundational questions and that followed directly from quantum theory, without any interpretational baggage attached." And Zurek's work was received very differently from Zeh's—unsurprising, given the difference in their approaches and the large changes that had happened in physics in the previous decade. While Zeh had found it difficult to publish his ideas at all, Zurek's work was published in a prominent physics journal without much trouble. And Zurek also had a powerful sponsor in Wheeler—again, unlike Zeh, whose work on decoherence had painfully ruined his relationship with his mentor, Jensen. In addition to serving as a sounding board and encouraging Zurek's work, Wheeler helped secure invitations for Zurek to meetings on quantum foundations, which would not normally have been accessible to such a junior researcher. Zurek's talks and ideas were received well at these conferences, which further convinced him to spend much of his professional efforts on the foundations of quantum physics. "I had been under the impression that quantum foundations was a kiss of death to a physicist's career," Zurek recalled. "During my student days, this was the message I had gotten from essentially everyone, with the notable exception of Wheeler. So getting invited to meetings on the basis of my foundations research was solid evidence that times were a-changin'." Zurek published half a dozen more papers on decoherence over the next five years, as well as several other papers on quantum foundations—none of which visibly hampered his career, which took him from Texas to Caltech, and ultimately to Los Alamos.

The success of Zurek's papers convinced Zeh the time was right to start working on decoherence again. He took on a promising young student, Erich Joos, and wrote several papers with him on decoherence. But Zeh didn't want his dissent from Copenhagen to impact Joos. "A young man should not immediately ruin his career by talking about Everett," Zeh told Joos when they first started working together. "So we'll write this paper without ever mentioning that." Zeh deliberately avoided talking about Everett altogether for several years after Zurek's papers appeared, in a vain attempt to safeguard Joos's career. But despite the excellent work that Zeh, Joos, Zurek, and others were doing on decoherence, Zeh's colleagues in Heidelberg were still unconvinced that it was real physics—when they were aware of it at all. "[In 1990] I had the idea that I might suggest Joos for his habilitation [the 'second PhD' required in Germany for anyone lecturing at a university]," Zeh recalled. "I suggested that to some of the people who could have influenced that. Well, the answer was, 'What has he done?' I said 'Decoherence.' [They replied] 'Decoherence? What is that?' In 1990!"

Decoherence finally found a broader audience of physicists in 1991, when Zurek wrote an article on the subject for *Physics Today*, the magazine of the American Physical Society. But Zurek's article made some controversial statements about decoherence—in particular, it came close to implying that decoherence could solve the measurement problem single-handedly. "In spite of the profound nature of the difficulties, recent years have seen a growing consensus that progress is being made in dealing with the measurement problem," Zurek wrote. "Macroscopic systems are never isolated from their environments. . . . The resulting 'decoherence' cannot be ignored when one addresses the problem of the [collapse] of the quantum mechanical wave packet." And toward the end of the article, he stated bluntly that "decoherence destroys superpositions."

A flurry of letters to *Physics Today* pushed back on Zurek, pointing out that decoherence couldn't solve the measurement problem in the absence of an interpretation to go along with it. For a small object in a Schrödinger's cat–like superposition of states put in contact with its environment, decoherence wouldn't destroy the superposition—it would make it worse. Instead of merely having the object in a superposition, the bigger system of object and environment would itself be in a superposition. And, without an interpretation to explain what that superposition means, the measurement problem would remain: Why don't we ever see dead-and-alive cats in the real world? Why does the Schrödinger equation work so well for small objects yet appears to fail so miserably for the objects of everyday life?

Zeh, unsurprisingly, agreed that "environment-induced decoherence by itself does not solve the measurement problem"—he contended that Everett's many-worlds interpretation was needed to complete the picture. And Zurek, despite his *Physics Today* article, did agree that decoherence was not the whole solution. He had been much clearer about this in his very first paper on decoherence, where he said explicitly that decoherence could not address the question of "what causes the collapse of the system-apparatus-environment combined wave function?" But Zurek's views on many-worlds weren't like Zeh's. Instead, they were somewhat reminiscent of those of his mentor, Wheeler—Zurek tried to find a diplomatic way to reconcile Everett's many worlds with Bohr's Copenhagen interpretation, just as Wheeler had attempted on his ill-fated trip to Copenhagen in 1956.

Unfortunately, many physicists took Zurek's diplomatic approach as a sign that decoherence somehow vindicated the Copenhagen interpretation. To them, decoherence was, like the Copenhagen interpretation itself, a magic phrase that could be invoked to dispel the specter of the measurement problem and the rest of the halo of weirdness surrounding quantum theory. Experiments probing decoherence in the late 1990s only added fuel to the fire: as the quantitative predictions of decoherence were confirmed, some physicists concluded that the measurement problem had finally been laid to rest. Among many others, Philip Anderson—the same physicist who had accepted Bell's theorem for publication, apparently on the erroneous understanding that Bell had ruled out Bohm's pilot-wave theory—fell victim to this error. In 2001, he stated that "'decoherence' . . . describes the process that used to be called 'collapse of the wave function.' The concept is now experimentally verified by beautiful atomic beam techniques quantifying the whole process." Anderson's misunderstanding about the nature of decoherence, like his misunderstanding of Bell's result, was definitely not due to any serious deficiency as a physicist on his part—Anderson won the Nobel Prize in 1977 for his seminal contributions to solid-state physics; he is also one of the architects of the modern Standard Model of particle physics. His errors were merely a sign of the times: quantum foundations had become so complex so quickly that it was difficult for even the best physicists to speak intelligently about the subject if they didn't specialize in it—and because Copenhagen-fueled preconceptions about quantum physics remained so ingrained, it was difficult for physicists to see that this was the case. Jeff Bub, Bohm's former student and a philosopher of quantum physics, lamented this state of affairs in 1997. "The 'new orthodoxy' appears to center now on the idea that the original Copenhagen interpretation has been vindicated by the recent technical results on environmental decoherence," wrote Bub. He argued "that there is no real advance here with respect to Einstein's qualms about the Copenhagen interpretation. It is still a 'gentle pillow for the true believer,' perhaps now with that added attraction of a rather fancy goose-down comforter."

Zeh, for his part, had been worried about this outcome from the start. "I expect that the Copenhagen interpretation will some time be called the greatest sophism in the history of science," he wrote to Wheeler in 1980, "but I would consider it a terrible injustice if—when some day a solution should be found—some people claim that 'this is of course what Bohr always meant,' only because he was sufficiently vague."

During his time in Texas, Wheeler was also one of the animating forces behind a new set of ideas about quantum foundations. Quantum interpretations were proliferating wildly in the 1980s and 1990s, with suggestive new ideas taking flight alongside revitalized old ones—and the most prolific class of new interpretations was based on information theory. Taking inspiration from the work being done in quantum computation and cryptography, these interpretations proposed using the theoretical underpinnings of computer science to solve the difficult problems at the heart of quantum foundations. Wheeler was among the earliest proponents of this approach. He summed up the concept as "it from bit": find a way to ground reality itself, as described in quantum physics, in the notion of information.

The motivation behind the information-theoretic interpretations was relatively simple: if the wave function is information of some sort, rather than being a physical object, then many of the puzzles at the heart of quantum physics seem to melt away. In particular, the measurement problem seems much easier to explain if the wave function is information—your information changes when you make a measurement, so it's no surprise that wave functions change dramatically when measurements occur. And the EPR experiment and Bell's theorem might become less puzzling too. When two photons with entangled polarization go flying off in opposite directions and we measure the polarization of one of them, we do instantly learn the polarization of the other one—but there's nothing mysterious or nonlocal about that, any more than there's something mysterious or nonlocal about being able to instantly infer the time in Buenos Aires by looking at a clock in Beijing. And since there's nothing nonlocal about this, there's no longer any puzzle about why entanglement can't be used for faster-than-light communication.

Except that can't quite be right, as any advocate of information-theoretic interpretations would point out. Bell's theorem explicitly states that photon polarizations can't be like clocks, nor like Bertlmann's socks.

If wave functions are information rather than objects in themselves, they must be information of a rather peculiar sort. "*Whose* information?" demanded John Bell. "Information about *what?*" To resolve the measurement problem, information-theoretic interpretations had to answer these questions. The most immediate and Copenhagen-friendly answers were "my information" and "information about my observations"—but to Bell, such answers were profoundly inadequate. Placing observation at the center of physics smacked of positivism, a philosophy that Bell had entertained and rejected during his college days, concluding that it led inevitably to solipsism. Solipsism—the idea that you are the only person, and everyone and everything else is merely a hallucination of some kind in your own mind—was a problem that had haunted positivism from the start. Information-based interpretations of quantum physics ran the risk of collapsing into solipsism as well. If the information that the wave function represented was your information, what makes you so special? And how could different observers agree on the same information? How could your information appear to be an objective fact in the world, something capable of creating interference patterns plain for all to see?

Some physicists tried to address these questions about information-theoretic interpretations by stating that the wave function was information about an unseen world underlying quantum physics, one that obeyed different and as yet undiscovered laws. But such a world would have to be nonlocal to satisfy Bell's theorem—in which case, much of the appeal of the information-theoretic interpretations was lost. (Wheeler himself misunderstood the Bell experiments as ruling out determinism, rather than locality.) Others attempted to get around Bell's theorem by altering the laws of probability or breaking one of the other handful of assumptions that entered into Bell's proof—but each of these solutions came with its own strange and difficult problems.

None of these problems meant that information-theoretic interpretations couldn't work. These were challenges that would have to be met or convincingly dismissed, and physicists and philosophers interested in information-theoretic interpretations worked on doing just that. But, for some physicists, the simple idea of the wave function as "information" held the same allure as decoherence: the promise of a quick and easy way to dismiss niggling doubts about the measurement problem. Wheeler said that his inspiration for "it from bit" was Bohr's approach to quantum physics; some took this to mean that this was what Bohr himself had meant all along, that the Copenhagen interpretation was and always had been merely stating that the wave function was information (flatly refusing to answer what it was information about), that this was the One True Way to "understand" quantum physics.

Bell, of course, knew that there was nothing in quantum physics or in his own theorems that led inexorably to the Copenhagen interpretation. He had been promoting pilot-wave theory for decades to illustrate precisely this point. "Why is the pilot wave picture ignored in text books?" Bell asked in 1982. "Should it not be taught, not as the only way, but as an antidote to the prevailing complacency? To show that vagueness, subjectivity, and indeterminism are not forced on us by experimental facts, but by deliberate theoretical choice?" But not long after Bohm returned to pilot-wave theory, Bell took up the banner of one of the newer ideas being developed at the time: *spontaneous-collapse theory*.

Rather than interpreting the existing mathematics of quantum physics, as Bohm and Everett had done, spontaneous-collapse theory actually modifies the equations of quantum physics to solve the measurement problem. It does so in a subtle way—as it would have to, since quantum physics is spectacularly successful in predicting the outcomes of experiments. But spontaneous-collapse theory manages to leave most of the predictions of standard quantum physics intact, while altering them enough to solve the measurement problem.

In spontaneous-collapse theory, the quantum wave function is real, but it doesn't obey the Schrödinger equation perfectly. Instead, sometimes the wave function collapses. But this collapse has nothing to do with observation or measurement—the collapse happens entirely at random, for no reason at all, whether or not anyone is looking. Think of the wave function as playing a collapse slot machine (Figure 10.3a): every time the wave function hits the jackpot, it collapses. It pulls the handle millions of times a second, but the collapse jackpot only comes up once every 10 *million billion billion* times or so—a one with 25 zeros after it—so it takes hundreds of billions of years for the wave function to collapse. This means that subatomic particles can almost always go down two paths at once, just like our nanometer Hamlet in the Introduction—but every once in a very great while, they're forced to a single path. (Just how great a while is a matter to be resolved by experiment, but it must be at least tens of thousands of years, otherwise the theory would contradict existing experiments.)

Yet this still leaves us with the question from the Introduction: if subatomic particles can behave so strangely and we and the objects in our everyday lives are composed of such particles, why don't we see such strange behavior on a regular basis? According to spontaneous-collapse theory, the answer lies in two key facts: entanglement and the vast number of particles that comprise the objects of our everyday experience. Though a single-particle wave function might not collapse on average until a billion years have passed, the solid objects of our everyday lives, like this book, are generally composed of at least 10 million billion billion individual particles. If each one of those particles' wave functions is compulsively pulling the handle of its own slot machine (Figure 10.3b), then, on average, at least one of them will hit the collapse jackpot every millionth of a second. But because the particles in this book are all continually interacting with each other, they're all entangled—which means they all share a single wave function. So when one of them hits the jackpot, the wave function for the entire book collapses, meaning this book can't be in two places at once for much longer than a microsecond or so—a hundred thousand times faster than a blink of an eye. As Bell put it, in spontaneous-collapse theory, Schrödinger's cat "is not both dead and alive for more than a split second." This neatly resolves the measurement problem: all objects, large and small, obey the same laws, with no special role played by measurement. Wave function collapse happens at random to everything, all the time, without any need for intervention from an observer.

Spontaneous-collapse theory isn't really just one theory; it's a collection of related theories, which were developed by a small handful of people dissatisfied with the Copenhagen interpretation over the years. The one described above—the one that caught Bell's attention and through

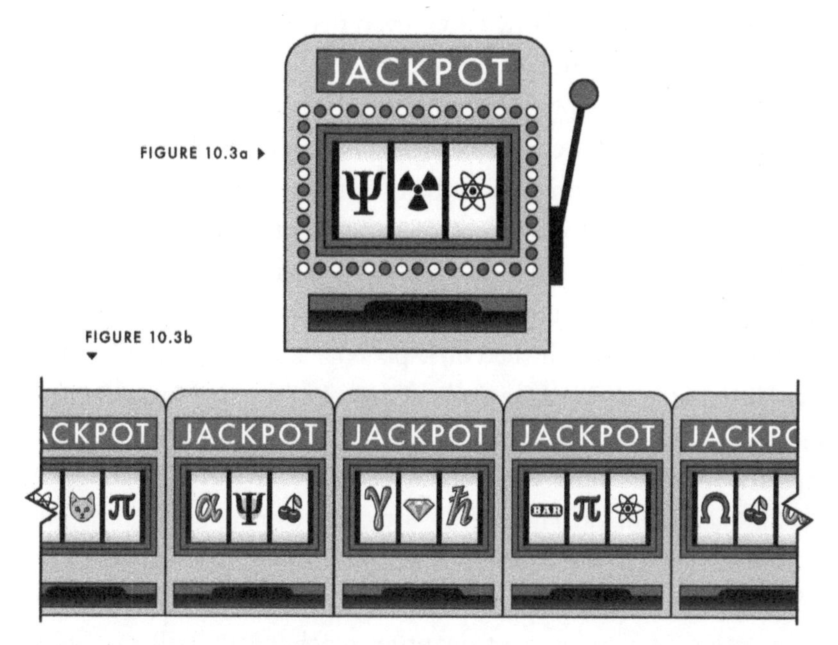

Figure 10.3. Spontaneous-collapse theory. (a) A single-particle wave function only has one slot machine and is unlikely to hit the collapse jackpot for millions or billions of years. (b) A wave function shared by many entangled particles has many slot machines and is likely to hit a collapse jackpot much sooner.

him the attention of many other physicists—was developed in 1985 by a trio of physicists working in Italy, GianCarlo Ghirardi, Alberto Rimini, and Tullio Weber. It came to be known as the "GRW model" after their initials. "I see the GRW model as a very nice illustration of how quantum mechanics, to become rational, requires only a change which is very small (on some measures!)," Bell wrote soon after the GRW paper first appeared. Bell's paper on GRW brought it to the attention of many other physicists, including Philip Pearle, who had been working on similar ideas since the early 1970s. (Ten years earlier, Pearle's work had also singled him out for the unfortunate distinction of an interview with a sociologist studying "social deviance" among physicists.) Pearle wrote to Bell for more information on GRW, and Bell arranged for Pearle to spend a sabbatical with Ghirardi, where the two men attempted to adapt GRW for relativistic quantum field theory. But the common refrain that had greeted Bohm, Everett, and others for decades hit GRW and Pearle doubly hard: quantum theory works astonishingly well. Why fix what is manifestly not broken? Why do we even need a different interpretation, much less a different theory altogether?

Bell framed his reply as a moral issue. "It is not right to tell the public that a central role for conscious mind is integrated into modern atomic physics. Or that 'information' is the real stuff of physical theory. It seems to me irresponsible to suggest that technical features of contemporary theory were anticipated by the saints of ancient religions . . . by introspection." Bell felt an urgent need to solve the problems at the core of quantum physics—but he had little patience for purported solutions that were little more than vague declarations of faith. He wanted something more definite, something that wasn't a professional embarrassment, a true theory that didn't shy away from questions about what happened during measurement. His relentlessly clear writing gave no quarter to anyone hoping to take solace in the comforting platitudes of the Copenhagen interpretation. "Surely, after 62 years, we should have an exact formulation of some part of quantum mechanics?" Bell said in 1989. "[Measurement devices] should not be separated off from the rest of the world into black boxes, as if [they] were not made of atoms and not ruled by quantum mechanics." In a talk he gave in Geneva in January 1990, Bell acknowledged that the task at hand was difficult and that his own theorem proved that there was some radical change necessary, something that physics would have to come to terms with. "I think you're stuck with the nonlocality," Bell told the small group who had come to hear him speak that day. "I don't know any conception of locality which works with quantum mechanics."

Eight months later, Bell died suddenly of a massive stroke at the age of sixty-two. Memories and tributes poured in from his colleagues and friends. "He was one of the most rigorously honest men ever, and I never met anything like it, myself. He was awesome," Abner Shimony recalled. "Bell proved Bell's Theorem, and no one else did, because of his character. . . . Of course he had a tremendous intellect. But what he had to such a superlative degree was the honesty, the tenacity to push through his questions." "John Bell had a consuming commitment to wresting an *understanding* of the natural world from the great theories of physics," wrote Mermin and Kurt Gottfried (a physicist who had sparred with Bell over the Copenhagen interpretation several times). "He held that a theory that merely succeeded in accounting brilliantly for data, without providing a satisfactory understanding of what it described, should be subject to stringent critical scrutiny, and if such an understanding was found to be unattainable the theory should be expected to crumble, its superficial triumphs notwithstanding. . . . John was truly unique in the world of physics, as a personality and as an intellect—at once scientist, philosopher, and humanist. He was a person to whom deep ideas mattered deeply. Fate has been most cruel to steal him from us when he was still so brimful of vitality."

Bell had spent a quarter century fighting the overwhelming dominance of the Copenhagen interpretation. "Trust me, you could not do what [Bell] has done without a strong personality," said Gisin. "He would have been destroyed." Instead, Bell thrived—not only weakening the Copenhagen interpretation more than anyone since Einstein but also discovering a profound new truth about nature in the process. "Nonlocality, I think, was [Bell's] great discovery," said Bertlmann. "I think it [is] one of the greatest discoveries in the last century, that there is nonlocality in nature." Yet Bell, a modest man, did not receive the recognition and accolades he deserved in his lifetime for his work. Several years before Bell's death, he and Bertlmann were having tea in the outdoor cafeteria at CERN, enjoying the views of the Alps and the Jura in the afternoon sun, when Bertlmann mentioned that he thought Bell had been sold short. "I just suddenly said to him, John, I think you deserve the Nobel Prize," Bertlmann recalled. "He was surprised, and asked 'Why?' [I said] because of Bell's theorem!" Bell pointed out that the experimental tests of his theorem had not shown any deviation from quantum physics, which was not worthy of a Nobel Prize. And besides, he added, "I don't think I deserve it because I stick to Nobel's original rules, and I do not see how my inequalities could contribute to the benefit of mankind." (Alfred Nobel had originally stipulated that the prizes were to be given to people who had contributed most to the benefit of humanity in the previous year through work in their field.) Bertlmann protested, "I replied, no I don't agree. I think it is nonlocality [that is worthy of the prize]. . . . Then on the one hand he was pleased, but on the other hand, he was saying, a bit disappointed and sadly, 'Who cares about this nonlocality?' . . . So he felt that the community did not realize this enough, or did not appreciate this enough. I mean it was clear at CERN, he was highly appreciated there as a particle physicist. But his work in quantum physics was not appreciated." Unbeknownst to Bell, he was shortlisted for the Nobel Prize the year before he died and might well have won it had he lived longer—but the Nobel Prize isn't awarded posthumously (another stipulation of Alfred Nobel's will).

Yet Bell's legacy was assured, though he did not live to see it. "The great boom came in the 90s with quantum information," said Bertlmann. "It is a new community now, which was not there in the 80s. . . . So actually he could not see the fruits of his work." Bell, through his profound physical insights and his pellucid, urgent prose, had managed to shift the mindset of physics as a whole and had unintentionally conjured the entirely new field of quantum information processing out of thin air at the same time. And his contributions to "quantum engineering"—his work on particle physics and accelerator design at CERN—was of the very first rank.

Bell also left behind a research program in quantum foundations. The year before he died, at a conference in the mountain village of Erice on the western tip of Sicily, Bell "delivered what came close to being the most spell-binding lecture I have ever heard," Mermin later recalled. "What exactly qualifies some physical systems to play the role of 'measurer'?" Bell asked, dripping with sarcasm. "Was the wave function of the world waiting to jump for thousands of millions of years until a single-celled living creature appeared? Or did it have to wait a little longer, for some better qualified system . . . with a PhD?" Bell then pointed out the flaws in how quantum physics was commonly taught (taking several specific textbooks to task for their lapses in explanation) and finally laid out the two approaches to quantum physics that he thought were the most promising: the pilot-wave interpretation and spontaneous-collapse theory. He concluded his talk with a challenge. "The big question, in my opinion, is which, if either, of these two precise pictures can be redeveloped in a [way consistent with special relativity]."

But Bell also entertained a third option, beyond pilot waves and spontaneous collapse, though he did not mention it in his talk at Erice. "The 'many world interpretation' seems to me an extravagant, and above all an extravagantly vague, hypothesis," Bell said in 1986. "I could almost dismiss it as silly. And yet. . . . It may have something distinctive to say in connection with the 'Einstein Podolsky Rosen puzzle,' and it would be worthwhile, I think, to formulate some precise version of it to see if this is really so. And the existence of all possible worlds may make us more comfortable about the existence of our own world . . . which seems to be in some ways a highly improbable one." Though the number of physicists interested in pilot waves and spontaneous collapse grew after Bell's death, the many-worlds interpretation gained far more popularity and notoriety in the last decades of the twentieth century. And much of the reason had nothing to do with Bell's work, or any work in quantum physics at all. Instead, many-worlds came roaring back thanks to research in an entirely different field of physics, a field that studied not the ridiculously small but the unthinkably huge: the entire universe itself.

11

Copenhagen Versus the Universe

"If a poll were conducted among physicists," Bryce DeWitt wrote in 1970, "the majority would profess membership in the [Copenhagen] camp, just as most Americans would claim to believe in the Bill of Rights, whether they had ever read it or not." DeWitt had managed to convince the editor of *Physics Today*, the monthly magazine for members of the American Physical Society, to publish an article on the foundations of quantum physics. In a sign of the times, the editor, Hobart Ellis Jr., had not been particularly difficult to convince. "For a long time I personally have been dissatisfied with the apparent contradictions that physicists appear to be ready to live with in quantum mechanics and its interpretation," he wrote to DeWitt. "I think a general review of different interpretations of quantum mechanics without special emphasis on any one would be of interest."

DeWitt's article, "Quantum Mechanics and Reality," did review several interpretations. But DeWitt made his own opinions quite clear. "The Copenhagen view promotes the impression that the collapse of the [wave function], and even the [wave function] itself, is all in the mind," he wrote. "If this impression is correct, then what becomes of reality? How can one treat so cavalierly the objective world that obviously exists all around us?" When dealing with a system in a quantum superposition, like Schrödinger's cat, DeWitt said most physicists "conceive the [measurement device] to have entered a kind of schizophrenic state in which it is unable to decide what value it has found for the system," a living cat or a dead one. This problem, he concluded, was not resolved by the Copenhagen interpretation. And other interpretations, like Bohm's, added hidden variables to quantum physics, a move that DeWitt thought was unnecessary. "What if we assert that the [Schrödinger equation] is all, that nothing else is needed?" he wrote in *Physics Today*. "Can we get away with it? The answer is that we can."

DeWitt used the rest of the article to advocate Hugh Everett's "relative-state" interpretation of quantum physics, which DeWitt had subscribed to since his correspondence with Everett in 1957. Everett had never talked explicitly about many worlds, but DeWitt ventured where Everett hadn't bothered to tread, rechristening the idea as the "many-worlds" interpretation. "The universe is constantly splitting into a stupendous number of branches, all resulting from the measurementlike interactions between its myriads of components," DeWitt wrote. "Moreover, every quantum transition taking place on every star, in every galaxy, in every remote corner of the universe is splitting our local world on earth into myriads of copies of itself." DeWitt knew that this idea was dazzlingly strange:

> I still recall vividly the shock I experienced on first encountering this multiworld concept. The idea of 10^{100+} slightly imperfect copies of oneself all constantly splitting into further copies, which ultimately become unrecognizable, is not easy to reconcile with common sense. Here is schizophrenia with a vengeance.

Nonetheless, DeWitt argued, many-worlds "has a better claim than most to be the natural end product of the interpretation program begun by Heisenberg in 1925." He pointed out that the interpretation does not require wave functions to ever collapse and claimed that it didn't require anything else at all.

Many readers of *Physics Today* were unconvinced by DeWitt's arguments. "The idea of infinitely many multiplying, noninteracting worlds should be taken somewhat less seriously than the Ptolemaic [earth-centered system's] epicycles," wrote one physicist in reply. "At least

Ptolemy's theory 'explained,' in some sense, the one observable world without invoking infinitely-many unobservable worlds." The many-worlds interpretation "would also imply the (happy!) feeling that if an airline passenger were in an aircraft about to crash, he need not really worry, for in another world, this same aircraft . . . will land at home, safe and sound," wrote another. "I ask whether it is really necessary to go to such extreme lengths of straining physical sensibilities (admitting that here I speak for myself) to resolve the logical difficulties of the quantum theory."

But DeWitt remained convinced, and some of his readers were swayed as well. Everett's interpretation had spent over a decade in deep obscurity. Now, "one of the best kept secrets of this century," as DeWitt called it, was finally out.

DeWitt's enthusiasm for the many-worlds interpretation was not merely driven by a desire to resolve the mysteries of quantum physics. Replying to his critics in *Physics Today*, DeWitt wrote that many-worlds "is the only conception that, within the framework of presently accepted [equations and mathematics], permits quantum theory to play a role at the very foundations of cosmology." At the time DeWitt was writing, cosmology was more established as a research field than the foundations of quantum physics was—but that wasn't saying very much. The idea that the universe as a whole was a suitable subject for scientific investigation was difficult for some physicists to swallow. General relativity, Einstein's theory of gravity and warped spacetime that underpins cosmology, was a theoretical backwater, an accepted theory that was nonetheless considered to be useless. Einstein's theory only differs appreciably from Newtonian gravity when dealing with extremely massive objects, objects at least as big as stars. But these massive objects that it dealt in were considered to be too far removed from everyday experience to be relevant to physics, and opinion was split over whether the theory's implications for cosmology should be taken seriously at all. In 1962, a young physics student named Kip Thorne had just finished his undergraduate degree at Caltech and was about to go off to study general relativity at Princeton under John Wheeler. One of his Caltech professors attempted to dissuade him. "General relativity has little relevance for the real world," Thorne recalled his professor telling him. "One should look elsewhere for interesting physics challenges."

General relativity didn't merely deal with abstruse situations—it was also written in abstruse mathematics. The theory is very mathematically complex, far more so than quantum mechanics. Einstein famously had to enlist the help of a mathematician friend, Marcel Grossman, just to learn the differential geometry necessary to formulate and understand his own theory. This combination of unfamiliar subject matter and obscure mathematics made it difficult to be sure of what the theory was saying and led many physicists to be suspicious of its conclusions. Even Einstein himself had trouble accepting the consequences of his own theory after he initially developed and published it in 1915. He realized that general relativity implied the universe as a whole must be either contracting or expanding, a conclusion he found troubling, at odds with all known data at the time. So he put in a "cosmological constant," a fudge factor to keep the universe at a static size. But, in 1929, the astronomer Edwin Hubble discovered that distant galaxies appeared to be receding at a rate proportional to their distance—exactly what you would expect to see in an expanding universe. Einstein readily dismissed his ad hoc cosmological constant—he had never liked it anyhow, calling it "gravely detrimental to the beauty of the theory"—and accepted the cosmological picture presented by general relativity as correct. But not everyone was convinced—including Edwin Hubble himself. Hubble and others thought that distant galaxies only appeared to be receding and that the universe was in fact static. Others agreed the universe was expanding, but they proposed modifications to the laws of physics that would ensure the universe looked basically the same at all times in the past and future, despite its expansion. This came to be known as the "steady-state" theory of the universe. For decades afterward, the steady-state theory was considered a reasonable scientific theory, and many physicists found it quite a bit more reasonable than the expanding universe of general relativity. After all, such a universe must have once been in a fabulously hot, dense, and small state, expanding rapidly—what steady-state theorist Fred Hoyle termed a "Big Bang." General relativity, Hoyle and others held, was not necessarily to be trusted under such strange conditions, nor when applied to the universe as a whole.

Meanwhile, confusion persisted about the implications of general relativity even for objects much smaller than the entire universe, like stars. In 1938, Robert Oppenheimer and his student George Volkoff at Berkeley, along with Richard Tolman at Caltech, used a very early forerunner of a computer to calculate that supermassive stars, far larger than our Sun, must end their lives by collapsing down to a fabulously dense object from which nothing, not even light, can escape. The notion of these "collapsed stars," as they were known then, generated intense debate. The forbidding mathematical edifice of general relativity, along with the computational intensity and unusual (at the time) tools required for the Oppenheimer-Volkoff calculation—not to mention the sheer strangeness of their result—made it difficult for other physicists to take the idea of collapsed stars seriously.

This mathematical complexity also made it hard for Einstein to understand the implications of his own theory. "Together with a young collaborator, I arrived at the interesting result that gravitational waves do not exist," Einstein wrote to his old friend Max Born in 1936. Gravitational waves—ripples in spacetime, formed by colliding hyperdense stars and similarly intense events, racing out at the speed of light from their violent births—were a unique prediction of general relativity, not found in Newton's theory of gravity. But the strange mathematics of the new theory led Einstein and his collaborator Rosen astray. They published a paper claiming to prove that gravitational waves aren't physical objects and are merely mathematical fictions of the theory. Einstein was later set straight by the American physicist Howard Percy Robertson, but Rosen remained unconvinced that gravitational waves were real, and his paper with Einstein was never retracted, leading to a great deal of confusion about the reality of one of the fundamental predictions of general relativity, a confusion that persisted for decades.

The mathematical difficulty of the theory, the confusing arguments surrounding its predictions, and the remoteness of its predictions from the realm of experiment left general relativity off to the side as physics boomed after World War II. The new sources of scientific funding that flowed from the military-industrial complex largely ignored general relativity. But, in the late 1950s, the field slowly began to blossom. Several conferences were held on the subject, and a professional community of relativistic astrophysicists and cosmologists began to form. One of the most important of these conferences was the 1957 Chapel Hill conference, organized by Bryce DeWitt and Cécile DeWitt-Morette, a talented physicist who had studied with de Broglie in France (and who was married to Bryce DeWitt).

A variety of physics luminaries descended on Chapel Hill for the conference. Aside from the DeWitts, John Wheeler was there, as was his student and Everett's friend Charles Misner. Feynman was there too, but he registered under a pseudonym, "Mr. Smith," to protest the sorry state of research in the field. At the conference, Feynman and physicist Hermann Bondi presented closely related ironclad arguments that finally convinced the physics community that gravitational waves must be real if general relativity was correct, clearing up a major embarrassment for the still-nascent research field. (This also kicked off a sixty-year search for gravitational waves, ultimately culminating in the first successful detection of gravitational waves at LIGO, a pair of four-kilometer-long laser-powered gravitational wave observatories, in 2015—an accomplishment that earned a Nobel Prize for Thorne and two of his colleagues in 2017.) And Wheeler, meanwhile, pushed his agenda of "radical conservatism," taking the predictions of established theories seriously even in wildly strange, untested, and remote domains—for example, the small, hot, dense period in the universe's history, immediately after the still-controversial Big Bang, when both general relativity and quantum physics would be necessary for understanding the behavior of the universe.

In the 1960s, the new field accelerated significantly. New mathematical techniques led to the realization that collapsed stars—or as John Wheeler dubbed them in 1968, "black holes"—must be real. And in 1964, two physicists at Bell Labs, Arno Penzias and Robert Wilson, stumbled upon a radio hiss coming from all directions in the sky, and realized that they had discovered the CMB radiation, the oldest light in the universe and an echo of the Big Bang. Within fifteen years, the steady-state theory had lost all credibility, the Big Bang model was accepted as basically correct, and Penzias and Wilson had shared a Nobel Prize. There was still a great deal of disagreement about basic questions like the rate at which the universe was expanding, but relativistic cosmology was finally off and running, with a shared model of how the universe as a whole behaves.

But the rise of cosmology made the inadequacy of the Copenhagen interpretation more acute. How can you draw a line between the observer and the observed system, as Bohr had required, when the system in question is the entire universe? "Quantum gravity is certainly going to be important in the early moments of the universe, then you are driven to the notion of a wave function for the universe, and how do you interpret a thing like that when there's no observer outside?" DeWitt said. "The Everett point of view was the only way of doing this." In the late 1960s, as Clauser and others were first discovering Bell's theorem and devising ways to test it, DeWitt started spreading the gospel of Everett among cosmologists and astrophysicists. "[I] felt that Everett had been given a raw deal," DeWitt said. He gave a talk on Everett's theory in 1967 at a Seattle conference on relativistic astrophysics and cosmology that Wheeler and DeWitt-Morette organized together. He managed to get *Physics Today* to publish his article on the subject, which he later said "was deliberately written in a sensational style." He uncovered Everett's original longer version of his thesis, significantly easier to understand than the edited-down version that Wheeler had insisted on, and published it, along with other works of Everett's, similar work from others, and responses from other physicists, in a volume he edited along with his student Neill Graham, published in 1973. The classic science-fiction magazine *Analog* even ran an article on the many-worlds interpretation, mostly based on DeWitt's *Physics Today* article, in December 1976. And in 1977, DeWitt and Wheeler asked Everett himself to give a seminar on his interpretation. Everett accepted and drove from his sleepy Virginia suburb down to Austin with his wife and two teenaged kids to give his first talk on quantum physics in fifteen years.

On January 2, 1971, a White House courier and two US air marshals took a red-eye flight from Washington, DC, to Los Angeles. They were carrying classified information to National Security Advisor Henry Kissinger: routine work but sensitive. Needless to say, the courier and marshals were startled when a portly middle-aged man with a goatee snapped a picture of them with a small camera as he walked past their seats. Their alarm only intensified when they questioned him immediately afterward; he would only say that he had taken the picture "for my files." He smelled of gin and Kent cigarettes. When the plane landed, the marshals lost sight of the mysterious man in the crowd, but an inquiry with the airline identified him as one Hugh Everett III. The marshals reported the incident with Everett to the FBI, who sent an agent to Everett's airport hotel room several hours later. By then, Everett had sobered up, and sheepishly admitted to the FBI agent that he had merely been playing a practical joke on the courier and marshals. He had inferred their profession when he overheard their chatter in the Dulles airport bar. The FBI agent, finally satisfied that no harm had been done and Everett was merely a man with a strange sense of humor, let Everett off with a warning and left him alone in his hotel room. Both the marshals and the FBI agent remained blissfully unaware that Everett himself had a security clearance that far outstripped their own.

Everett had done reasonably well for himself over the previous fifteen years since leaving Princeton and John Wheeler's tutelage. After eight years working directly for the Pentagon, he had struck out on his own, founding a statistical consulting firm and contracting for his former employers. His optimization algorithm had made his reputation at the Pentagon, and his work provided him with enough income for the finer things in life he enjoyed. He spent his days gaming out various scenarios for nuclear apocalypse and his evenings eating, smoking, and womanizing; he and his wife had agreed to an open marriage in the middle of the 1960s, though by then Everett had been conducting various short-lived affairs for years. By the time DeWitt and Wheeler invited him to the Austin conference in 1977, Everett was spending some of those evenings parked in front of his television with an early VCR, watching *Dr. Strangelove,* his favorite movie, on an endless loop, drink in hand.

Everett was pleased that DeWitt had brought new attention to his ideas. And it tickled him to see his own theory discussed in the pages of the same science-fiction magazines he had been reading all his life. "I certainly approve of the way Bryce DeWitt presented my theory," Everett wrote, "since without his efforts it would never have been presented at all." But it's not clear that Everett himself truly believed that the many worlds implied by his interpretation were literally real, as DeWitt did. Shortly after leaving Princeton, Everett corresponded with Philipp Frank, a founding member of the positivist Vienna Circle. Their letters reveal that the two men shared similar philosophical inclinations. "I find that you have expressed a viewpoint which is very nearly identical with the one which I have developed independently in the last few years, concerning the nature of physical theory," Everett wrote to Frank in 1957. Everett's dissatisfaction with Copenhagen had less to do with any commitment to realism and more to do with its irrational and inconsistent usage of the Schrödinger equation. Even from a positivist perspective, the measurement problem appeared grave to Everett. When does collapse happen? Why does the Schrödinger equation apply at some times but not others? Frank was clearly troubled by this too, as he wrote in his reply to Everett: "I have always disliked the traditional treatment of 'measurement' in Quantum Theory according to which it seems as if measurement would be a type of fact which is essentially different from all other physical facts." Rather than saving realism, as Bohm and Shimony and others wanted to do, Everett had simply wanted to fix this hole in physics and to have a good time while doing it. "He wanted a quick short thesis project and punking the measurement problem made him laugh," Everett's biographer, Peter Byrne, said.

Everett had kept up with developments in the foundations of physics over the years, but he had never published anything on the subject after finishing his PhD thesis. He had certainly never spoken about it publicly—he hated public speaking—and he rarely even spoke about it with friends or colleagues. When Don Reisler, a physicist with a PhD in quantum foundations, applied to work at Everett's company, Everett shyly asked if he was aware of the relative-state interpretation. Reisler immediately thought, "Oh my God, you are *that* Everett, the crazy one." He said he had heard of the theory. They became great friends, but they never spoke of quantum physics again. And even as awareness of Everett's ideas spread, they were still often met with ridicule and withering contempt. Writing about "cognitive repression in contemporary physics," the physicist-turned-philosopher Evelyn Fox Keller said that the many-worlds interpretation "demonstrates remarkable ingenuity," in the solutions it provides for the measurement problem and other quantum paradoxes. But, she concluded, "a price has been paid—namely the price of seriousness." More criticism of Everett was yet to come, and not from some unknown quarter but from an old ally.

Shortly after Everett's Austin seminar, Wheeler received a draft of a paper criticizing the many-worlds interpretation, which the paper termed the "Everett-Wheeler interpretation." Wheeler hastily replied, pointing out that "Everett's PhD thesis was on a topic entirely conceived by him and ought to be called the Everett Interpretation, not the Everett Wheeler Interpretation." Ever the scientific diplomat, Wheeler had tried to maintain his commitment to the ideas of Bohr, his late mentor, without explicitly denouncing the ideas of Everett, his former student. This wasn't a difficult position for him to hold while Everett's work languished in obscurity and remained cloaked in the language of the "relative-state formulation." But now DeWitt was calling Everett's view the "many-worlds" interpretation and saying that Wheeler was partly responsible for it—and the fact that it was showing up in science-fiction magazines didn't help either. So Wheeler publicly distanced himself from Everett's work and DeWitt's spin on it. "[Everett's] infinitely many unobservable worlds make a heavy load of metaphysical baggage," Wheeler wrote in 1979. Though Wheeler had always strongly supported Everett's physics career—and was still interested in Everett returning to academia after his twenty years in industry—he claimed that he had never supported Everett's ideas. "Wheeler told me that he was always implacably opposed to the theory—what he supported was Everett," said David Deutsch, who was a young researcher at Austin when Everett came to speak there. Shortly thereafter, Wheeler started promoting his own ideas about an information-based interpretation of quantum physics, ideas that he saw as compatible with the Copenhagen interpretation.

But Deutsch and many of the younger attendees at Everett's seminar in Austin were enthusiastic about the many-worlds interpretation.

Deutsch sat next to Everett at lunch in a beer garden after the talk. Everett was "full of nervous energy, high-strung, extremely smart, very much in tune with the issues of the interpretation of quantum mechanics," Deutsch recalled. "He was extremely enthusiastic about many universes, and very robust as well as subtle in its defense, and he did not speak in terms of 'relative states' or any other euphemism." Several years later, in his seminal paper on quantum computing, Deutsch claimed that only the many-worlds interpretation could explain the fabulous increase in speed that quantum computers offered. "The Everett interpretation explains well how the [quantum] computer's behaviour follows from its having delegated subtasks to copies of itself in other universes," Deutsch wrote. "When the [quantum] computer succeeds in performing two processor-days of computation, how would the conventional interpretations explain the presence of the correct answer? *Where was it computed?*" Other interpretations of quantum physics would also prove capable of explaining the power of quantum computers. Nonetheless, Deutsch's enthusiasm was infectious, and many-worlds soon gained in popularity within the new field of quantum information processing.

Many-worlds also continued to gain in popularity among physicists who took cosmology seriously—and even inspired new interpretations. "Measurements and observers cannot be fundamental notions in a theory that seeks to discuss the early universe when neither existed," wrote Murray Gell-Mann and James Hartle in 1990. Gell-Mann had won a Nobel Prize in 1969 for suggesting the existence of quarks; Hartle, his former student, had done work on quantum cosmology with Stephen Hawking. Both had long been convinced that the Copenhagen interpretation had to be wrong. "The fact that an adequate philosophical presentation [of quantum physics] has been so long delayed is no doubt caused by the fact that Niels Bohr brainwashed a whole generation of theorists," Gell-Mann wrote in 1976. Gell-Mann and Hartle combined Everett's interpretation with the work on decoherence done by Zeh, Joos, and Zurek, along with ideas from Roland Omnès and Robert Griffiths, to develop what they called the "decoherent-histories" interpretation of quantum physics. Despite the fact that their interpretation had a single world, Gell-Mann and Hartle acknowledged their intellectual debt to Everett and saw their ideas as an extension of his work.

But Everett didn't live to see Gell-Mann's work, or Deutsch's. On July 19, 1982, Everett died of a heart attack at the age of fifty-one. In accordance with his wishes, his family had him cremated, and left his ashes out with the trash.

Within a decade of Everett's death, cosmology entered a golden age. For much of the previous century, the field had been driven primarily by theoretical advances, such as the development of general relativity in the first place. But in the 1990s, the Hubble Space Telescope, the Cosmic Microwave Background Explorer, and other space-based observatories, complemented by a new generation of enormous terrestrial telescopes, flooded cosmologists with data. Around the same time, the advent of high-speed computing made it possible not only to process this data but to simulate the entire universe, testing out different theories about its composition and behavior. Cosmology rapidly went from guessing some of the most fundamental properties of the universe to nailing them down with shocking precision. In 1996, estimates for the age of the universe ranged from 10 to 20 billion years, much as they had for the three decades since Penzias and Wilson had discovered the CMB. By 2006, the age of the universe had been pinned down to 13.8 billion years, give or take a percent.

With that precision came new pictures of the universe. The Wilkinson Microwave Anisotropy Probe (WMAP), a space telescope launched in 2000, created a detailed map of the tiny differences in intensity in the CMB, which is uniform to about one part in 100,000. That map lent support to a theory of the extremely early universe at the time of the Big Bang, known as "inflation." Inflation, first proposed by the physicist Alan Guth in 1981 and refined by Andreas Albrecht and Andrei Linde shortly thereafter, says that the very early universe expanded extraordinarily quickly for a minuscule fraction of a second—increasing in size by a factor of about 100 trillion trillion in about a billionth of a trillionth of a trillionth of a second—then resumed expanding much more slowly. This rapid expansion was driven by hypothetical "inflatons," high-energy subatomic particles, which decayed into normal matter at the end of inflation. Crucially, the theory dictated that tiny quantum fluctuations in the density of inflatons were blown up by the inflation process, and then led to the tiny fluctuations in the density of normal matter in the small, hot universe immediately after inflation. Those fluctuations led, in turn, to the fluctuations in the CMB—and ultimately seeded the formation of all structure in the universe today, including our own galaxy and Earth itself. In short, inflation suggested that we are all the products of quantum fluctuations that occurred in the very early universe—and WMAP's data suggested inflation was right. "WMAP's data supports the notion that galaxies are nothing but quantum mechanics writ large across the sky," Brian Greene said in 2006. "This is one of the marvels of the modern scientific age."

But the Copenhagen interpretation couldn't explain how the early universe worked—and the mathematics of quantum physics couldn't handle such situations either. The early universe was fabulously small, suggesting that quantum physics was needed, but also fabulously dense, requiring the forbidding mathematical machinery of general relativity. Unfortunately, a theory unifying general relativity with quantum physics hadn't been found, despite having been sought for decades by an army of physicists, including Einstein. As late as the 1960s, some people had suggested that no such unification was necessary: Léon Rosenfeld had said (like a good positivist) that because quantum gravitational effects could never be seen, there was no need to develop a theory to address such unobservable phenomena. But, as general relativity became more reputable, the need to unify it with quantum field theory had become more pressing. By the 1990s, ideas like Rosenfeld's were as far outside the mainstream of physics research as the old claims that cosmology shouldn't be taken seriously. A theory of quantum gravity—now nicknamed a "theory of everything"—was widely considered to be the single most important unsolved problem in all of physics. The most promising candidate was string theory, whose abstruse mathematics yielded glimpses of elegant connections between quantum physics and general relativity. By the early 2000s, combining string theory and inflation seemed like the best hope for a theory of the early universe.

Surprisingly, both string theory and inflation, which were developed quite independently, seem to point to a common conclusion: the existence of a multiverse, an enormous number of multiple independent universes. According to inflation, the universe is unable to escape "eternal inflation": as inflation ends in one part of the universe, it continues in others, and "bubbles" of noninflating universe continually appear in the inflating region. We live in one of these bubbles; other bubbles would be their own universes, cut off from all the others, and each might have its own laws of physics and assortment of fundamental particles. And because inflation is eternal, there would be an infinity of these bubbles—an infinite multiverse of inflation. String theory, meanwhile, doesn't describe a single universe but instead describes a "string landscape," a phenomenally huge number of possible universes—10^{500} or more.

The similarities to the many-worlds interpretation's multiverse were not lost on quantum cosmologists. The appearance of multiverses independent of Everett's interpretation made its strange profusion of worlds downright appealing. Some physicists even proposed that all three of these multiverses—Everettian many-worlds, eternal inflation, and the string landscape—were in fact a single multiverse, and the three theories were simply describing the same reality in different ways. In any case, the many-worlds interpretation was (mostly) no longer laughed out of the room without serious consideration. Indeed, by the start of the twenty-first century, many-worlds had become the most popular rival to the Copenhagen interpretation itself among physicists, with particular popularity among cosmologists. But with wider consideration came the realization of a new problem, one that any theory with an infinite multiverse faced: the problem of probability.

At its core, the measurement problem asks when wave functions obey the deterministic harmony of the Schrödinger equation and when they undergo the random process of collapse. The many-worlds interpretation gets around the measurement problem by denying that wave function collapse happens at all. In the multiverse of many-worlds, the wave function of the universe always obeys the Schrödinger equation, splitting off into an endless series of branches that constitute the many worlds. But there's a problem with this picture. It's not clear how randomness and probability enter our quantum physics experiments if the universal wave function really does obey the Schrödinger equation at all times, since that equation is completely deterministic, with no element of chance. After all, one thing that absolutely everyone agrees on, no matter what interpretation (or incoherent pseudo-interpretation) they subscribe to, is that the outcomes of quantum physics experiments have an element of randomness. In general, the mathematical machinery of quantum physics only lets us predict the probabilities of particular experimental outcomes, rather than stating with certainty that a particular thing will happen. But if the entire universe is deterministically obeying a single equation, how can probability enter into physics at all?

Normally we think of probability as being like rolling a die: there are six outcomes, you can only get one, so the odds of any particular outcome are one in six (unless the die is weighted). The probability of rolling an odd number on a die is three in six, because there are three ways to get that outcome out of six total possible outcomes (Figure 11.1a). But probability in the many-worlds interpretation can't work that way. With Schrödinger's cat, there are two possible outcomes—the cat is either alive or dead—so it's tempting to think the probability of either outcome is one in two: 50 percent. But say we had set up the experiment slightly differently—say we hadn't left the cat in there quite so long, out of some belated sense of mercy, and the probability of a radioactive decay (and thus a dead cat) was only 25 percent, rather than 50 percent. Now we have a problem: there are still two possible outcomes, but quantum physics dictates that they have unequal probabilities. There's a 75 percent chance that the cat is alive and a 25 percent chance that the cat is dead—but there's still just two branches, each inhabited by nearly identical versions of you. Is the copy of you in the dead-cat branch somehow "less real" than the copy of you in the living-cat branch? How do we make sense of this?

It gets worse. This is just one experiment, and it's a big universe out there—in fact, in any reasonable understanding of Everett's interpretation, there's an infinity of branches on the universal wave function. How can we make sense of probabilities when there's an infinite number of copies of ourselves? The only reason we can calculate probabilities for our

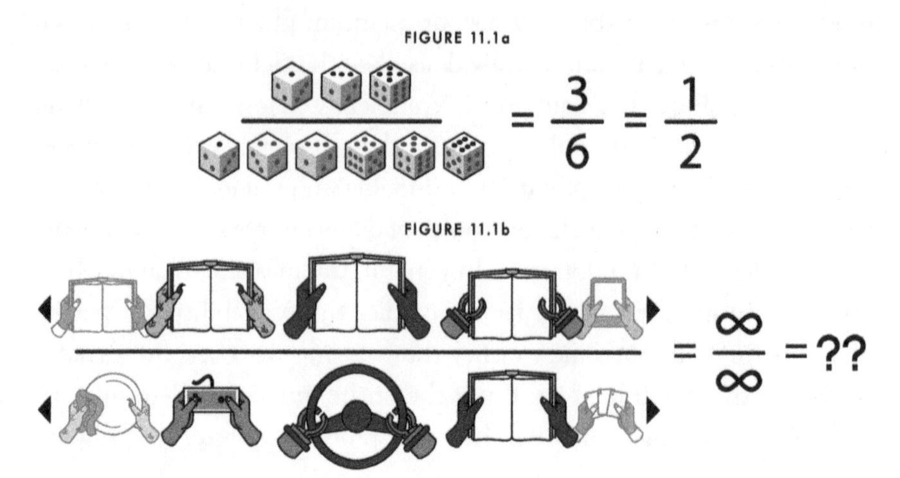

Figure 11.1. (a) Probabilities are relatively easy to calculate for dice and other situations where there are only a finite number of outcomes. The odds of rolling an odd number with an ordinary six-sided die are three out of six, or one-half. (b) Probabilities are much more difficult to calculate in an infinite multiverse. In the many-worlds interpretation, what are the odds that a randomly chosen version of you is reading this book right now?

dice throws is that we can count the number of possible outcomes. This approach doesn't work in an infinite multiverse, because the numbers involved are always infinite. If we want to know the number of branches in which a particular event occurs—say, the number of branches of the universal wave function where you're reading this book right now—the answer will always be infinite. And the number of branches in which you are *not* reading this book is also infinite. So what's the probability, in the multiverse, that a randomly chosen version of you would be reading some version of this book? What does the fraction infinity over infinity come out to? (Figure 11.1b). There are entire branches of mathematics devoted to dealing with infinity, and they tell us that such fractions can equal nearly anything at all—zero, some finite number, or even another infinity. So how do we handle this? How can we recover the fabulously accurate probabilistic predictions of quantum physics in the totally deterministic universe of the many-worlds interpretation? How do we measure the infinite fraction of infinite outcomes in which you're reading this book? And what can it even mean to talk about probability in a world where literally anything that is physically possible actually occurs somewhere?

The answer, or at least an answer, is that probability shows up in the many-worlds interpretation because we're hopelessly lost. Although the universal wave function obeys the Schrödinger equation and splits in a deterministic way, we don't know where in that huge and complicated wave function we are. We know that we're in only one branch of the universal wave function—but which one? After all, there are many copies of each of us scattered across a multitude of quantum worlds, each only slightly different, so it's not immediately obvious which world we're in. In particular, after conducting a quantum experiment, we know that we're in one and only one of the several worlds that the universe split into after the experiment concluded. But we can't know which of those worlds we're in without looking at the outcome of the experiment—we can't tell just by looking around us, because all of those universes look identical otherwise. The best we can do is to use the mathematics of quantum physics to say how likely, how probable, it is that we're currently in a particular branch of the wave function—which means we're assigning a probability to seeing a particular outcome of the experiment when we do look. So probability is still an essential part of quantum physics in the many-worlds interpretation; it's just that the probability isn't, strictly speaking, about outcomes of experiments, but rather about where you find yourself in the universe right now.

It's not clear whether this explanation actually works, though—it may be guilty of dualism, the idea that your mind is a nonphysical entity somehow separate from your body. Nor is it clear whether this explanation also yields the specific probabilistic predictions of quantum physics. But it is a promising idea, one among several attempts to explain this problem. Figuring out how to calculate probabilities in an infinite multiverse is one of the most pressing problems both in modern inflationary cosmology and among those who subscribe to the many-worlds interpretation. There are a variety of proposals for solving it, none of which have been widely accepted. (Some of them invoke Everett's other great mathematical love, game theory, to answer the question.) As with any open problem in science, there are no easy answers yet. But the consensus seems to be that the problem is not unequivocally solved—but that it probably can be solved, and either one of the existing promising solutions will turn out to be correct or another will be found, hopefully soon.

Despite the challenge that probability poses to the many-worlds interpretation and other multiverse theories, the most frequent objection to the idea of a multiverse (be it quantum, cosmological, or stringy) is simply its profusion of worlds. "It is hard to imagine a more radical violation of Occam's razor, the law of parsimony which urges scientists to keep entities to a minimum," complained Martin Gardner, the writer and recreational mathematician. But parsimony is in the eye of the beholder—advocates of Everett's view point out that their interpretation of quantum physics requires far fewer assumptions than any other. And arguments from simplicity, by themselves, can easily lead science astray. There are many complicated scientific theories that are inarguably correct. "Here's a 'multiverse' that basically everyone believes in," says David Wallace, a philosopher and advocate of the many-worlds interpretation. "Think about the planets of the stars of distant galaxies. Pretty much everyone thinks that there are indeed planets around the stars in distant galaxies, and that there are rocks on the surface of those planets. . . . That's not an infinite multiverse, but [ten thousand billion billion] solar systems is quite a lot to be getting on with. And the reason you take that seriously is not really because we can observe it. . . . It's more that it's a completely unavoidable consequence of a theory that we think is just rock-solid."

Physicists attacking many-worlds (or inflation, or string theory) usually have a more serious objection to the idea of a multiverse: they denounce it as a prime example of "unfalsifiability." This unwieldy word, a ghost of philosophy past, comes from the work of Karl Popper. Popper was a celebrated mid-twentieth-century philosopher of science who spent most of his career at the London School of Economics. Popper had once held truck with the logical positivism of his native Vienna but ultimately took an iconoclastic stance of his own. Rather than championing a verification theory of meaning, as the Vienna Circle had done, Popper promoted a scientific worldview based on *falsification*. Theories that could be proven false, Popper declared, were potentially scientific theories—and theories that could not be proven false were not scientific at all.

Popper's views became unusually popular among practicing scientists, and, by the end of the twentieth century, many physicists believed that falsifiability was a vital acid test that any potential theory must pass. Viewed through this lens, any multiverse theory appears quite suspect. If the other universes are not accessible and can never directly influence our own universe, then what possible experimental data could falsify the theory that we live in a multiverse? And if no data could ever show the theory wrong, then how can it be an acceptable scientific theory? "As the philosopher of science Karl Popper argued: a theory must be falsifiable to be scientific," wrote prominent cosmologists George Ellis and Joe Silk in a *Nature* editorial in 2014. "These unprovable hypotheses [many-worlds, string theory, and the inflationary multiverse] are quite different from those that relate directly to the real world and that are testable through observations—such as the standard model of particle physics and the existence of dark matter and dark energy. As we see it, theoretical physics risks becoming a no-man's-land between mathematics, physics and philosophy that does not truly meet the requirements of any." Straying from Popper's dictum, they warned, was a "drastic step" with potentially dire consequences. "This battle for the heart and soul of physics is opening up at a time when scientific results—in topics from climate change to the theory of evolution—are being questioned by some politicians and religious fundamentalists. Potential damage to public confidence in science and to the nature of fundamental physics needs to be contained by deeper dialogue between scientists and philosophers."

Yet if Ellis and Silk had bothered to engage in such a dialogue before writing their editorial, they would have learned that Popper's work hasn't been taken seriously by philosophers of science for decades, and with good reason. The idea that falsifiability marks the boundary of science is vulnerable to the same arguments that made the verification theory of meaning untenable, which we saw in Chapter 8. Just as individual beliefs can't be verified, as Quine pointed out in "Two Dogmas of Empiricism," individual theories can't be falsified either, for much the same reason. Say Karl Popper's remote control isn't turning his television on, and he theorizes that the batteries in the remote are dead. So Popper runs down to the corner store, buys new batteries, and pops them into his remote. But the remote still doesn't turn the TV on. "Aha!" Popper exclaims.

"My theory has been falsified!" But it ain't necessarily so. Despite the fact that the remote still isn't working, it's still possible that the old batteries were dead. Maybe the new batteries are dead too. Maybe a rat gnawed through the TV's power cable while Popper was at the corner store. Maybe the laws of physics actually change depending on where you are, and while Popper was at the store, the Solar System moved through its orbit around the center of the Milky Way and entered a patch of space where the laws of electromagnetism that govern the behavior of batteries in remote controls are different. The problem is that Popper's "dead battery theory" of his remote control doesn't actually make any predictions on its own: it only makes predictions in conjunction with a huge number of other basic assumptions that Popper has made about the functioning of the world. So Popper is wrong: his theory was not falsified. When the remote continued not to work, he could have rejected his theory about the dead batteries, but he could just as easily have rejected any of his other assumptions about the world. As Quine said, our beliefs about the world can only be tested against the world as a group, not individually, and this holds for falsification just as much as for verification. No theory, in isolation, is falsifiable.

The history of science bears this out: when an experimental or observational result doesn't match a theoretical prediction, often one of the auxiliary assumptions used to generate the prediction is discarded, rather than the "main" theory itself. In 1781, John Herschel discovered Uranus, and astronomers of the time immediately set about predicting its motion using the cutting-edge science of Isaac Newton's gravitation and laws of motion. Over the course of the next few decades, as more observations poured in and calculations were refined, several astronomers realized that Uranus was not, in fact, moving as Newton's law of universal gravitation said it should be. But rather than throwing out Newtonian gravity as falsified by observation, they theorized that there was another as yet unseen planet, even farther out than Uranus, that was causing the anomalies in Uranus's motion. One of these astronomers, Urbain Le Verrier, calculated exactly where to find this planet, and, in 1846, a group of German astronomers found Neptune exactly where Le Verrier said it would be. So, rather than being falsified, Newtonian gravity lived to fight another day. And when, several years later, Le Verrier and others noticed that Mercury, the closest planet to the Sun, was not moving as it should, they once again did not throw out Newtonian gravity and instead postulated another new planet, so close in to the Sun that it was perpetually lost in its glare. They named this hypothetical scorched planet "Vulcan," after the Roman god of the forge, and immediately set about looking for it. They hunted for it during solar eclipses, when the Sun was masked by the Moon. Several teams, including one led by Le Verrier himself, claimed to have found the elusive Vulcan, but the planet was never conclusively pinned down. Finally, in 1915, Albert Einstein proved Vulcan was a phantom: his new theory of general relativity explained the motion of Mercury perfectly, without invoking a new planet. Newtonian gravity had been wrong all along—but it took a new theory to displace it, rather than an alleged "falsification."

Even Popper understood that falsification couldn't be the acid test for scientific theories: he admitted that no theory could be falsified in isolation but suggested that good scientists would reject their own theories rather than auxiliary hypotheses. But as the saga of Neptune and Vulcan shows, it is far from obvious when a theory should be discarded in light of conflicting evidence, rather than rejecting some other assumption used to make the prediction. Claiming, then, that multiverse theories are unscientific because they are unfalsifiable is to reject them simply because they do not live up to an arbitrary standard that no scientific theory of any kind has ever met. Claiming that no data could ever force the rejection of a multiverse theory is merely stating that a multiverse theory is just like any other theory. And claiming that there could never be any observable evidence in favor of a multiverse theory is to forget Einstein's admonition that "it is the theory which decides what we can observe." Just as Grover Maxwell said in Chapter 8, what is considered observable can and does change over time, as scientific theories change. Atomic theory was once considered impossible to falsify, and atoms were once thought to be unobservable in principle. Evidence of a multiverse could share the same fate. Ultimately, arguments against a multiverse purportedly based on falsifiability are really arguments based on ignorance and taste: some physicists are unaware of the history and philosophy of their own field and find multiverse theories unpalatable. But that does not mean that multiverse theories are unscientific.

If scientific theories don't need to be falsifiable, what do they need to do? They need to give explanations, unify previously disparate concepts, and bear some relationship with the world around us. That's vague, of course, but science, like the people who do it and the world it describes, is complicated. Pat answers to complex questions, like Popper's cry of "falsifiability!", should always be suspicious: as H. L. Mencken once said, "There is always a well-known solution to every human problem—neat, plausible, and wrong."

What, then, is the right solution to the human problem of the Copenhagen interpretation? For, despite everything—despite pilot waves and many worlds, despite Bell and Bohm and Everett, despite the rise of quantum computing and the fall of logical positivism—Copenhagen is still dominant within physics. Copenhagen is still the view taught in every single popular introductory quantum physics textbook. There are still plenty of physicists who not only prefer the Copenhagen interpretation but think any other view is unscientific; some even claim that Bell's theorem proves that the Copenhagen interpretation is the only possible consistent view. Quantum foundations is a much more respectable field than it once was, but it is still a small field, and there are still many physicists who have disdain for it. It's hard to find any jobs in quantum foundations, albeit not as hard as it was for John Clauser fifty years ago. And while the many-worlds interpretation is generally known to most physicists, many other views, such as pilot-wave theory, are all but unheard of.

How did we get here? Or rather, why are we still here? It's a good question. David Albert—the graduate student who nearly got kicked out of Rockefeller University back in Chapter 9 for having the temerity to question the Copenhagen interpretation—is now a professor of philosophy at Columbia University, and has spent the past forty years working in quantum foundations. "This is a really bizarre story," he said, summing up the history of his field. "You have, at the same time, the following two wildly contradictory things going on: the 20th century outstripped every other century . . . for the number of smart people who were interested in physics and working actively on physics. That same century witnessed the longest period of psychotic denial of this deep logical problem right at the center of this whole project!"

"Psychotic" might be overstating it. But it's pretty bizarre. Now that you've seen the whole story—now that you know how we got here—let's see just how weird things are now.

12

Outrageous Fortune

In the wooded foothills of the Austrian Alps, on the outskirts of Vienna, there is a hut in a vineyard with a small mirror in the window. The vineyard has been there for centuries; it was already old when one of the founders of the Vienna Circle, Otto Neurath, met with Einstein and other scientists on the neighboring hill in 1920 to discuss his idea of an *International Encyclopedia of Unified Science*. The mirror is a newer addition. It was placed there in 2011, by graduate students at the Institute for Quantum Optics and Quantum Information of the University of Vienna, as part of a long-distance quantum encryption network. The students, working for Professor Anton Zeilinger, shot photons at the mirror, one at a time, from their lab on the top floor of their building in the heart of Vienna, five kilometers away. On the roof of that building, a specially equipped telescope—named after Hedy Lamarr, the Viennese movie star and cryptography pioneer—remained trained on the mirror in the vineyard, carefully collecting the reflected packets of light after their passage through the turbulent air over Vienna.

This feat, unimaginable outside of thought experiments for the founders of quantum theory, was merely a test. Right now, Zeilinger and his students are using this same equipment to exchange photons with a specially designed satellite in low Earth orbit, attempting to enable quantum-encrypted communication between Vienna and Yunnan Astronomical Observatory in China, where physicist Jianwei Pan, a former student of Zeilinger, has a similar setup already in place. And if the past is any guide, they are likely to succeed: Zeilinger is an experimental master of photon manipulation. Zeilinger's group has already demonstrated that it can send and receive single photons over distances much longer than the 10 kilometers from the lab to the mirror in the vineyard and back. In 2012, they successfully sent entangled photons over 143 kilometers, between La Palma and Tenerife in the Canary Islands. And Zeilinger has also spent decades conducting improved versions of Aspect's Bell experiments, verifying the existence of quantum nonlocality with enormous experimental precision.

Yet despite his intimate knowledge of the most bizarre aspects of the quantum world, Zeilinger has no qualms about the Copenhagen interpretation. "The quantum state, as Heisenberg says, is a mathematical representation of our knowledge," said Zeilinger. "It tells us the set of possible future measurement results, together with their probabilities." Measurement, for Zeilinger, plays a central role in quantum physics. "There is no measurement problem," he claims. "Measurement results live in the classical world, and the quantum state is what we call a quantum world, which is only a mathematical representation, according to Heisenberg. . . . What you can talk about with your classical language, these are the objectively existing objects of the universe, these are the classical objects. And that's it. That's what can be talked about. The rest is mathematics." In other words, there are two worlds: a world of actually existing everyday objects obeying classical prequantum physics, and a quantum "world" that isn't real in the same way, just as Heisenberg said. But Zeilinger doesn't think that there is a true boundary between them, some line beyond which quantum physics does not apply. "There is no fundamental boundary," he said. "There is a transition from classical to quantum, but that is not a boundary." It's not surprising that Zeilinger would say this: almost no physicists still believe that there is any such fundamental boundary, and some of the most convincing work against such an idea came from Zeilinger himself. Back in 1999, Zeilinger and his collaborators managed to coax a buckyball—a collection of sixty carbon atoms in the shape of a soccer ball—to interfere with itself, like a photon in a double-slit experiment. Finding quantum effects in an object so much larger than an individual subatomic particle (though still about a billion times smaller than the objects in our everyday lives) might have been shocking to some of the founders of quantum physics. But Zeilinger, through his experimental work, has been determined to show that there is no limit to the validity of quantum physics.

But this leaves a question: If only classical things exist objectively, yet quantum physics applies to everything, what is classical? More generally, how can we account for the world we see around us? According to Zeilinger, our everyday world is classical—but quantum physics must also accurately describe what we see in our everyday lives, because there's no boundary to its validity. How can we form a coherent picture of reality from this version of the Copenhagen interpretation? Zeilinger's answer to this question is unexpectedly simple. "I don't know what you mean by that," he said. "I don't think that you can even define it precisely."

What the hell is going on here?

Not all physicists agree with Zeilinger. "The Copenhagen interpretation assumes a mysterious division between the microscopic world governed by quantum mechanics and a macroscopic world of [measurement] apparatus and observers that obeys classical physics," said Steven Weinberg, winner of the 1979 Nobel Prize in Physics. "This is clearly unsatisfactory. If quantum mechanics applies to everything, then it must apply to a physicist's measurement apparatus, and to physicists themselves. On the other hand, if quantum mechanics does not apply to everything, then we need to know where to draw the boundary of its area of validity. Does it apply only to systems that are not too large? Does it apply if a measurement is made by some automatic apparatus, and no human reads the result?" Gerard 't Hooft, winner of the 1999 Nobel Prize in Physics, struck a more conciliatory note. "I go along with everything [Copenhagen] says, except for one thing, and the one thing is you're not allowed to ask any questions," he says. "Or to be more precise, there are certain questions you should not ask. I say, no, I'm going to ask them anyway. You don't want me to ask questions? Sorry. I have the strong impression that there is much more to say, and that asking questions is useful." And Sir Anthony Leggett, winner of the 2003 Nobel Prize in Physics, said that he must make "an awful confession: If you were to watch me by day, you would see me sitting at my desk solving Schrödinger's equation . . . exactly like my colleagues. But occasionally at night, when the full moon is bright, I do what in the physics community is the intellectual equivalent of turning into a werewolf: I question whether quantum mechanics is the complete and ultimate truth about the physical universe. In particular, I question whether the superposition principle really can be extrapolated to the macroscopic level in the way required to generate the quantum measurement paradox. Worse, I am inclined to believe that at *some* point between the atom and the human brain it not only may but *must* break down."

Yet Weinberg, 't Hooft, and Leggett are exceptions among physicists. Views like Zeilinger's are far more common. There have been multiple informal surveys in the past twenty years asking physicists about their preferred interpretation of quantum physics. In most of these, the Copenhagen interpretation wins a commanding plurality of votes. And there is good reason to believe this seriously understates the support of Copenhagen among physicists, since these surveys are usually conducted at conferences on quantum foundations, creating a massive sample bias in the results. There are still a significant number of physicists who consider such conferences a waste of time, because they think that the Copenhagen interpretation sorted out all these problems long ago.

Strangely, though, Zeilinger had difficulty recommending a reference that laid out the Copenhagen interpretation clearly. "Maybe I, or someone else, should write a clear paper about quantum mechanics," he said. Part of the problem is that Bohr, the obvious choice, is fabulously (and famously) opaque. But there's a deeper reason underlying this difficulty. "Copenhagen is no longer the dominant interpretation," said physicist-turned-historian Sam Schweber (who bailed out David Bohm back in Chapter 5). In the original old-school versions of the Copenhagen interpretation, classical objects like measurement devices couldn't be described with quantum physics, even in principle. But today, Schweber points out, nearly all physicists agree with Zeilinger that there are no such limits on quantum physics. Why, then, do so many physicists still think they subscribe to the Copenhagen interpretation? How can so many of them run blithely past the edge of the quantum cliff, like Wile E. Coyote, not realizing just how far they have to fall? "That's a different story," Schweber said.

Part of the problem is that there is no single "Copenhagen interpretation" and never really was. "The name 'Copenhagen interpretation' has gotten pretty slippery," said Nina Emery, a philosopher of physics at Mt. Holyoke. "The semantic confusion makes it easy for physicists to avoid dealing with those flaws directly. For instance, when you push them on the idea that measurements cause collapse . . . they shift and start talking about some kind of Bohrian view or about the [mathematics of the theory]. And if you point out the issues with those views (e.g. who knows what the former is; and the latter isn't even a complete interpretation), they go back to talking about [measurements causing collapse]." The flexibility afforded by these contradictory positions makes it easy to fend off any attacks on "the" Copenhagen interpretation. Physicists can simply hop from one position to another—sometimes without even realizing that they've done it.

But if you're an instrumentalist—if you think that science is merely a tool for predicting the outcomes of experiments and nothing more—then this kind of hopping around isn't a problem, because questions of interpretation are pointless and unscientific anyhow. It doesn't matter whether you hold a single consistent position about the meaning of quantum theory. All that matters is what you can directly observe. These kinds of positivist-like ideas are still very popular among physicists, especially when the subject of quantum physics comes up. Zeilinger has suggested that the "message of the quantum" is the fairly positivist idea that "the distinction between reality and our knowledge of reality, between reality and information, cannot be made." And the eminent physicist Freeman Dyson, like Rosenfeld before him, has suggested that it may be impossible to observe any consequences of a theory of quantum gravity, which he claimed "would imply that theories of quantum gravity are untestable and scientifically meaningless," in true positivist style.

Yet philosophers have known for more than half a century that the positivism underpinning statements like these is fundamentally flawed.

And philosophers of physics today almost unanimously reject the Copenhagen interpretation. (Logical empiricism of a sort has made a comeback since 1980, but scientific realism is still the standard position among philosophers of physics—and even the most staunch defenders of empiricism today agree that the sort of naive positivism deployed in standard defenses of Copenhagen doesn't work.) How have physicists failed to get the memo from philosophers after all this time? Part of the problem is that physicists generally don't know much about philosophy. There is a massive asymmetry between the two fields: while philosophers usually take physics very seriously indeed—philosophers of physics are mathematically conversant in physics and often have advanced degrees in both fields—physicists are rarely trained in philosophy at all. Despite their ignorance of philosophy (or more likely because of their ignorance), some physicists are openly contemptuous of the subject. "Philosophy is dead," declared Stephen Hawking in 2011. "Philosophers have not kept up with modern developments in science. Particularly physics." And according to Neil deGrasse Tyson, studying philosophy "can really mess you up." "Pretty much after quantum mechanics . . . philosophy has basically parted ways from the frontier of the physical sciences," Tyson claimed. "I'm disappointed because there is a lot of brainpower there, that might have otherwise contributed mightily, but today simply does not." Physicist Lawrence Krauss surmised that this antagonism between physics and philosophy comes from envy on the part of the philosophers, "because science progresses and philosophy doesn't," he claimed. "Philosophy is a field that, unfortunately, reminds me of that old Woody Allen joke, 'those that can't do, teach, and those that can't teach, teach gym.' And the worst part of philosophy is the philosophy of science. . . . It's really hard to understand what justifies it."

These are breathtakingly ignorant claims. Yet Hawking, Tyson, and Krauss are certainly not stupid people—why do they know so little about philosophy? Their attitudes are even stranger when put in historical perspective. Just a few generations ago, at the birth of quantum physics, all physicists received some schooling in philosophy. Einstein read Mach, Bohr read Kant. But the shifts in research funding and physics classrooms after World War II also led to broader shifts in university curricula. For Einstein and Bohr's generation, philosophy was part of the core educational curriculum in central Europe. But in postwar America, it was (and is) relatively easy for an intelligent student to go all the way from kindergarten to a PhD in physics at a top-tier university without ever darkening the door of a philosophy classroom.

This is not a plea for the good old days. This problem isn't especially new. Even Einstein complained about this, and how it helped keep the Copenhagen interpretation entrenched. "This state of affairs will last for many more years," he wrote in 1951, "mainly because physicists have no understanding of logical and philosophical arguments." And in most ways, education and access to education are far better now than they ever have been before. But with the massive increase in knowledge and information in the last century, education became unavoidably specialized. There's nothing wrong with that, but with specialization comes boundaries on knowledge, and good specialists understand that. Indeed, it's unlikely that Hawking, Tyson, or Krauss would issue strong pronouncements in many areas where they have no background—say, parasite ecology or best practices in industrial sheet metal production. So why do they feel so comfortable saying so much about philosophy? Why is philosophy held in such contempt by many physicists (and other scientists of all stripes)?

Philosophy has an image problem. Philosophers are thought to be mystics, religious figures, bullshit artists—anything divorced from reality. The discipline as a whole is seen as millennia of people chasing down big questions—What is the meaning of life? Why is there suffering?—and coming back without any good answers. Philosophers of physics, and most other philosophers, are far removed from this picture: they work on well-defined questions with logical rigor and with input from the most recent developments in science and from the immediate experiences of the senses. How the practice and the image of philosophy have diverged so wildly is a subject for an entirely different book, but one part of the answer probably lies in the split between the two major branches of modern Western philosophy, *analytic* and *Continental* philosophy. (Those names are largely historical accidents and not relevant to their work.) How the two branches split is a long and complex story (it has to do with the debate between the positivists and the German idealists, which we touched on in Chapter 8), but while most philosophers of physics are analytic, most of the philosophers from the past seventy years that you've heard of are probably Continental. Continental philosophers like Sartre, Camus, Foucault, Derrida, and Žižek have become public figures, while very few analytic philosophers have done the same. And Continental philosophers tend to be much more suspicious of scientific claims about knowledge and truth than are their analytic colleagues. Yet the distinction between the two kinds of philosophy is not apparent from a distance—most scientists have never heard of the analytic-Continental divide. So, given that most of the highly visible philosophers in the public sphere today are Continental, and given the attitude that some (not all) Continental philosophers have toward science, it's not terribly surprising that scientists often have disdain for *all* philosophers, and sometimes even think that they can do philosophy better than the philosophers can.

But there's more to it than this. Not all physicists who support the Copenhagen interpretation are merely ignorant of philosophy. Zeilinger has spent a great deal of time at conferences on quantum foundations with philosophers of physics and is certainly aware of the history of positivism in his native Vienna. And it's not as if a deep commitment to positivism among a wide swath of physicists is driving support of the Copenhagen interpretation—if anything, it's the opposite. We physicists all learn some form of the Copenhagen interpretation in school, and many of us adopt it. And once you have that Copenhagen mindset, you'll probably be more favorably inclined toward positivism and related views. So perhaps it's not that physicists are eager to adopt positions that absolve them of responsibility for talking about reality—perhaps it's merely that such positions become appealing once you've adopted Copenhagen's viewpoint. Which leaves us back where we started: What's so appealing about the Copenhagen interpretation?

"If I were forced to sum up in one sentence what the Copenhagen interpretation says to me," wrote the physicist David Mermin in

1989, "it would be 'Shut up and calculate!'" Mermin followed his summary with a quick rejoinder—"But I won't shut up." Yet the phrase "shut up and calculate" took on a life of its own after Mermin set it to paper, and rapidly became the catchphrase of the Copenhagen interpretation among physicists. It was misattributed to Richard Feynman, and eventually even Mermin himself forgot where it came from, only to rediscover, years later, that he was the source of the phrase.

"Shut up and calculate!" certainly doesn't sound appealing if you're not mathematically inclined. But, even if you're a physicist, what's the virtue in shutting up and calculating? Mermin himself provided the answer in his 1989 article. "It is a fact about the quantum theory of paramount importance which ought to be emphasized in every popular and semi-popular exposition, that it permits us to calculate measurable quantities with unprecedented precision." Quantum physics works. The calculations enabled by the theory are astonishing in their range of applicability and the accuracy of their results. Quantum physics tells us how long it will take to heat up your frying pan to cook your eggs and how large a dying white dwarf star can be without collapsing. It reveals the exact shape of the double helix at the core of life, it tells us the age of the immortal cattle on the rock walls at Lascaux, it speaks of atoms split beneath the stone heart of Africa eons before Oppenheimer and the blinding light of Trinity. It predicts with uncanny accuracy the precise darkness of the blackest night. It shows us the history of the universe in a handful of dust. If shutting up is the price of doing these calculations, then pass the ball gag and break out the graph paper.

But why is that the price? Why does Copenhagen require that you shut up in order to calculate? For that matter, how does the Copenhagen interpretation allow you to calculate at all? The measurement problem is so centrally tied to the core of quantum physics that, without some answer to the problem, it's impossible to use the theory. Some interpretation must guide you in the use of the mathematics—and Copenhagen, as we've seen many times over, offers no such solution and is not a true interpretation. So how can shutting up allow you to calculate anything?

The form of the Copenhagen interpretation usually found in physics textbooks says (explicitly or otherwise) that measurement is a fundamentally different process from any other found in nature and that "measurement" is defined as "any time a large object encounters a small one." Large objects are simply assumed to obey classical physics, even as quantum physics is presented to the student as a more fundamental theory underpinning classical physics. In short, the student is implicitly asked to accept as part of the basic structure of quantum physics that there are two worlds, the classical and the quantum, just as Bohr taught. Yet, at the same time, they're being told that quantum physics is the fundamental theory from which classical physics emerges. So quantum physics students are asked to swallow a contradiction: on the one hand, they're told that the idea of a classical object is logically prior to the idea of quantum physics, since the idea of a classical object is needed in order to figure out when a measurement has happened; but, on the other hand, they're told that quantum physics is logically prior to classical physics, the latter emerging out of the former. These ideas can't both be right. And, in practice, the version of the Copenhagen interpretation most commonly found in textbooks and "in the wild" values the first idea over the second. Some objects are just classical, and it's interaction with those objects that is defined as measurement for the purposes of quantum physics, "solving" the measurement problem well enough to allow calculations to be performed. Certainly most physicists (including yours truly) also believe that quantum physics underpins classical physics, but when actually performing quantum physics calculations, this fact is conveniently forgotten, and some objects are simply treated as exempt from the Schrödinger equation. Hence the desperate need to shut up while calculating.

There have been physicists who have tried to bring the idea that quantum physics is fundamental into their calculations. To do so, they had to give up Copenhagen's solution to the measurement problem and perform conceptual work to solve the problem a different way. In other words, these people, people like David Bohm and Hugh Everett, had to develop new interpretations of quantum physics, because the Copenhagen interpretation doesn't take quantum physics seriously. It requires that we abandon the idea that quantum physics can be used to deal with everything in the universe and restricts it to a limited domain. Today, most physicists agree with Zeilinger that there is no limit to the validity of quantum physics—but the manner in which quantum physics is generally taught and used betrays that ideal.

Nonetheless, the appeal of Copenhagen makes some sense, seen in this light. Quantum physics drove much of the technological and scientific progress of the past ninety years: nuclear power, modern computers, the Internet. Quantum-driven medical imaging changed the face of health care; quantum imaging techniques at smaller scales have revolutionized biology and kicked off the entirely new field of molecular genetics. The list goes on. Make some kind of personal peace with Copenhagen, and contribute to this amazing revolution in science . . . or take quantum physics seriously, and come face-to-face with a problem that even Einstein couldn't solve. Shutting up never looked so good.

There's more at play here than a simple pragmatic desire to do physics or a clash between physics and philosophy. Ultimately, this is a story about people. "The case of the measurement problem," said David Albert, "was something very painful for the [physics] community. A lot of careers were destroyed. This whole business was a real trauma, in the psychological sense of the word, for physics." The history of quantum foundations is soaked in personalities. If David Bohm had held more palatable political convictions, if Hugh Everett hadn't hated public speaking, if Einstein had had Bohr's charisma, the story told in this book likely would have been dramatically different. So many of the key events were driven by political or social or interpersonal interactions, not by scientific considerations. This suggests another reason the Copenhagen interpretation is so popular: not because it is somehow better or more suited to the needs of physicists but simply because it was first.

From a naive perspective about science—from the view that science is merely a mechanism for deducing the One True Answer from the available clues, like a Sherlock Holmes story—this idea is disturbing. (Indeed, this whole book is probably disturbing to someone with such a view.) If these extraneous factors could have such a profound influence on fundamental physics, what part of science could possibly remain untouched? And, indeed, this isn't limited to quantum foundations: all of science is vulnerable to human biases and to influences from all the other spheres of human endeavor—politics, history, culture, economics, art—that some of those biases spring from. Most scientists, by and large, will agree to this. But agreeing with the abstract existence of these nonscientific biases within science is different from being faced with a concrete example. The idea that something as pervasive and central as the Copenhagen interpretation might be dominant for "accidental" nonscientific reasons can be scary, especially for people who have devoted their entire lives to physics. Once you give up Copenhagen, "there's more than one option on the table, and if there's more than one option on the table then how do you decide?" asks Doreen Fraser, a philosopher of physics at the University of Waterloo. "Is it because you have certain prejudices about what's interesting and what's not interesting? Actually that's a large part of it, but that's kind of uncomfortable." This discomfort, this fear, is yet another reason why it's tempting, as a physicist, to "shut up and calculate." But giving in to that fear just makes it harder for us to see our biases.

These biases include a lot of the factors we've seen directly at play in this book: political considerations, funding models, the milieu of ideas in particular places and times, even simple interpersonal disputes. There are also many biases that have been at work throughout this book, but not in the foreground. A woman, Grete Hermann, found the problem with von Neumann's proof thirty years before Bell did, but nobody noticed at the time. It's hard to imagine that her gender had nothing to do with the reception of her work in 1935, a time when women were still generally not allowed to teach at universities. And, given that specialization in the foundations of quantum physics was a professional strike against any physicist, it's also not far-fetched to imagine that few women and nonwhite physicists would be attracted to work in that area, since their identities were already a strike against them within science and academia as a whole, because of systemic bias. This would explain the otherwise remarkable fact that hardly any women or people who aren't white appear anywhere in this story at all. Prejudices, against certain kinds of people as well as certain kinds of ideas, are pervasive throughout science.

But the mere existence of these biases doesn't mean that science is identical to all those other spheres of human endeavor or that scientific truth isn't different from ill-informed opinions with no connection to experiment or reality. Our biases don't fully determine the content of our best scientific theories—reality pushes back, and we want it to push back as hard as we can allow it. That pushback constrains the possible hypotheses we entertain as scientists. There's a wide middle ground between "science is Pure and Perfectly Rational" and "science is just some bullshit somebody made up." There's still plenty of room for humans to interfere in that middle ground, as we've seen throughout this book. But that doesn't mean science isn't to be trusted—that's as naive as the Sherlock Holmes view of science.

That being said, the story of quantum foundations does seem to call into question how science works. We've seen how it doesn't work—it's not about verification or purely empirical statements, as the positivists thought; it's not about falsifiability, as Popper thought; nor is it about being completely independent from the complex historical forces that have buffeted and buoyed the characters we've met over the course of this book. So how does science work? Echoing the end of Chapter 11, that's a fabulously complex question. The long answer would take another book. But the short answer is that science involves a combination of experiment, mathematical and logical reasoning, unifying explanations, and biases that scientists bring to the table from their own lives and the cultures they live in. We work to reduce those biases; we don't always succeed, but the explicit attempt to account for and reduce those biases is an important part of the process, properly conducted. The whole edifice of science is geared toward this goal. And, given the phenomenal explanatory power and predictive success of science, it would be foolish in the extreme to give scientific truths no more credence than idle speculation, religious articles of faith, or deeply held cultural values. Science, done right, works hard to respect absolutely no authority at all other than experience and empirical data. It never succeeds entirely, but it comes closer and has a better track record than any other method we apes have found for learning about the world around us, a world we never made.

The story of the search to understand quantum physics is an emphatically scientific one. Yet the cultural and historical forces at work over the course of this book, while par for the course, are still troubling. How can we distinguish a controversy like the one over quantum foundations—a legitimate scientific controversy if ever there was one—from a manufactured pseudo-controversy about science, like the "debates" over evolution, global warming, and homeopathy? It might seem tempting to compare them, after all: from the perspective of someone who believes (erroneously) that climate change isn't real, evolution didn't happen, and homeopathy works, these are all stories of an overwhelming scientific consensus fought by a scrappy minority of independent thinkers dedicated to the truth at all costs. Yet even that apparent similarity is an illusion. The debates about evolution, global warming, and homeopathy have been explicitly manufactured and funded by a variety of corporate, religious, and political entities from outside science, who are not in the least interested in the quest to divorce our human biases from our understanding of the world. They are not committed to taking the science seriously and are instead devoted to taking their own aims and giving them a thin patina of scientific respectability, enough to justify their claims to equal or greater validity than the existing and overwhelming scientific consensus. These groups are not interested in examining the data and happily reject it when it does not meet their preordained conclusions, inventing new "data" to suit their purposes. In the cases of global warming and evolution, these "controversies" were invented to push back against a perceived political agenda on the part of science and scientists. And the forces behind intelligent design and climate change denial were not wrong about that—science is political, and always has been, in that it informs decisions about the best policies in the public sphere, as well it should. And science certainly is a threat to the institutions pushing these antiscience agendas. Science will always be a political threat to some institutions, simply by virtue of its attempts to respect no authority other than data and logic. So much the worse for those institutions. And that's another sign that these "debates" are not like the debate over quantum foundations—because those working against the scientific consensus are allied with (and often funded by) groups that are simply against the idea of science itself, like some fundamentalist religious groups.

By contrast, in the debate over quantum foundations, everyone agrees that science works, otherwise there wouldn't be much to debate. Despite the deep and sometimes bitter conflict over the Copenhagen interpretation, none of the physicists mentioned in this book doubted that quantum physics was correct, or at least an excellent approximation of some underlying theory. Nobody doubted the veracity of the experimental data that originally inspired quantum physics, nor the data that further supported its predictions once the theory was developed by Heisenberg and Schrödinger and the rest. Nor was there a concerted, organized effort to maintain Copenhagen's dominance. There was no conspiracy or any corporate or political interests at stake in this debate—just a dispute among physicists about the meaning of a theory that they all agreed was correct. Indeed, the debate over quantum foundations is, at its heart, a debate over just how seriously one should take quantum physics—and the dissenters from Copenhagen were the ones arguing that quantum physics should be taken very seriously indeed, as a theory of the entire world.

But there is one way in which quantum foundations does come into contact with the public debates between science and pseudo-science. The Copenhagen interpretation, with its vagueness, its seeming promise of a fundamental role for human consciousness, and its bevy of internal contradictions, has turned quantum physics into a wellspring of purported scientific support feeding a constant river of New Age nonsense and junk pseudo-science. The TV show *Futurama* skewered this fairly accurately, showing a physics professor in the year 3008 claiming that "as Deepak Chopra taught us, quantum physics means anything can happen at any time for no reason." Chopra does in fact claim that consciousness arises from quantum entanglement and that "quantum healing" allows the mind to heal the body through sheer willpower. "Our bodies ultimately are fields of information, intelligence, and energy," he said. "Quantum healing involves a shift in the fields of energy information, so as to bring about a correction in an idea that has gone wrong." Chopra is hardly alone in his spurious assertions about the fabulous implications of quantum physics for medicine. There are endless "quantum" health-care scams, claiming that their products can channel your thoughts to restructure your body on a quantum level, whatever that might mean. And perhaps most odiously, best sellers like *The Secret* have made fantastical claims about the power of quantum physics and have been so successful that they have inspired knockoff books like *Why Quantum Physicists Cannot Fail* and *Why Quantum Physicists Don't Get Fat.* (I can attest, from personal experience, that both of these claims are false.) These books breathlessly inform you that you can simply achieve what you want by wishing hard enough, reshaping your own reality, since quantum physics "proves" there is a fundamental role for conscious observers in creating the universe around us.

There is a massive irony here: critics of non-Copenhagen interpretations of quantum physics often say that concerns about Copenhagen come from a desire to keep the world sensible and "normal" as it was in classical physics. Yet the Copenhagen interpretation harks back to a far older and more comfortable vision of the world than any proposed in any of the other interpretations. Copenhagen puts humans, indeed the self, at the very center of the universe, more important than anything else, just as the ancients had it, with everything else revolving around us. This is why quantum physics holds such appeal in "alternative" circles. Rather than giving a humbling and strange vision of the universe, the Copenhagen interpretation makes physics familiar and comfortable. If we are to have any hope of understanding the universe, we must dare to imagine a world that is not bounded by our limited perspective.

But why does any of this matter? If shutting up and calculating works—and it does—then why do physicists need anything else? And why should any of this matter to anyone who isn't a physicist?

It's certainly true that we'll get the same answers when doing quantum-mechanical calculations whether we prefer the Copenhagen interpretation, the many-worlds interpretation, the pilot-wave interpretation, or anything else. Even alternatives to quantum physics, like spontaneous-collapse theories, will give the same answers in nearly every situation. Some people have argued, as Wolfgang Pauli did to Bohm, that, because the different interpretations don't make new predictions, we should just stick with Copenhagen—a silly argument, since you could say "we should just stick with many-worlds" or any other interpretation with that same reasoning.

Others claim that the alternatives to Copenhagen are driven by a desire to make things less weird than Copenhagen makes them out to be, and we should instead embrace the weirdness, that any discomfort with the Copenhagen interpretation is merely a sign of our limited human ability to understand the world of the quantum. This argument would carry more weight if there were no viable alternatives to the Copenhagen interpretation, if its conclusions were forced upon us. But there's another problem with it as well. "All of the proposals we have for solving the measurement problem are weird in one way or another," said David Albert. "Bell's theorem proves they have to be weird. . . . [But] there's a huge difference between being weird and being incoherent or unintelligible." And, Albert added, many physicists still do not seem to appreciate this point. "Physicists will say, 'Yes, the Copenhagen interpretation is weird, but so is everything else.' And you sort of want to slap them and say, 'No! The Copenhagen interpretation is not *weird*, it's gibberish, it's unintelligible.'"

And some physicists argue, like good positivists, that because no experiment can tell the difference between the different interpretations, it's meaningless to draw any distinctions between them—that even if Copenhagen is inconsistent, it doesn't matter which, if any, alternative to it we adopt. And this is simply not true. If we want to go beyond our current theories to devise a new theory, to find new physics and explain new experimental results, our interpretations matter. Ask two physicists, a pilot-wave theorist and a many-worlds theorist, what kind of theory they expect to see going beyond quantum physics, and you'll get two very different answers. Richard Feynman pointed out that although there's no experimental way to tell the difference between two mathematically equivalent theories (i.e., two different interpretations of the same math), subscribing to one theory or the other makes a huge difference in how you think about the world. That difference, in turn, affects the new ideas and new theories we develop. For example, the sixteenth-century astronomer Tycho Brahe had a theory in which the Earth was at the center of the universe, the Sun and Moon orbited the Earth, and the rest of the planets orbited the Sun. His theory was mathematically equivalent to the heliocentric model of Copernicus—it gave identical predictions about the movements of lights in the sky—but the idea that the Earth wasn't at the center of the universe led to an entirely different set of ideas about how the universe works. Similarly, we could develop an interpretation of quantum physics in which wave functions are powered by invisible unicorns, which obey aggregate laws of herding and flocking that lead to the Schrödinger equation. But (I hope) we can agree that this is a bad idea, significantly worse than any other interpretation. Experimental results are not the only things that enter into the formulation and evaluation of scientific theories, nor could they be. The full content of our theories—not only the mathematics but the claims about the nature of the world that come along with the mathematics—is important to the work of science.

And the worldviews we get from our best scientific theories also make their way to the public and inform how we see ourselves, as I argued in the Introduction. This has already happened with the Copenhagen interpretation—that is, after all, where that quantum healing nonsense is coming from. (Though, to be sure, Chopra and his ilk would probably find some other way to package their work if the Copenhagen interpretation didn't exist, and other interpretations would probably be misunderstood in some ways. Misappropriations of science are inevitable. It's just that Copenhagen seems particularly ripe for it.) In the past, new physics has opened up new horizons for human imagination, new ways of thinking about our own existence, new ideas in fields as wildly disparate as biology and art, geology and religion. If Copernicus hadn't unseated the Earth from its place at the center of the cosmos, it's hard to imagine that Darwin could have had the audacity to suggest that humans were not wholly unique creations but instead descended from apes—and without both of those insights, Kubrick certainly couldn't have filmed *2001*. Science and culture form an undivided whole, now more than ever, in our world whose every corner has been reshaped by human activity. If the past is any guide at all, finding the answer to the puzzle of quantum physics, and finding the next theory beyond it, will ultimately affect the daily lives of every human being, not just the professional lives of physicists.

The deep problems at the boundaries of physics—quantum gravity chief among them—have not yielded solutions for decades. These challenges are so profound that a handful of physicists have turned to quantum foundations for guidance and inspiration. Some have suggested that the structure of spacetime itself is built out of quantum entanglement, bridging far-distant points with wormholes. Others have argued that the multiverses of eternal inflation and string theory are actually the same as the multiverse of the many-worlds interpretation, that all three theories are merely different ways of arriving at the same fundamental truth about the cosmos. There's also work that explicitly takes quantum nonlocality as a starting point and tries to cook up a theory of quantum gravity that actually breaks Einstein's relativity, since nobody's had unequivocal success in building a theory of quantum gravity that doesn't break relativity.

And we've hardly done justice to the profusion of ways to interpret quantum physics that have been proposed to date. While the different possibilities laid out in this book are the most significant historically, and they are all mostly still around in various forms (minus Wigner's consciousness-based proposal, which has been dismissed as needlessly speculative and vague, and in danger of collapsing into solipsism), many, many more have been proposed in the past thirty years. There are retrocausal interpretations, which suggest that subatomic particles can affect their own pasts, taking quantum nonlocality to an extreme. There are interpretations that try to get around Bell's theorem by altering the axioms of probability itself, though it's not clear if they can succeed. 't Hooft is developing his own interpretation of quantum theory that takes an unusual avenue through the obstacle course of weirdness set up by Bell's results. His theory is "superdeterministic," a local hidden-variables theory that has deep prearrangements between subatomic particles and experimental setups. Many physicists and philosophers reject this kind of approach out of hand as a sort of cosmological conspiracy theory that would preclude the very possibility of doing science. Yet 't Hooft believes that he can find a way to do this without sacrificing science itself, and he may not be wrong. Roger Penrose, one of the foremost mathematical physicists alive, believes that wave function collapse is real and that the Schrödinger equation must be modified, as in spontaneous-collapse theories. But rather than collapse being entirely random, he believes that it is caused by gravity, marrying general relativity and quantum physics in an unexpected and novel fashion. There are even interpretations that are a kind of hybrid of existing interpretations, like the many-interacting-worlds interpretation, which has features of both the pilot-wave and many-worlds interpretations.

There are also challenging issues in the interpretation of quantum field theory, the theory that combines quantum mechanics with special relativity and describes the intricate high-energy physics seen at particle accelerators. QFT shares some of the problems of regular quantum theory—the measurement problem and nonlocality are still there—but it also has some new and strange foundational problems of its own. Getting some of the existing interpretations of quantum theory to work with QFT, such as the pilot-wave interpretation, is an ongoing challenge. (Others, such as many-worlds, have no problem with QFT, which is arguably a mark in their favor.) And there are so, so many other ideas and open problems in quantum foundations, all fascinating. Despite decades of discouragement and indifference from the rest of physics, the field of quantum foundations is healthy and burgeoning. John Bell, were he alive, would be astonished by what he has wrought.

So what is real? Pilot waves? Many worlds? Spontaneous collapse? Which interpretation of quantum physics is the right one? I don't know. Every interpretation has its critics (though the proponents of basically every non-Copenhagen interpretation are usually agreed that Copenhagen is the worst of the lot). Somehow, something is going on in the world that is related to the mathematics of quantum physics. There is a correct interpretation, though it may not be any of the ones that we have yet. Simply dismissing the quantum world as a convenient mathematical fiction means we aren't taking our best theories of the world seriously enough, and we are hobbling ourselves in the search for a new theory. Stating that the conclusions of the Copenhagen interpretation are "inevitable" or "forced upon us by the mathematics of the theory" is simply wrong. It is not true that it's meaningless to talk about reality existing independently of our perceptions, that we must think of the world solely as the subject of our observations. Solipsism and idealism are not the messages of quantum physics.

Instead, we physicists should learn the different interpretations available and keep them all in mind while working. Hold on to them loosely, not dogmatically, and keep a fresh perspective on the work we do. I'm not saying that interpretations of quantum physics are something that every physicist should work on, any more than every physicist should be working on any other specific open problem in physics, like quantum gravity or high-temperature superconductivity (an unexpected mystery worthy of its own book). But all physicists should be aware of the problem and passingly familiar with the field. We have a wildly successful theory, an embarrassment of interpretations, and a major challenge in moving past our theory to the next one. Pluralism about interpretations might be the right answer, pragmatically, while we face that challenge. Or if not pluralism, at least humility. Quantum physics is at least approximately correct. There is something real, out in the world, that somehow resembles the quantum. We just don't know what that means yet. And it's the job of physics to find out.

This is the great enterprise. This is what everyone in this sprawling tale fought for, in their ways: Bell with his scathing critic's pen, Bohm with his stubborn disregard for the status quo, Everett with his prankster's style. But their ideas aren't all that matters—their stories matter too. The history behind the physics can guide us in our pursuits, just as a new interpretation of a theory does. The path that led us here can give hints about the way forward. Demonstrating this, if nothing else, has been the aim of this book. I'll give the last word on the subject to someone far more qualified:

> So many people today—and even professional scientists—seem to me like somebody who has seen thousands of trees but has never seen a forest. A knowledge of the historic and philosophical background gives that kind of independence from prejudices of his generation from which most scientists are suffering. This independence created by philosophical insight is—in my opinion—the mark of distinction between a mere artisan or specialist and a real seeker after truth.
>
> —**Albert Einstein**

Appendix: Four Views of the Strangest Experiment

In 1978, not long after moving to the University of Texas, John Wheeler proposed a thought experiment that, according to him, "gets at the core of what fueled the Bohr-Einstein debate." In fact, he suggested that "this experiment may have something to tell us about the very machinery of the universe." He called it the delayed-choice experiment (Figure A.1).

The experiment has two configurations; we'll start with the simpler one, on the left (Figure A.1a). A laser beam (i.e., a beam of photons) enters from the lower left-hand corner and enters a beam splitter, which (as the name suggests) splits the beam into two equal parts: one gets bounced up and one passes straight through to the right. The two beams each hit one more mirror, setting them on a course to cross paths again. Each beam hits a photon detector, and that's it.

Now, consider the same experiment, but with a twist (Figure A.1b, on the right). At the point where the two beams cross in the upper right-hand corner, before they hit the detectors, place a second beam splitter. Each of our two beams splits again: now half of each beam goes off to the right, toward detector 2, and half of each beam goes up, to detector 1. But the beam splitter is constructed in a funny way, so the two combined half-beams don't behave the same way in both directions. The two half-beams going up are in sync: their peaks and valleys line up with each other, reinforcing their combined wave. This is constructive interference, like the bright bands in the double-slit experiment in Chapter 5. But the two half-beams going off to the right are perfectly out of sync:

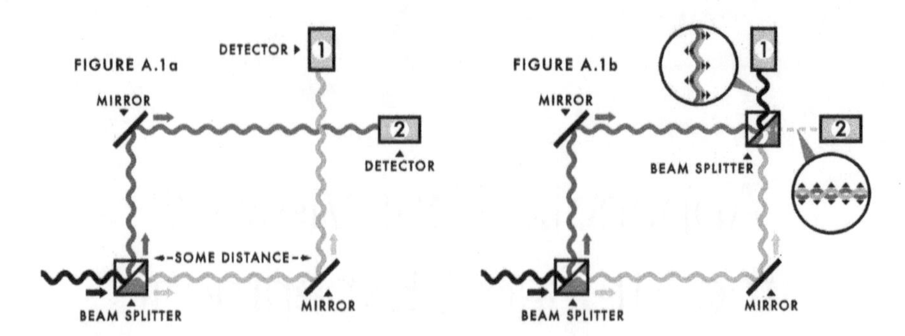

Figure A.1. Wheeler's delayed-choice experiment. (a) Without the second beam splitter, a single photon has a 50-50 chance of arriving at either detector. (b) With the second beam splitter, a single photon will interfere with itself, guaranteeing that it never arrives at detector 2.

the peaks of one line up with the valleys of the other, so they totally cancel out. This is destructive interference, like the dark bands in the double-slit experiment. The upshot is that no light reaches detector 2 at all, because the two beams heading toward it destructively interfere with each other—and the light reaching detector 1 is just as bright as the original laser beam that entered the first beam splitter in the lower left-hand corner of the experiment.

So far, so good. Aside from the existence of the laser in the first place, everything we've described so far is just classical physics. Now let's get quantum. Make the laser beam really dim—as dim as possible, in fact, so we're just sending one photon at a time through this experiment. If we don't have the second beam splitter in the upper right-hand corner, this is still pretty simple. The photon is detected at either detector 1 or detector 2, and we can tell which path the photon took through the setup by seeing which detector it arrived at. And if we send lots of photons through, one at a time, about half will show up at each detector.

But, Wheeler says, things become much more complicated when we put the second beam splitter back into the experiment. If we do that, a photon will never arrive at detector 2 because it will interfere with itself, just as in the double-slit experiment. Send as many photons as you like through the experiment, one at a time, and they'll all show up at detector 1. This, Wheeler says, is because each photon took both paths and interfered with itself, preventing it from arriving at detector 2. By inserting the second beam splitter, says Wheeler, "we make the whole idea of following a single path meaningless."

This is not terribly different from the double-slit experiment—indeed, this is the double-slit experiment, just with a slightly different geometry. And as with the double-slit experiment, it's tempting to say that the photon can tell whether the second beam splitter is there before it starts on its journey through the experiment. If there's only one beam splitter, the photon travels down only one path. But if the second beam splitter is in place, the photon travels down both paths, so it can interfere with itself.

But Wheeler introduced one more twist into this experiment: the delayed choice. There's some distance between the beam splitter and the mirror on the lower right-hand side (Figure A.1a). Let's make that distance big—several kilometers, say. In that case, it'll take the photon, traveling at the speed of light, about a dozen microseconds to get from the beam splitter to the detectors. That gives us plenty of time to direct a computer to insert (or remove) the second beam splitter *after* the photon leaves the first beam splitter. In other words, we can delay the choice of which experiment we want to do—Figure A.1a or Figure A.1b—until after the photon is already on its way through the experiment. Yet, if we do this, our results are no different. When the second beam splitter is in place, the photon will never arrive at detector 2. And when the second beam splitter is removed, the photon will arrive at each detector about half of the time.

These results are quite strange—yet they have been confirmed by actual experiments. Clearly, this happens. But how can the photon "decide" whether to travel down just one path after it's already passed through the first beam splitter? We can make the apparent paradox even worse by increasing the distance the photon travels. In principle, there's no reason this experiment can't be done using a setup a light-year long, or even billions of light-years. It's as if a photon can edit its own past in addition to sometimes being in two places at once—or as if our own choices about our experimental setup can change the distant past. And indeed, Wheeler endorsed this view, saying that "we must conclude that our very act of measurement not only revealed the nature of the photon's history on its way to us, but in some sense *determined* that history. The past history of the universe has no more validity than is assigned by the measurements we make—now!"

But this is just one view of this experiment, using Wheeler's version of the Copenhagen interpretation. What, after all, is a measurement? And how does it work? Wheeler never explains this, other than to insist that it has nothing to do with consciousness or life. Beyond that, he merely states that a measurement "is an irreversible act in which uncertainty collapses to certainty." Measurement, collapse—we're in familiar territory here, and Wheeler is in the usual trouble of needing to define what measurement is and how it happens, yet refusing to do exactly that. (Wheeler also states that the "essence" of quantum physics, "as the delayed-choice experiment shows, is *measurement*." But this is not particularly helpful for determining what actually constitutes a measurement.) There are, of course, other ways to look at this experiment—ways that are significantly different from Wheeler's ill-defined and questionably coherent ideas. Here are three of them.

Pilot-wave interpretation: A photon enters the beam splitter. Its pilot wave splits and travels down both paths, while the photon takes only one path (though we don't know which). If the second beam splitter isn't in place, the pilot wave reaches both detectors, carrying the particle along with it to one or the other.

If the second beam splitter is in place, the pilot wave interferes with itself when it arrives there, and never reaches detector 2. This prevents the photon from reaching detector 2 as well, no matter which path it traveled down.

It doesn't matter whether the second beam splitter is put in place before or after the photon goes through the first beam splitter—all that matters is whether the second beam splitter is there when the pilot wave arrives.

Many-worlds interpretation: A photon wave function enters the first beam splitter, splits, and travels down both paths. If the second beam splitter isn't in place, the photon wave function hits both of the detectors and entangles with the detectors' wave functions. Due to the enormous number of particles involved in this giant entangled wave function, decoherence occurs rapidly and the wave function branches. In one branch, the photon arrived at detector 1; in the other, it arrived at detector 2.

If the second beam splitter is in place, the photon wave function interferes with itself destructively when it gets there, ensuring that it never reaches detector 2. Thus, the photon hits only detector 1, and the world does not branch.

It doesn't matter whether the second beam splitter is put in place before or after the photon goes through the first beam splitter—all that matters is whether the second beam splitter is there when the wave function arrives.

Spontaneous-collapse theory: A photon wave function enters the first beam splitter, splits, and travels down both paths. If the second beam splitter isn't in place, the photon wave function hits both of the detectors and entangles with the detectors' wave functions. The enormous number of particles involved in this giant entangled wave function all but ensures that one of them will hit the collapse jackpot almost instantly, forcing the photon into one detector or the other entirely at random.

If the second beam splitter is in place, the photon wave function interferes with itself destructively when it gets there, ensuring that it never reaches the right-hand detector.

It doesn't matter whether the second beam splitter is put in place before or after the photon goes through the first beam splitter—all that matters is whether the second beam splitter is there when the wave function arrives.

In short, Wheeler's conclusions are, at best, not forced upon us. (At worst, they are logically incoherent.) Nor is this experiment particularly odd, seen from these other viewpoints—certainly not as strange as the Bell experiments. There are versions of this experiment that combine it with aspects of the Bell experiments, but those are similarly possible to explain in all of these interpretations (though the explanations do get quite a bit more complex).

One final note: although pilot waves are in general nonlocal, in this case, everything is entirely local using the pilot-wave interpretation. So there's a sense in which Wheeler was right—this is exactly the kind of thing that was at the heart of the Einstein-Bohr debates, because it can in theory be explained locally, yet adherents of Copenhagen insist on giving a nonlocal explanation for it!

Acknowledgments

Over forty physicists, philosophers, and historians were kind enough to take the time to speak with me on the record for this book. I will not list them all here (there's a list in the references) but I will single out a few of them. David Albert, Shelly Goldstein, Tim Maudlin, Roderich Tumulka, and Nino Zanghì took the time to speak with me on the record for my book proposal, before any of us knew if this would ever see the light of day. Dieter Zeh invited me into his home for lunch and lovely conversation. Mary Bell not only spoke with me at length, but also let me come back the next day. And Sam Schweber was very generous with his time, but sadly did not live to see this book come to print.

Conversations at Cornell with David Mermin and Dick Boyd nearly fifteen years ago set me down the path that led to this book. They are not responsible for its content—indeed, I know they would both disagree with some of what I've written here—but they surely deserve part of the credit for anything worthwhile herein. And though I suspect he also disagrees with much of what I've written here, Dragan Huterer's unflagging support during my time at Michigan led me here too.

If Peter Aldhous hadn't taken a chance on an entirely green writer with no clips whatsoever, this book wouldn't exist at all. Peter also introduced me to Anil Ananthaswamy, who walked me through the process of finding an agent, encouraging me every step of the way and beyond.

Peter Tallack, my agent (and Anil's), made sure this book landed in the right hands: those of my editor, T.J. Kelleher at Basic Books. T.J.'s editing, enthusiasm, and patience made this book far better than I had any right to expect. Hélène Barthelemy, formerly at Basic, gave helpful comments and suggestions on the structure of the first part of this book, solving many difficult problems in one stroke. And Melissa Veronesi and Carrie Watterson patiently shepherded this book through the copyediting process, despite my efforts to make needless changes at the last minute.

David Baker, Peter Byrne, Olival Freire, Benj Hellie, Nikki Huhn, David Kaiser, Colin Nichols, and Elizabeth Seiver all read significant chunks of this manuscript and gave me valuable feedback on it. All infelicities and errors that remain in this book are there despite their best efforts. And Andrew McNair took on the task of poring over the entire manuscript shortly before it went into production; he never let me down.

Doron Weber, Eliza French, and Josh Greenberg at the Alfred P. Sloan Foundation helped me make this book my full-time work for the time it took to get it right. Chip Sebens let me use his Mathematica code, pointed me toward the UCSC summer school, and was generous with his time on top of that. And Olival Freire's book *The Quantum Dissidents* easily halved the time I spent on research for this book; Olival himself provided helpful information and feedback throughout my research.

John Clauser provided access to his own private letters, and Robert Crease gave me the original audio recordings of his interviews with John Bell. David Wick and Andrew Whitaker also graciously gave me access to private archival materials. Jeremy Bernstein, Troels Petersen, Gerald Holton, and David Cassidy provided prompt and helpful replies to my questions. And Chris Fuchs gave me a second chance over email.

Nick James turned my scribbles and ramblings into beautiful illustrations. Adrienne Grant transcribed the lion's share of the interviews I conducted for this book and let me steal her friends. Lipa Long picked up where Adrienne left off (with the interviews, not the friends). Andy Schwarzkopf has been fielding last-minute optics questions from me for fifteen years and has generally been willing to indulge my crazy ideas, including this one. Daniel Jordan understands that Copenhagen delenda est. And Lisa Grossman was always there, even though Boston is stupidly far away.

Lunches at the 2013 UC Santa Cruz Summer School on Philosophy of Cosmology led me back to this project after setting it aside. Hurried conversations at the 2015 Emergent Quantum Mechanics conference in Vienna and more leisurely ones at the 2016 International Summer School in Philosophy of Physics in Saig were all invaluable during the writing. The UC Berkeley Office for History of Science and Technology gave me an academic home during the writing and editing. And the AIP archives, Niels Bohr archives, and CERN archives all graciously granted me access for my research.

There are many, many more people who deserve thanks for this book, including (but not limited to) Gordon Belot, Celeste Biever, Anne Brown, Glen Chiaccheri, Sarah Cobey, Peter Coles, Alex Demasi, Jonathan Dugan, Lucas Dunlap, Jared Emerson-Johnson, Nina Emery, Amanda Gefter, Louisa Gilder, Kate Hanley, Melissa Hogenboom, Parker Imrie, Rob Irion, Victoria Jaggard, Cagliyan Kurdak, Tom Levenson, Chris Lintott, Mike Marshall, Katy Meadows, Robin Meadows, Alyssa Ney, Emily Nichols, Robert Ochshorn, Pierangelo Pirak, Michael Polashenski, Ari Rabkin, Ryan Reece, Stefan Richter, Laura Ruetsche, Jim Sethna, Larry Sklar, Arfon Smith, Kimberly Smith, Jonah Waissman, Brian Wecht, Alex Zani, and the Weirs.

Thanks also to my parents, and the rest of my family, for their continual encouragement, and for putting up with decades of being asked more questions than they could reasonably have been expected to answer.

Thanks to Copernicus, for being the quietest and furriest coworker.

And finally, thanks to Elisabeth for her patience, and for everything.

Permissions

Figure 2.1: Courtesy of the Niels Bohr Archive, Copenhagen.
Figure 3.3: Photograph by Paul Ehrenfest, courtesy and © AIP Emilio Segrè Visual Archives.
Figure 4.1: Based on Figure 2 of Kaiser 2012.
Figure 5.1: Library of Congress, New York World-Telegram and Sun Collection, courtesy AIP Emilio Segrè Visual Archives. (NB: The Library of Congress lists the date of this photograph as 1940; however, the headline on the paper reveals that it must have been taken in May 1949, very shortly after Bohm testified to HUAC, possibly later that same day.)
Figure 6.1: Photograph by Alan Richards, courtesy AIP Emilio Segrè Visual Archives and © Princeton University Library.
Figure 6.3: Courtesy Institute for Advanced Study and Princeton University Library.
Figure 7.1: Courtesy Ruby McConkey and Dorothy Whiteside.
Figure 8.1: Courtesy of University of Chicago Library Special Collections Research Center and Harvard University Library.
Figure 9.2: Photo courtesy Lawrence Berkeley National Laboratory. © 2010 The Regents of the University of California, through the Lawrence Berkeley National Laboratory.
Figure 10.1: © CERN. Licensed under a Creative Commons Attribution 4.0 international license: https://creativecommons.org/licenses/by/4.0/.
Figure 10.2: © CERN.

Excerpts from "A Farewell to String and Sealing-Wax," in *From Hiroshima to the Moon* by Daniel Lang, © 1959 by Daniel Lang. Reprinted with the kind permission of Frances Lang, Helen Lang, and Cecily Lang.
Excerpt from David Bohm's undated letter to Arthur Wightman c. 1952, © Bohm Estate. Copy letter held at Niels Bohr Archive, Copenhagen. Reprinted with the kind permission of Basil Hiley.

Excerpts from the letters of John Wheeler, © Wheeler Estate. Reprinted with the kind permission of James Wheeler.

Excerpts from the letters of Hugh Everett III, © Everett Estate. Reprinted with the kind permission of Peter Byrne, Jeffrey Barrett, and Mark Oliver Everett.

Notes

Introduction

5 **as the objects in our everyday lives do:** Werner Heisenberg 1958, *Physics and Philosophy*, Harper Torchbooks ed. (Harper and Row), p. 129.

6 **"shut up and calculate":** N. David Mermin 1990, *Boojums All the Way Through: Communicating Science in a Prosaic Age* (Cambridge), p. 199.

7 **"The theory decides what we can observe":** Werner Heisenberg 1971, *Physics and Beyond* (HarperCollins), p. 63.

7 **three centuries after Galileo's discovery:** See Stanley L. Jaki 1978, "Johann Georg von Soldner and the Gravitational Bending of Light, with an English Translation of His Essay on It Published in 1801," *Foundations of Physics* 8 (11/12): 927–950. The experiment would have been possible decades before Einstein—and was in fact proposed a century before Einstein, by Johann Soldner, as a test of Newtonian physics. But nobody cared until Einstein proposed a rival theory to Newtonian gravity that could be tested in this manner.

Prologue

9 **"but I *knew* it was *rotten*":** Jeremy Bernstein 1991, *Quantum Profiles* (Princeton University Press), p. 20. Emphasis on "knew" is Bell's, according to Bernstein; emphasis on "rotten" is inferred from the context in Bernstein: "Bell pronounced the word 'rotten' with a good deal of relish."

10 **"got on with more practical things":** John S. Bell 2004, *Speakable and Unspeakable in Quantum Mechanics*, 2nd ed. (Cambridge University Press), p. 160.

10 **"the impossible done":** Bell 2004, p. 160.

10 **"waiting for me":** Charles Mann and Robert Crease 1988, "Interview: John Bell." *OMNI*, May, 90.

Chapter 1

14 **"only an abstract quantum physical description":** Max Jammer 1974, *The Philosophy of Quantum Mechanics* (John Wiley & Sons), p. 204. But see also N. David Mermin 2004a, "What's Wrong with This Quantum World?," *Physics Today*, February, pp. 10–11.

14 **"is impossible":** Heisenberg 1958, p. 129.

14 **"produce the results of measurement":** Jammer 1974, p. 164. Also note that Jordan's position contradicts Bohr's, and Heisenberg's may not be compatible with either. There are in fact many contradictory schools of thought that all go under the name "Copenhagen interpretation," though they claim to be the same. For more on this, see Chapter 3.

14 **"exceedingly intelligent paranoiac":** Letter from Einstein to D. Lipkin, July 5, 1952, as quoted in Arthur Fine 1996, *The Shaky Game*, 2nd ed. (University of Chicago Press), p. 1.

14 **"damned little effect on me":** Kaiser 2011, *How the Hippies Saved Physics: Science, Counterculture, and the Quantum Revival* (W. W. Norton), p. 8.

14 **"epistemology-soaked orgy":** Fine 1996, p. 94.

14 **greatest mathematical genius alive:** Max Born 2005, *The Born-Einstein Letters: Friendship, Politics and Physics in Uncertain Times* (Macmillan), p. 140.

15 **anything he proved was correct:** Richard Rhodes 1986, *The Making of the Atomic Bomb* (Simon and Schuster), pp. 108–109.

15 **even aware of this proof:** See Fine 1996, p. 42n3, where there's a long discussion of this.

15 **"silenced the objectors":** As quoted in Mara Beller 1999b, *Quantum Dialogue: The Making of a Revolution* (University of Chicago Press), pp. 213–214.

15 **whole proof was flawed:** Jammer 1974, pp. 273–274; also see an English translation of the relevant part of Hermann's paper here: http://mpseevinck.ruhosting.nl/seevinck/trans.pdf, accessed September 20, 2017.

15 **because she was a woman:** See N. David Mermin 1993, "Hidden Variables and the Two Theorems of John Bell," *Reviews of Modern Physics* 65 (3): 805. "Grete Hermann pointed out a glaring deficiency in the argument, but she seems to have been entirely ignored. Everybody continued to cite the von Neumann proof." For more on Hermann, see the presentation slides by M. P. Seevinck on Grete Hermann (2012). Also see the resources at http://web.mit.edu/redingtn/www/netadv/PHghermann.html, accessed September 20, 2017.

15 **hole at its heart:** See Jammer 1974, p. 247: "In spite of the opposition to Bohr's views by some leading physicists like Einstein and Schrödinger the vast majority of physicists accepted the complementarity [i.e. Copenhagen] interpretation in general without reservations, at least during the first two decades after its inception."

16 **a number for every point in space:** For specialists: I'm just using position-space wave functions for single-particle stationary states as an example. I'll get to more complex stuff later.

16 **it'd display 0.02:** Sometimes the Wave-Function-O-Meter™ might display an imaginary number, like the square root of negative one. But put that complication aside for now.

17 **the *probability* that the electron is in that place:** Technically, it's the square of the wave function that gives you the probability, but the idea is the same.

19 **"does that change the [quantum] state of the universe?":** Walter Isaacson 2007, *Einstein: His Life and Universe* (Simon and Schuster), p. 515.

19 **"highly qualified measurer—with a Ph.D.?":** Bell 2004, p. 117.

Chapter 2

21 **"at greater length":** Heisenberg 1971, p. 62.

22 **came to be known as "quantum physics":** Bohr's model of the atom was not the origin of the term "quantum physics." The term slowly came into use over the first decade of the twentieth century, as various phenomena were discovered that involved the absorption or emission of discrete packets of electromagnetic radiation, starting with Planck's law of black-body radiation. This period in the history of physics that I'm quickly breezing through—from Planck's discovery in 1900 to the theories that Heisenberg and Schrödinger developed in 1925, which are described in this chapter—is worthy of a book in its own right. Many have been written; see Manjit Kumar 2008, *Quantum: Einstein, Bohr, and the Great Debate About the Nature of Reality* (Icon Books/Norton); or David Lindley 2007, *Uncertainty: Einstein, Heisenberg, Bohr, and the Struggle for the Soul of Science* (Anchor) for good examples.

23 **"countless arithmetical errors":** Heisenberg 1971, p. 61.

23 **"so generously spread out before me":** Ibid.

23 **"generally my severest critic":** Ibid., p. 64.

23 **"once again possible to move forward":** Kumar 2008, p. 193.

25 **"purely a thought-thing":** Isaacson 2007, p. 84.

26 **"profound influence on me":** Albert Einstein 1949a, "Autobiographical Notes," in *Albert Einstein: Philosopher-Scientist*, edited by Paul Arthur Schilpp (MJF Books, 1949), p. 21.

26 **he was not a dogmatic Machian:** See Don Howard 2015, "Einstein's Philosophy of Science," in *The Stanford Encyclopedia of Philosophy*, Winter ed., edited by Edward N. Zalta, http://plato.stanford.edu/archives/win2015/entries/einstein-philscience/. Also see Chapter 8 for more on the influence Einstein had among Mach's followers and their reactions upon learning his true philosophical views.

26 **introduced by the physicist Max Planck:** Gerald Holton 1998, *Thematic Origins of Scientific Thought*, rev. ed. (Harvard University Press), p. 70.

26 **"invariant theory":** Ibid., p. 130.
26 **"essentially untenable":** Einstein 1949a, p. 21.
26 **"It can only exterminate harmful vermin":** Isaacson 2007, p. 334.
27 **"determining what *is*":** Kumar 2008, p. 262. Emphasis in original.
28 **a few holdouts:** The last holdouts were finally forced to accept the reality of photons as a side effect of the Bell experiments conducted by John Clauser in the 1970s. See Chapter 9.
28 **"one-horse cart":** Kumar 2008, p. 35.
28 **prejudices accumulated by the age of eighteen:** Lincoln Barnett 1949, *The Universe and Dr. Einstein* (Victor Gollancz), p. 49.
28 **"In Göttingen they believe in it. I don't":** Isaacson 2007, p. 331.
28 **"such strange assumptions":** Heisenberg 1971, p. 62.
29 **"restrict myself to these":** This was likely a post hoc justification Heisenberg gave for his own work. His true motivation in ignoring orbits was probably that they had proven largely useless over the previous decade in explaining new experimental results. See Beller 1999b, Chapters 2 and 3, and especially pp. 52–58.
29 **"It is the theory which decides what we can observe":** Ibid., p. 63.
29 **"what do you yourself think about it?":** Ibid., p. 64.
30 **"get you into hot water":** Ibid., pp. 65–66.
30 **go work with Bohr:** Kumar 2008, p. 227.
30 **"such pleasure by his mere presence":** Ibid., p. 131.
30 **"in a kind of hypnosis":** Ibid., p. 132.
31 **"wisest of living men":** Mara Beller 1999a, "Jocular Commemorations: The Copenhagen Spirit." *Osiris* 14, p. 266.
31 **"Erasmus and Lincoln":** Ibid., p. 257.
31 **"the birds whisper their secrets to you":** John L. Heilbron 1985, "The Earliest Missionaries of the Copenhagen Spirit," *Revue d'histoire des sciences* 38, nos. 3–4, pp. 195–230. doi:10.3406/rhs.1985.4005, p. 223.
31 **"whose authority not many dared to challenge":** Beller 1999a, p. 258.
31 **"married after no more than two years":** Ibid., p. 271n54.
32 **"intoxicated with the heady spirit of Platonic dialogue":** Ibid., pp. 258–259.
32 **"while the visitor's interpretation was wrong":** George Gamow 1988, *The Great Physicists from Galileo to Einstein* (Dover), p. 237.
33 **Bohr did not publish a single paper alone:** Beller 1999a, p. 261.
33 **"symbolized by Planck's quantum of action":** Niels Bohr 1934, *Atomic Theory and the Description of Nature* (Cambridge University Press), p. 53, in the published version of his Como speech (which originally appeared in English, in *Nature*, as Bohr explains in the introduction to that volume).
33 **"stumbled over incomplete sentences":** Beller 1999a, p. 256.
33 **"less intelligible the more important the subject became":** Ibid., p. 257.
33 **for hours or days on end:** Beller 1999a, p. 257.

34 **"loved him without limits":** Ibid., p. 252.

34 **"transition probabilities, energy levels, and the like":** David Cassidy 1991, *Uncertainty: The Life and Science of Werner Heisenberg* (W. H. Freeman), p. 214.

34 **"Schrödinger has now come to our rescue":** Ibid., p. 213.

34 **"deepest form of the quantum laws":** Ibid.

34 **a problem that had been outstanding for more than seventy years:** Beller 1999b, p. 29.

36 **"in other words it's crap":** Kumar 2008, p. 212.

36 **"open questions of atomic theory":** Ibid., p. 222.

36 **"a fact that marked their every utterance":** Heisenberg 1971, p. 73.

36 **"sorry I ever got involved with quantum theory":** Ibid., p. 75.

37 **"But you must surely admit that . . . ":** Ibid., p. 76.

37 **"we were on the right track":** Ibid.

37 **"probability of measuring the particle in that location":** See Chapter 1.

39 **"observed without disturbing them appreciably":** Bohr 1934, p. 53.

39 **"agency of observation not to be neglected":** Ibid., p. 54.

39 **"indispensable for a description of experience"**: Ibid., pp. 56–57.

40 **Bohr then echoed Mach:** He may have been echoing Kant instead. Or he may have been doing something else entirely; Bohr's difficult writing has led to differing opinions on the subject.

40 **other interpretations of quantum physics are possible:** Several other interpretations are described in detail from Chapter 5 onward. Also note that it doesn't particularly matter whether any of these other interpretations are correct—because Bohr is claiming that it's impossible to describe a quantum world without complementarity, the mere logical *possibility* of other interpretations of quantum mechanics blows Bohr right out of the water.

40 **"[Complementarity] doesn't provide you any with any equations":** Paul Dirac, interview by Thomas S. Kuhn, May 14, 1963, Cambridge, England, courtesy of the Niels Bohr Library & Archives, American Institute of Physics, College Park, MD, USA, https://www.aip.org/history-programs/niels-bohr-library/oral-histories/4575-5, Part 5.

41 **"Bohr's principle will not change the way we do physics.":** *Discussion Sections at Symposium on the Foundations of Modern Physics: The Copenhagen Interpretation 60 Years after the Como Lecture*, 1987, p. 7.

Chapter 3

43 **origin myth for quantum physics:** Pieces of this fable have been written down. Popular physics books by physicists often contain a version of it; for example, it appears in Stephen Hawking 1988, *A Brief History of Time* (Bantam Dell), p. 56, as well as Stephen Hawking 1999, "Does God Play Dice?," http://www.hawking.org.uk/does-god-play-dice.html, accessed

March 18, 2016. This narrative stems primarily from a handful of histories of quantum physics, especially Jammer 1974; and Max Jammer 1989, *The Conceptual Development of Quantum Mechanics*, 2nd ed. (Tomash) (see, e.g., p. 374, in Jammer 1989). It also appears in the recollections of this period written by Bohr and Heisenberg decades later (Niels Bohr 1949, "Discussion with Einstein on Epistemological Problems in Atomic Physics," in Schilpp 1949; Heisenberg 1971). However, this narrative is contradicted by the materials available from the actual period during which quantum physics was developed (e.g., the proceedings of the Fifth Solvay conference, contained in Guido Bacciagaluppi and Antony Valentini 2009, *Quantum Theory at the Crossroads: Reconsidering the 1927 Solvay Conference*, arXiv:quant-ph/0609184v2, as well as the contemporary letters of Einstein, Schrödinger, Bohr, and others) and must therefore be considered unreliable. For more on this (aside from the present work), see Don Howard 2004, "Who Invented the 'Copenhagen Interpretation'? A Study in Mythology," *Philosophy of Science* 71 (5): 669–682; Don Howard 2007, "Revisiting the Einstein-Bohr Dialogue," *Iyyun: The Jerusalem Philosophical Quarterly* 56:57–90; Fine 1996; Beller 1999b; James Cushing 1994, *Quantum Mechanics: Historical Contingency and the Copenhagen Hegemony* (University of Chicago Press); Olival Freire Jr. 2015, *The Quantum Dissidents: Rebuilding the Foundations of Quantum Mechanics* (Springer-Verlag); and Jean Bricmont 2016, *Making Sense of Quantum Mechanics* (Springer International).

44 **"God does not play dice":** Letter from Albert Einstein to Max Born, December 4, 1926, reprinted in Born 2005.

45 **"lifted a corner of the great veil":** Kumar 2008, p. 150.

46 **Pauli remained unsatisfied:** Bacciagaluppi and Valentini 2009, pp. 242–244.

46 **for reasons related to Kramers's objection:** Ibid., pp. 254–255.

46 **"no longer susceptible of any modification":** Ibid., p. 435.

46 **the same object at the same time:** See the end of Chapter 2. We don't actually know what Bohr said—rather than submitting a copy of his comments to the conference proceedings, he asked that a copy of his Como lecture be substituted. But notes taken during the conference suggest that the content was largely similar. See Bacciagaluppi and Valentini 2009 for more on this.

47 **"strove for precision":** Beller 1999a, p. 268.

48 **antimaterialist culture of interwar Weimar Germany:** Paul Forman 1971, "Weimar Culture, Causality, and Quantum Theory: Adaptation by German Physicists and Mathematicians to a Hostile Environment," *Historical Studies in the Physical Sciences* 3:1–115.

48 **nonsensical:** We'll see more of the logical positivists in Chapter 8.

48 **"the Wrath of God":** Kumar 2008, p. 157.

48 **"publish more quickly than you can think":** Ibid., p. 160.

48 **"it is not even wrong":** Ibid.

48 **"what Mr. Einstein said is not so stupid":** Ibid.

48 **"how many angels are able to sit on the point of a needle":** Born 2005, p. 218.

48 **"their interaction with other systems":** First half, Jammer 1974, p. 204; second half, Bohr 1934, pp. 56–57.

49 **"world of potentialities or possibilities":** Heisenberg 1958, p. 186.

49 **observed in uncontrollable ways:** Wolfgang Pauli 1994, *Writings on Physics and Philosophy,* edited by Charles P. Enz and Karl von Meyenn, translated by Robert Schlapp (Springer-Verlag), p. 33.

49 **a myth:** To their credit, Heisenberg, Jordan, and others didn't say that there was a unified interpretation—at least, not at the time. Jordan spoke of a "Göttingen-Copenhagen spirit" in 1927, and Heisenberg mentioned a "Copenhagen spirit of quantum theory" three years later in a similar context, but the phrase "Copenhagen interpretation" was first used in 1955, by Heisenberg. See Chapter 4 for more on this, as well as Howard 2004.

49 **"what we can say about nature":** Jammer 1974, p. 204; but see also N. David Mermin 1985, "Is the Moon There When Nobody Looks? Reality and the Quantum Theory," *Physics Today* 38 (4): 38–47.

50 **"a pitying smile":** Albert Einstein 1949b, "Reply to Criticisms," in Schilpp 1949, p. 667.

50 **"[to be is to be perceived]":** Ibid., p. 669.

51 **the problem is one of *locality*:** Einstein had been concerned about locality in quantum physics for several years at this point; even before Heisenberg's matrix mechanics, Einstein had recognized that the statistics of photons implied some kind of nonlocality. See Howard 2007. Einstein also knew that the idea of photons, taken together with locality, implied a serious revision of Maxwell's laws of electromagnetism, as early as 1909. See Bacciagaluppi and Valentini 2009.

52 **"contradiction with the principle [i.e., the special theory] of relativity":** Bacciagaluppi and Valentini 2009, p. 487.

52 **"Mr. de Broglie is right":** Ibid., p. 487.

52 **"No doubt it is my fault":** Ibid., pp. 487–488. Bacciagaluppi and Valentini make this point themselves: "Einstein's argument is so concise that its point is easily missed, and one might well dismiss it as arising from an elementary confusion about the nature of probability" (p. 195).

52 **confused about the nature of probability:** After all, if the wave function is simply a statement about the probability that a single electron will be registered at a location on the film, then it's logically impossible for one electron's wave function to lead the film to register two electrons in two different locations. But that argument is circular reasoning—literal

question begging—because it already assumes that the wave function is merely a probability distribution. In other words, that argument already assumes the conclusion that Bohr and company wish to reach. See ibid., p. 195, for more on this.

53 **Einstein hoisted by his own petard:** See, for example, Kumar 2008 for a "traditional" account of this encounter, in which Einstein is laid low by his own theory, wielded by a victorious Bohr.

54 **"for a totally different purpose":** Don Howard 1990, "'Nicht sein kann was nicht sein darf,' or the Prehistory of EPR, 1909–1935: Einstein's Early Worries About the Quantum Mechanics of Composite Systems," in *Sixty-Two Years of Uncertainty: Historical, Philosophical, and Physical Inquiries into the Foundations of Quantum Mechanics*, edited by Arthur I. Miller, 61–111 (Plenum Press). Quote appears on p. 98.

54 **missed the point:** Even if Einstein's concern really had been the uncertainty principle, Bohr's invocation of general relativity is alarming rather than ironic. The logical consistency of quantum physics shouldn't rely on the existence of general relativity, as the two theories are not only independent but famously incompatible. There is a resolution of the paradox that Bohr attributed to Einstein that does not involve anything other than quantum physics, but that resolution was not provided by Bohr, and indeed was not provided by anyone at all for decades. See Howard 1990; Howard 2007; and Bricmont 2016, pp. 238–241, for more on all of this.

54 **"Can Quantum Mechanical Description of Physical Reality Be Considered Complete?":** Reprinted in Wheeler and Zurek 1983, p. 138.

55 **"Not 'Complete' Even Though 'Correct'":** *New York Times* 1935, "Einstein Attacks Quantum Theory," Science Service, May 4, 1935.

55 **"EINSTEIN ATTACKS QUANTUM THEORY":** *New York Times* 1935, "Statement by Einstein," May 7, 1935.

55 **"the secular press":** Fine 1996, p. 35.

55 **"couldn't care less":** Ibid., p. 38.

56 **"smothered" in the EPR paper:** Writing many years later, Einstein said this explicitly: "The [EPR] paradox forces us to relinquish one of the following assertions:

(1) the description by means of the [wave]function is complete

(2) the real states of spatially separated objects are independent of each other [locality]" (Einstein 1949b, p. 682).

56 **"spooky action at a distance":** Born 2005, p. 155.

56 **"by a more complete and direct one":** Ibid., pp. 169–170.

56 **"Einstein proved that it does not work":** Kumar 2008, p. 313.

57 **reply for publication:** Ibid., p. 307.

57 **"clear up such a misunderstanding at once":** Wheeler and Zurek 1983, p. 142.

57 **"an astonishing speed" for Bohr:** Ibid., p. 143.

57 ***"future behavior of the system":*** Ibid., p. 148. Emphasis in original.
58 **whether he thought quantum physics was nonlocal:** Jammer (1974) thinks he did; Bell isn't sure (John Bell 1981, "Bertlmann's Socks and the Nature of Reality," *Journal de Physique*, Seminar C2, suppl., 42 (3): C2 41–61, reprinted in Bell 2004). Bell 2004, pp. 155–156.
58 **"inefficiency of expression":** Bohr 1949, p. 234.
58 **crucial part:** I.e., the part quoted in the previous paragraph, which Bohr himself identified as crucial in ibid.
58 **"frequently nebulous and obscure":** Born 2005, p. 207.
58 **Few actually read what Bohr had written:** Kumar 2008, p. 313.
58 **most other physicists didn't:** See Jammer 1974 for a sampling of contemporary and later reactions.
58 **"dogmatic quantum mechanics":** Fine 1996, p. 66.
58 **dubbed this connection "entanglement":** One of these papers, in an attempt to explain the weirdness of the Copenhagen approach to the measurement problem, laid out the famous "Schrödinger's cat" thought experiment, explained in the Introduction.
59 **"'it is very hot in Florida'":** Fine 1996, p. 74.
59 **disagreed about where exactly that was!:** Jammer 1974, p. 187.
59 **clockwork deterministic universe:** Most famously, Max Born thought Einstein's problems with quantum physics had to do with determinism, until Pauli set him straight. See Born 2005, and also Mermin 1985 for more on this. Despite Pauli clearing up Born's confusion in 1954, the misunderstanding has persisted to the present; see Jammer 1974, p. 188, and Hawking 1988, p. 56, for two prominent examples.
59 **"avoids reality and reason":** Jammer 1974, p. 188.
59 **"hobgoblin of the naive":** Einstein to Schrödinger, June 19, 1935. Translation by Don Howard 1985, "Einstein on Locality and Separability," *Studies in History and Philosophy of Science* 16:178.
59 **"the details of a physicist's philosophy":** Beller 1999b, p. 4.
60 **"Bohr's Sunday word of worship":** As quoted in Beller 1999a, p. 257.

Chapter 4

61 **"exist objectively in the same sense as stones or trees":** Heisenberg 1958, p. 129.
61 **"the mind of the observer":** Ibid., pp. 54–55.
62 **"consistent interpretation of quantum theory":** Ibid., p. 43.
62 **"criticize the Copenhagen interpretation and to replace it":** Ibid., p. 128.
63 **meet with new heads of state:** Whether this was truly the reason for Planck's visit is a matter of some controversy. There are also varying accounts of how this meeting played out; see Ball 2013 for more on this.
63 **"we need their scientific work":** Ibid., p. 62.
63 **"do without science for a few years!":** Kumar 2008, p. 293.

63 **"no choice except to fall silent and leave":** Ball 2013, p. 62.
63 **fewer impediments to Jews advancing in the sciences:** Rhodes 1986, p. 188.
63 **what had been the unrivaled center of the physics world:** Ball 2013, p. 72; Rhodes 1986, p. 185.
64 **"You will never see it again":** Isaacson 2007, p. 401.
64 **"like the end of the world":** Max Born 1978, *My Life: Recollections of a Nobel Laureate* (Scribner's Sons), p. 251.
64 **"political unreliability":** J. J. O'Connor and E. F. Robertson 2003, "Erwin Rudolf Josef Alexander Schrödinger," http://www-groups.dcs.st-and.ac.uk/~history/Biographies/Schrodinger.html, accessed September 25, 2017.
64 **"great duplicity":** Ibid.
65 **"leave Italy as soon as possible":** Laura Fermi 1954, *Atoms in the Family: My Life with Enrico Fermi* (University of Chicago Press), p. 120.
65 **"My heart aches at the thought of the young ones":** Born 2005, p. 111.
65 **emigrated from the European continent to the United States and the UK:** Rhodes 1986, pp. 195–196.
66 **wearing a three-piece pinstriped suit:** Marina Whitman (von Neumann's daughter), interview by Gray Watson. January 30, 2011, https://web.archive.org/web/20110428125353/http://256.com/gray/docs/misc/conversation_with_marina_whitman.shtml.
66 **"We know it will":** Eugene Wigner, interview by Charles Weiner and Jagdish Mehra on November 30, 1966, Princeton, NJ, USA (courtesy of the Niels Bohr Library & Archives, American Institute of Physics, College Park, MD, USA), http://www.aip.org/history-programs/niels-bohr-library/oral-histories/4964, accessed April 6, 2016.
66 **"brilliant men all lived elsewhere":** Rhodes 1986, p. 106.
66 **"could imitate them perfectly":** Ibid., p. 109.
67 **"requirements of mathematical rigor":** John von Neumann 1955, *Mathematical Foundations of Quantum Mechanics*, translated by Robert T. Beyer (Princeton University Press), p. ix.
67 **"discontinuous, non-causal, and instantaneously acting":** Ibid., pp. 349–351.
68 **"it requires the [collapse of the wave function]":** Ibid., p. 420.
69 **"a physical quantity has a certain value":** Ibid.
69 **claimed that Bohr's work supported this "dual description":** Ibid.
71 **impossible to build a bomb out of U-238:** Hitting U-238 with a slow neutron sometimes creates plutonium-239, an entirely different element. P-239 can be split by slow neutrons in much the same way as U-235, but producing P-239 from U-238 requires a good source of slow neutrons in the first place—and the best source of slow neutrons is a controlled nuclear chain reaction. So getting P-239 from U-238 is a lot easier if you already have some U-235.

71 **"turn the United States into one huge factory":** Rhodes 1986, p. 294.
72 **"it would all disappear":** Ibid., p. 275.
72 **"jaundice doesn't hurt really":** Wigner 1966, interview.
72 **"the rest and the detachment were wonderful":** Ibid.
72 **"'there will be a chain reaction'":** Ibid.
72 **"Hitler's success could depend on [nuclear fission]":** Rhodes 1986, p. 281.
73 **Top Policy Group:** Ibid., pp. 378 and 387.
73 **"he knew them so well":** Ibid., p. 381.
73 **reason to think so:** Daniel Lang 1953, "A Farewell to String and Sealing Wax," reprinted in *From Hiroshima to the Moon: Chronicles of Life in the Atomic Age*, by Daniel Lang (Simon and Schuster, 1959), p. 58.
74 **"Germany needs me":** David Cassidy 2009, *Beyond Uncertainty: Heisenberg, Quantum Physics, and the Bomb* (Bellevue Literary Press), p. 295.
74 **"machine gun practice in the Bavarian Alps":** Wheeler and Ford 1998, p. 32.
74 **"very casual about numbers":** Jeremy Bernstein 2001, *Hitler's Uranium Club: The Secret Recordings at Farm Hall*, 2nd ed. (Copernicus), pp. 35–36. Heisenberg's problems with numbers were well-known among his colleagues. (And, to be clear, Peierls worked with Heisenberg in the 1920s, not on the German bomb program; during the war and afterward, Peierls was in the UK.)
75 **escaped the attention of Heisenberg and his colleagues:** Although some people within the German program did seem to understand that purified graphite was a viable option, it is unclear how widely that information was shared, and those who were aware dismissed graphite purification as too expensive to pursue. See ibid., pp. 25–26.
75 **new power source for the Reich's war engine:** Cassidy 2009, p. 322.
75 **never led an experimental team before in his life:** Bernstein 2001, p. 40. To be fair, Oppenheimer (the scientific leader of the Manhattan Project) had never been an experimental physicist either—but he had many working for him, and Oppenheimer did not have Heisenberg's careless approach to experimental work. Oppenheimer respected experimental physics and knew his own limitations. Heisenberg, it would seem, did not.
75 **"make use of warfare for physics":** Cassidy 2009, p. 305.
75 **"agreed to sup with the devil":** Rhodes 1986, p. 386.
75 **"it would have been so beautiful if we had won":** Bernstein 2001, p. 43.
76 **more comforts than the average English family:** Cassidy 2009, p. 372.
76 **"a bit old fashioned":** Bernstein 2001, p. 78.
76 **In the express hope of provoking discussion:** Ibid., p. 78n7.
77 **"I don't believe it has anything to do with uranium":** Ibid., pp. 116–117.
77 **"Poor old Heisenberg":** Ibid., p. 116.
78 **One of them was Niels Bohr:** Bohr's mother was Jewish, which was enough for the Nazis to mark Bohr for execution.

79 **"'You have done just that'":** Rhodes 1986, p. 500.

79 **cost the nation nearly $25 billion:** All figures in this section are adjusted for inflation to 2016 dollars, using the CPI inflation calculator. The original figure is $1.9 billion.

79 **locations across the United States and Canada:** David Kaiser 2014, "History: Shut Up and Calculate!," *Nature* 505 (January 9): 153–155, doi:10.1038/505153a.

79 **$17 million:** Paul Forman 1987, "Behind Quantum Electronics: National Security as Basis for Physical Research in the United States, 1940–1960," *Historical Studies in the Physical and Biological Sciences* 18 (1): 149–229. Original figure is $1 million.

79 **$400 million:** Ibid. Original figure is $44 million.

79 **Atomic Energy Commission:** Kaiser 2014.

80 **"the war introduced me to the scientific life":** David Kaiser 2002, "Cold War Requisitions, Scientific Manpower, and the Production of American Physicists After World War II," *Historical Studies in the Physical and Biological Sciences* 33 (1): 138–139.

80 **faster rate of growth than any other academic field:** This trend was also seen in other fields, but not as dramatically as in physics—from 1945 to 1951, US PhDs awarded in all fields grew, but the annual growth rate in physics was twice the average, greater than any other field. In contrast, in the half century before the war, PhDs awarded in physics grew annually at 87 percent of the average growth rate across all fields in the United States. See Kaiser 2002.

80 **under the age of thirty:** Lang 1953, p. 217; Kaiser, private communication.

80 **"tools of war needed for the preservation of our freedom":** Henry D. Smyth 1951, "The Stockpiling and Rationing of Scientific Manpower," *Physics Today* 4 (2): 18, doi:10.1063/1.3067145.

80 **"maladjusted veterans of the Second World War":** Lang 1953, p. 216.

80 **"string-and-sealing-wax days":** Ibid.

81 **"atomic energy is on the agenda":** Ibid., pp. 216–217.

81 **"possible on a university campus":** Ibid., p. 239.

82 **"strangers to each other":** Ibid., p. 221.

82 **"almost certain to horrify the instructor":** David Kaiser 2007, "Turning Physicists into Quantum Mechanics," *Physics World* (May): 28–33. Much of this paragraph is based on this remarkable article.

82 **"efficient, repeatable means of calculation":** David Kaiser 2004, "The Postwar Suburbanization of American Physics," *American Quarterly* 56 (4): 851–888.

82 **"philosophically tainted questions":** Kaiser 2007.

83 **"musty atavistic to-do about position and momentum":** Ibid.

83 **Heisenberg's account of a meeting they had in 1942:** This meeting was later the subject of Michael Frayn's excellent play *Copenhagen*.

83 **friends and family members murdered by the Nazis:** M. Norton Wise 1994, "Pascual Jordan: Quantum Mechanics, Psychology, National Socialism," in *Science, Technology, and National Socialism*, edited by Monika Renneberg and Mark Walker (Cambridge University Press), pp. 251–252. Heisenberg and Pauli complied with Jordan's request, enabling Jordan to start a second career after the war as a far-right politician in West Germany, where he advocated for nuclear weapons to be deployed along the border between the Germanys.

84 **"puts nature on the fence and leaves it there":** Henry Margenau 1950, *The Nature of Physical Reality: A Philosophy of Modern Physics* (McGraw-Hill), p. 422.

85 **"legislates a difficulty into a norm":** Henry Margenau 1954, "Advantages and Disadvantages of Various Interpretations of the Quantum Theory," *Physics Today* 7 (10): 9, doi:10.1063/1.3061432.

Chapter 5

90 **Bohm and the reception of his ideas:** The story was apparently told by Dresden in May 1989 at an American Physical Society (APS) meeting. However, there was no official record of Dresden's comments at that meeting. The story appears in F. David Peat 1997, *Infinite Potential: The Life and Times of David Bohm* (Addison Wesley Longman), p. 133; but Peat did not actually record Dresden's comments in real time at the APS meeting, and, though Peat claims that Dresden repeated his story in a letter to him later on, Peat was unable to produce the letter when asked. A somewhat different account of the same story appears in Cushing 1994, pp. 156–157; although Cushing doesn't name Dresden, it is clearly very similar to the story recounted by Peat, and Cushing was on the same panel as Dresden at the 1989 APS meeting. Even if we grant that Peat and Cushing are accurately representing Dresden's comments from 1989, we are ultimately relying on a single person's recollection of events nearly forty years after they occurred. This story must, at best, be taken with a sizable grain of salt.

91 **no clear evidence:** A sampling (not a comprehensive list) of Peat's errors (1997):

- He claims Bohm harbored doubts about the Copenhagen interpretation during his early days at Berkeley; Bohm explicitly denies this in his interview with Wilkins, stating that he had no such doubts until he arrived at Princeton.
- He claims Feynman was one of Oppenheimer's PhD students at Berkeley with Bohm. Feynman never attended Berkeley.
- He claims that Fritz Zwicky supposedly spoke every language with an accent, even his native Russian. Zwicky was Swiss. That claim was originally made about George Gamow.

- He claims Einstein called Bohm's theory a "nursery rhyme" in a letter to Max Born. Einstein said no such thing; he was clearly referring to his own paper as a "nursery rhyme."
- He repeatedly claims Bohm testified in front of HUAC in 1950. That happened in 1949.

Moreover, Peat did not record any of the interviews he conducted with Bohm's friends and colleagues for his book; he simply talked with the people he interviewed and then, at a later date, wrote down his recollections of what they said, and presented those as direct quotes (Peat, personal communication).

92 **"a talent for being unhappy":** Peat 1997, p. 81.

92 **"more interested in competition and getting ahead":** David Bohm, interview by Maurice Wilkins, July 7, 1986, courtesy of the Niels Bohr Library & Archives, American Institute of Physics, College Park, MD, USA, http://www.aip.org/history-programs/niels-bohr-library/oral-histories/32977-3, accessed August 28, 2016, Part 3.

92 **"probably a little low":** Ibid.

93 **met Bohr and came to know him quite well:** Kai Bird and Martin J. Sherwin 2005, *American Prometheus: The Triumph and Tragedy of J. Robert Oppenheimer* (Vintage), p. 273.

93 **"Bohr was God and Oppie was his prophet":** Ibid., p. 169.

93 **"gave it a lot of weight in my mind":** Bohm 1986, interview, Part 3.

94 **"listen to what they said more sympathetically":** Bohm interview with Sherwin, June 15, 1979, New York, NY, USA. Atomic Heritage Foundation, "Voices of the Manhattan Project," http://manhattanprojectvoices.org/oral-histories/david-bohms-interview, accessed August 28, 2016.

94 **"The meetings were interminable":** Ibid.

94 **his association with Weinberg:** Bird and Sherwin 2005, p. 193.

94 **Bohm deserved his PhD:** Ibid.

94 **hired Bohm as an assistant professor:** Wheeler and Ford 1998, p. 216.

94 **"the ablest young theoretical physicists that Oppenheimer has turned out":** Russell Olwell 1999, "Physical Isolation and Marginalization in Physics: David Bohm's Cold War Exile," *Isis* 90 (4): 738–756.

94 **"stockpiling scientific manpower.":** See Chapter 4.

94 **"very status conscious":** Bohm 1986, interview, Part 3.

95 **research collaborations with several promising graduate students:** Chris Talbot, ed., 2017, *David Bohm: Causality and Chance, Letters to Three Women* (Springer), p. 4.

95 **"guaranteed by the First Amendment":** *Hearings Before the Committee on Un-American Activities, House of Representatives* 1949, Eighty-First Congress, First Session (March 31 and April 1) (Statement of David Bohm), p. 321.

95 **"the whole issue was dying away":** Bohm 1986, interview.

95 **"wasn't quite satisfied that I really understood it":** Ibid., Part 4.

96 **"put out notes and then finally a book":** Ibid.

96 **"very enthusiastic" response to his book from the notoriously harsh Wolfgang Pauli:** Ibid., Part 3. Note that here, Pauli's name is transcribed as "Pavvy." In Part 4, Bohm repeats some of this and the transcription is correct; from that, and from the context in Part 3, it's clear that Bohm is actually saying "Pauli" here.

97 **"he felt that it was incomplete":** Ibid., Part 4.

97 **"complete description of reality?":** Ibid.

98 **"of no practical interest":** Ibid., p. 125. Also Talbot 2017, p. 224.

99 **condition of the cat:** Measurements in Bohm's theory do have an effect on the systems measured, but this effect is well defined and easy to characterize for any given system. For more on this, see Chapter 7.

99 **"it contains the *only* mystery":** Richard Feynman, Robert B. Leighton, and Matthew L. Sands 1963, *The Feynman Lectures on Physics*, vol. 1 (Basic Books), ch. 37, section 37-1.

101 **must strike in only one spot:** This is exactly the behavior that Einstein objected to in Solvay in 1927 (see Chapter 3): stating that the photon is a wave until it hits the screen leads unavoidably to nonlocality. And if the photon is not a physical wave before it hits the screen, then what is it? Bohr and the others claimed that the photon was not a physical wave before it hit the screen, but they were remarkably vague about what, exactly, the photon *was* doing before it hit the screen.

103 **"accounting for the functions of the measuring instruments in purely classical terms":** See Beller 1999b, p. 163, for the quote source. Also Niels Bohr 2013, *Collected Works*, vol. 7, *Foundations of Quantum Physics II (1933–1958)*, edited by J. Kalckar (Elsevier), p. 311.

103 **"irritated rather than amused":** Beller 1999a, p. 263.

103 **"ordering and surveying human experience":** Mermin 2004a, pp. 10–11.

104 **"the need for new kinds of observations":** Wheeler and Zurek 1983, p. 392.

104 **"permit them to be observed directly":** Wheeler and Zurek 1983, p. 391.

105 **"first evidence for existence of atoms":** See Chapter 2 for the story of atoms and Brownian motion.

105 **"if this should help explain something":** Letter from David Bohm to Arthur Wightman, undated, c. 1952, while Wightman was visiting the Niels Bohr Institute. Courtesy of the Niels Bohr Archive, Copenhagen. Emphasis in original.

105 **"by a large number of modern theoretical physicists":** Wheeler and Zurek 1983, p. 391.

105 **"an analysis in classical terms of typical quantum phenomena":** Quoted in Bricmont 2016, p. 274.

106 **"what those little farts think":** Talbot 2017, p. 439.

106 **stamped valid for return only to the United States:** Bohm 1986, interview, Part 5.

106 **"reopening this whole dirty business again":** Freire 2015, *The Quantum Dissidents: Rebuilding the Foundations of Quantum Mechanics* (Springer-Verlag), p. 33.

106 **"or else nobody will take the trouble to read [my paper]":** Talbot 2017, p. 224.

107 **"a check that cannot be cashed":** Letter from Wolfgang Pauli to David Bohm, c. 1951, Pauli Archives at CERN, https://cds.cern.ch/record/80946.

107 **"artificial metaphysics":** Cushing 1994, p. 149.

107 **Bohr thought that Bohm's theory was "very foolish":** Talbot 2017, p. 147. Unfortunately, Wightman's original letter to Bohm reporting Bohr's impressions has been lost. We only have accounts of what Wightman said from letters that Bohm sent to other friends around the same time. See Bohm's letter to Wightman from the Niels Bohr Archive in Copenhagen, which is a reply to the lost letter from Wightman to Bohm; Bohm thanks Wightman for Wightman's report of Niels Bohr's impressions of Bohm's ideas.

Regarding Bohr's reaction to the pilot-wave interpretation, there is another legend about Bohm, which comes by way of the philosopher of science Paul Feyerabend. Feyerabend claimed that, while visiting Bohr's institute in Copenhagen in 1952, Bohr had a very different reaction to Bohm's work. "It seemed that, for him, the sky was falling in. . . . Bohr was neither dismissive nor shaken. He was amazed." When Feyerabend asked Bohr what was so amazing about Bohm's work, Bohr started to explain, only to be called away on other business—at which point Bohr's disciples swooped in and dismissed Bohm's ideas by invoking the all-powerful proof by von Neumann (Peat 1997, p. 129). But this story is another one told at a remove of nearly forty years; it is unclear whether it happened at all, much less whether it happened that way, especially since it seems to contradict the available contemporary evidence about Bohr's reaction to Bohm's ideas (i.e., Bohm's letter to Wightman).

107 **"consistent," and even "very elegant":** Talbot 2017, p. 247.

107 **encompass the quantum phenomenon of spin:** Ibid., p. 147.

108 **"the stone belongs to the second man?":** Freire 2015, p. 32.

108 **"elegant and suggestive":** David Bohm 1957, *Causality and Chance in Modern Physics*, Harper Torchbooks ed. (Harper and Row), p. xi.

108 **was a common thread in many strains of Marxist thought:** Marxism is probably more accurately described as a constellation of related ideologies, so it's hard to say anything about "Marxism" as a monolithic entity.

108 **"age of banishment of complementarity" in the USSR:** Freire 2015, p. 36.

109 **"scared people away from these problems":** Talbot 2017, p. 230.
109 **for ideological reasons:** Ibid., p. 178.
109 **"there is not the slightest controversial point about it":** Freire 2015, p. 36.
110 **prevent the publication:** Ibid., pp. 37–38.
110 **"mapped down to one part in twenty thousand":** Quoted in ibid., p. 39. The original is in French; the translation is my own and Alex Zani's. Rosenfeld omitted this sentence from the English translation of his review after several of his colleagues suggested that he had been a bit hard on Bohm.
110 **"Great honor for somebody so young":** Freire 2015, p. 38.
110 **"little to do with immediate physical reality":** Heisenberg 1958, pp. 131–132.
110 **"slays Bohm not only philosophically but physically as well":** Letter from Born to Einstein, November 26, 1953. Born 2005, p. 203.
110 **too politically motivated in his science:** Freire 2015, pp. 39–40.
111 **"It lay on the side":** Schweber interview with the author, September 7, 2016.
111 **"Newton, Einstein, Schrödinger, and Dirac all rolled into one":** Talbot 2017, p. 311.
112 **"calculations on a theory that is known to be of no use":** Ibid., p. 121.
112 **"anxiety about the verdict of history":** Freire 2015, p. 48.
112 **"In Portuguese, I would call Schrödinger 'um burro'":** Talbot 2017, p. 247.
112 **"That way seems too cheap to me":** Born 2005, Einstein to Born, May 12, 1952.
113 **quantum physics should agree with classical physics for large objects:** Einstein 1953, "Elementary Considerations on the Interpretation of the Foundations of Quantum Mechanics," translated by Dileep Karanth, http://arxiv.org/abs/1107.3701.
113 **"that [locality] will have to be abandoned":** Einstein to Born, April 1948, in Born 2005. Bell's theorem, of course, is just such a fact—but Bell's theorem didn't come for another fifteen years, by which time Einstein was dead. See Chapter 7 for more on this.
113 **"'music of the future'":** Born 2005, p. 199.
114 **they would do all of their work in "normal" quantum physics:** Yakir Aharonov, interview with the author, Vienna, October 24, 2015.
115 **"my interests began to turn in other directions":** Freire 2015, p. 54.
116 **"so as to get a better job":** Ibid., p. 56. Emphasis in original.

Chapter 6

117 **"more brain grease":** Otto Stern, interview by Thomas S. Kuhn on May 29 and 30, 1962, Berkeley, CA, USA, courtesy of the Niels Bohr Library &

Archives, American Institute of Physics, College Park, MD, USA, https://www.aip.org/history-programs/niels-bohr-library/oral-histories/4904.

117 **"When a mouse observes":** Isaacson 2007, p. 515.

118 **"strange difficulties created by himself for this purpose":** Peter Byrne 2010, *The Many Worlds of Hugh Everett III: Multiple Universes, Mutual Assured Destruction, and the Meltdown of a Nuclear Family* (Oxford University Press), p. 26.

118 **"the heretic":** Ibid., p. 30.

119 **a resolution he couldn't keep:** Ibid., p. 32.

119 **"probably *no* graduate student is his equal in native ability":** Ibid., p. 38.

119 **"make you stay there until he succeeded":** Ibid., p. 57.

119 **"favorite sport was one-upmanship":** Charles W. Misner 2015, "A One-World Formulation of Quantum Mechanics," *Physica Scripta* 90 (088014) 6pp., p. 1.

119 **"it was always friendly competition":** Byrne 2010, p. 57.

119 **"this is talent":** Ibid., pp. 57–58.

120 **classic in the field:** Ibid., p. 56.

120 **no exception:** Freire 2015, p. 87n46.

120 **reasonable field of research at the time:** See Chapter 11 for more on this.

121 **"encouraged to think about quantum gravity":** Byrne 2010, p. 132.

122 **wandering the Princeton campus with Petersen and Bohr:** Ibid., p. 89.

122 **dismiss the idea of a "quantum theory of measurement":** Ibid., p. 89.

122 **"don't you remember, Charlie?":** Everett-Misner "cocktail party" tape, 1977. In Jeffrey Alan Barrett and Peter Byrne, eds., 2012, *The Everett Interpretation of Quantum Mechanics: Collected Works 1955–1980 with Commentary* (Princeton University Press), p. 300.

122 **"he still feels that way a little bit":** Ibid., pp. 302–307.

122 **"a serious hearing":** Ibid.

123 **"obey perfectly natural continuous laws":** Letter from Everett to Jammer, September 19, 1973, in Barrett and Byrne 2012, p. 296.

123 **"other persons to whom it does not apply":** Barrett and Byrne 2012, p. 75.

123 **"denial of the same for the microcosm":** Letter from Everett to DeWitt, May 31, 1957, in Barrett and Byrne 2012, p. 255. Emphasis in original.

123 **"derive classical physics from it":** Letter from Everett to Petersen, May 31, 1957, in Barrett and Byrne 2012, p. 239.

127 **"an original man":** Byrne 2010, p. 91.

127 **"walks and talks under the beech trees":** J. A. Wheeler 1985, "Physics in Copenhagen in 1934 and 1935," in *Niels Bohr: A Centenary Volume*, edited by A. P. French and P. J. Kennedy (Harvard University Press, 1985), pp. 221–226.

128 **"contrary to Bohr's ideas":** Byrne 2010, p. 161.

128 **"need further analysis and rephrasing":** Wheeler to Bohr, April 24, 1956. Hugh Everett III Papers, UC Irvine, http://ucispace.lib.uci.edu/handle/10575/1195.

129 **"the words that are to be attached to the quantities of the formalism":** Wheeler to Everett, May 22, 1956 (letter 1), Hugh Everett III Papers, UC Irvine, http://ucispace.lib.uci.edu/handle/10575/1143.

129 **"How soon can you come?":** Ibid. Emphasis in original.

129 **"an INDEFINABLE interaction":** Letter from Alexander Stern to John Wheeler, May 20, 1956, Everett Papers, http://ucispace.lib.uci.edu/handle/10575/1160. Capitalization in the original.

130 **"get the bugs out of the words":** Second letter from Wheeler to Everett, May 22, 1956, Box 4, Folder 3, Correspondence from Wheeler to Everett and Others, Oct 1955–Dec 1957, Hugh Everett addition to papers, 1935–1991, American Institute of Physics, Niels Bohr Library & Archives, College Park, MD, USA, http://ucispace.lib.uci.edu/handle/10575/14608.

131 **"to accept it and *generalize* it":** John Wheeler to Alexander Stern, May 25, 1956, Everett Papers, http://ucispace.lib.uci.edu/handle/10575/1123. Emphasis in original.

131 **"this approach is insufficient":** Petersen to Everett, May 28, 1956, Everett Papers, http://ucispace.lib.uci.edu/handle/10575/1188.

131 **"read more carefully":** Letter from Everett to Petersen, June 1956 (draft, undated?), Everett Papers, http://ucispace.lib.uci.edu/handle/10575/1191.

132 **as Petersen had proposed:** Petersen to Everett, May 28, 1956, Everett Papers, http://ucispace.lib.uci.edu/handle/10575/1188.

132 **"revise the draft":** Wheeler and Ford 1998, p. 268.

132 **"told him what to say":** Bryce DeWitt and Cecile DeWitt-Morette, interview by Kenneth W. Ford, February 28, 1995, Austin, TX, USA, courtesy of the Niels Bohr Library & Archives, American Institute of Physics, College Park, MD, USA, http://www.aip.org/history-programs/niels-bohr-library/oral-histories/23199, accessed October 26, 2016.

132 **finally received his PhD:** Freire 2015, p. 111.

132 **judged "very good":** Ibid.

132 **"new and independent foundation":** John A. Wheeler 1957, "Assessment of Everett's 'Relative State' Formulation of Quantum Theory," in Barrett and Byrne 2012, p. 201.

132 **"some confusion as regards the observational problem":** Freire 2015, p. 114.

133 **"permits that neglect of quantum effects":** Letter from Petersen to Everett, April 24, 1957, in Barrett and Byrne 2012, p. 237.

133 **"when applied to any measuring processes!":** Letter from Everett to Petersen, May 31, 1957, in ibid., p. 240.

134 **"an infinity of possible worlds":** Byrne 2010, p. 182.

134 **"sympathetic to [their] point of view":** Weiner to Wheeler, April 9, 1957, in Barrett and Byrne 2012, p. 232.

134 **"given sacramental unction and expected to perform a redemptive act":** All in Margenau 1958.
134 **had not had time to read the thesis carefully:** Margenau to Wheeler and Everett, April 8, 1957, Everett Papers, http://ucispace.lib.uci.edu/handle/10575/1179.
134 **"I simply do *not* branch":** DeWitt to Wheeler, May 7, 1957, in Barrett and Byrne 2012, p. 246. Emphasis in original.
135 **"Do you feel the motion of the earth?":** Everett to DeWitt, May 31, 1957, in ibid., p. 254. Emphasis in original.
135 **"touché":** DeWitt and DeWitt-Morette 1995, interview.
136 **"relight his pipe seventeen times":** Cocktail party tape, Barrett and Byrne 2012, p. 307.
136 **"You had to lean close":** Byrne 2010, p. 221.
136 **"totally change his viewpoint":** Ibid., p. 221.
136 **"doomed from the beginning":** Ibid., p. 168.
136 **"could not understand the simplest things in quantum mechanics":** Freire 2015, pp. 114–115.
137 **"afford people the guidance they needed":** Bricmont 2016, p. 8.
137 **"new concepts by which to replace them":** Beller 1999b, p. 183.
137 **"had a cigarette *all the time*":** Byrne 2010, p. 221. Emphasis in original.
137 **"without things getting out in the newspapers":** Byrne 2010, p. 251.
138 **"non-denumerable [infinity] of alternative Furrys":** Xavier conference transcript, p. 95, http://ucispace.lib.uci.edu/handle/10575/1299. Also Byrne 2010, p. 255.

Chapter 7

141 **"[particle] theory group":** Bernstein 1991, p. 67.
142 **"quite intense discussions":** Ibid., p. 68.
142 **"carpenters, blacksmiths, laborers, farm workers, and horse dealers":** Ibid., p. 12.
142 **least expensive high school:** Free public secondary schools didn't become universal in the UK until several years later (ibid., p. 13).
142 **"did the first year of my college physics":** Ibid., p. 14.
142 **"Then the puzzles start":** Ibid., p. 50.
142 **"bookkeeping operation":** Mann and Crease 1988, p. 86.
142 **what was that information about?:** Bell 2004, p. 215.
143 **"we couldn't get all that clear":** Bernstein 1991, p. 51.
143 **"to epistemology, to philosophy, to humanity in general":** Ibid., p. 52.
143 **"perfectly obscure":** Ibid., pp. 52–53.
143 **"couldn't interpret quantum mechanics":** Ibid., p. 64.
143 **"a hole I wouldn't get out of":** Ibid., p. 53.
144 **"They did a fantastic somersault":** Ibid., p. 66.
144 **"von Neumann's unreasonable axiom":** Ibid., p. 65.

144 **his talk about accelerators:** Ibid., p. 67.

145 **"if the lift had gotten stuck between floors":** Mann and Crease 1988, p. 85.

145 **"not merely false but *foolish!*":** Ibid., p. 88. Emphasis in original.

145 **so-called hidden variables:** Gleason's proof actually didn't mention hidden variables. Gleason was a mathematician, not a physicist, and his proof had to do with certain features of Hilbert space, the mathematical structure underlying quantum physics. But Jauch and his colleague Piron pointed out to Bell that Gleason's proof had an obvious corollary that ruled out hidden variables—and that this corollary seemed far stronger than von Neumann's proof, or their own.

146 **an unjustified assumption at the foundation:** Actually, Bell found two different assumptions: one used by von Neumann and Jauch, and one by Gleason. Gleason's assumption is really what led Bell to contextuality. Von Neumann's assumption is related to Gleason's, but it's more specific than Gleason's, and rejecting it is easier. And it was von Neumann's assumption that Grete Hermann correctly recognized as unwarranted in the 1930s and that Bell himself later characterized as "silly."

146 **the same is true for evens and odds:** Technically, this isn't how real roulette wheels are constructed (though it is how the roulette wheel in Figure 7.2 is constructed). There are ten red odds and eight red evens on a real roulette wheel, and vice versa for black, so the colors aren't quite evenly split among odds and evens. And a real roulette wheel also has one or two slots that are neither red nor black, the 0 and 00 slots, which the house uses to win in the long run. But we'll imagine that Flo is at a roulette wheel like that pictured in Figure 7.2—in a better world than ours, one where the colors are evenly distributed between evens and odds and where players actually have a chance at beating the house in the long run.

148 **"We ourselves produce the results of measurement":** Jammer 1974, p. 164.

148 **"judo-like maneuver":** Mermin 1993, p. 811n23, quoting Abner Shimony.

148 **it's impossible to draw "any sharp distinction":** Bell 2004, p. 2.

148 **the electron has a definite position all the while:** Position does play a special role in the pilot-wave interpretation—although particles always have *positions*, there are other properties that are not always well defined outside of the context of a measurement apparatus. But all measurements of quantum properties ultimately end up being measurements of position, in the Bohmian view, so there is no real problem as long as position is always well defined. This is related to a problem known as the "preferred-basis problem," which is beyond the scope of this book but has to do with decoherence, a major subject in Chapters 9 and 10.

148 **"lack of imagination":** Bell 2004, p. 167.

149 **"moved a magnet anywhere in the universe":** Bernstein 1991, p. 72.

149 **a series of clerical errors:** Bell sent off his paper demolishing von Neumann's proof for publication in *Reviews of Modern Physics*, a widely read journal. Bell was asked to make some small changes to the paper before publication. Bell did so and sent the revised paper back. But once it arrived, it was misfiled and lost. Edward Condon, the editor of *Reviews of Modern Physics*, was eager to publish Bell's excellent paper and wrote back to Bell asking what had happened. But Condon wrote to Bell at SLAC, and by that time Bell was already back at CERN. Condon's letter was returned to him marked "Return to Sender—Addressee not in Directory." Eventually, Bell wrote to Condon asking when his paper would be published. Condon finally figured out what had happened, asked Bell to resend the paper with revisions, and published it immediately—two years after it had been originally submitted.

As a result of the delay in publishing this paper, Bell had already provided himself with an answer to his question by the time this paper was published. Thus, the published version includes not only the question but a reference to the answer provided by Bell himself in his later (and much more famous) paper. See Jammer 1974, p. 303.

149 **"leave everything local":** Bernstein 1991, p. 72.

149 **photons with entangled polarization:** Actually, Bohm's version involved electrons with entangled spin, but the idea is nearly identical, and photons are easier to deal with experimentally—and polarization is easier to think about than spin.

150 **"impossibility proof":** Bernstein 1991, p. 73.

151 **"the most profound discovery of science":** H. P. Stapp 1975, "Bell's theorem and world process," *Nuovo Cim B* 29 (2): 271, https://doi.org/10.1007/BF02728310.

152 **"the establishment of empirically testable laws":** Translated and quoted in Howard 1985, pp. 187–188.

153 **we'll need a whole casino:** The presentation of Bell's theorem in the following section owes much to a classic paper by Mermin 1985. There is also a vaguely similar presentation along the same lines, using slot machines rather than roulette wheels, in W. David Wick 1995, *The Infamous Boundary* (Copernicus), though I was not aware of Wick's version until after I had devised and written my own.

153 **Ronnie the Bear:** Apologies to Brad Neely.

153 **roulette wheels with numbers on them are illegal:** This is an actual law in the state of California. I do not know why it's true, but California roulette, as depicted in Figure 7.3a, is a real game played in casinos in the Golden State. The triple wheel in Figure 7.3b is Ronnie's innovation, though.

157 **perfectly matched outcomes:** This corresponds to the EPR experiment.

159 **another assumption in Bell's proof:** For example, David J. Griffiths 2005, *Introduction to Quantum Mechanics*, 2nd ed. (Pearson Education)

(a widely used textbook), claims this on pp. 423–426; Ernest S. Abers 2004, *Quantum Mechanics* (Pearson), does the same on pp. 192–195. Also appears on p. 244 of Freire 2015, and in dozens of older papers. As Travis Norsen 2007, "Against 'Realism,'" *Foundations of Physics* 37 (3): 311–340, doi:10.1007/s10701-007-9104-1, puts it, before about 1980, "[Bell's theorem] had been typically characterized as a constraint on local deterministic theories or local hidden-variable theories."

159 **[hidden variables are] not a *presupposition* of the analysis:** Bell 2004, p. 143. Emphasis in original. Bell was actually talking about determinism here, which is an equivalent (and equally irrelevant) assumption in this context. See Tim Maudlin 2002, *Quantum Non-locality and Relativity*, 2nd ed. (Blackwell), pp. 15–16, for more on exactly how irrelevant determinism is to the analysis here.

159 **commentators have almost universally reported:** Bell 2004, p. 157n10. Emphasis in original.

159 **especially popular claim:** See Norsen 2007 for numerous examples of this claim.

159 **quantum objects have well-defined properties before they're measured:** For example, see Michael A. Nielsen and Isaac L. Chuang 2000, *Quantum Computation and Quantum Information* (Cambridge University Press), p. 117.

160 **"we're stuck with nonlocality":** John Bell, Antoine Suarez, Herwig Schopper, J. M. Belloc, G. Cantale, John Layter, P. Veija, and P. Ypes 1990, "Indeterminism and Non Locality" (talk given at Center of Quantum Philosophy of Geneva, January 22), http://cds.cern.ch/record/1049544?ln=en; transcript: http://www.quantumphil.org./Bell-indeterminism-and-nonlocality.pdf.

161 **"embarrassed to ask them to pay for my article":** Bernstein 1991, p. 74.

161 **material from all subfields of physics:** Wick 1995, p. 289.

161 **"good way to avoid embarrassment":** Bernstein 1991, p. 74.

161 **"basically right":** Anderson to Wick, September 15, 1993, private collection.

162 **folded altogether:** Whitaker 2016, p. 210; Anderson to Wick, September 15, 1993.

Chapter 8

164 **"illegal to use photo-cells":** Niels Bohr interview by Thomas S. Kuhn, Aage Petersen, and Erik Rudinger, November 17, 1962, Copenhagen, Denmark, courtesy of the Niels Bohr Library & Archives, American Institute of Physics, College Park, MD, USA, http://www.aip.org/history-programs/niels-bohr-library/oral-histories/4517-5, accessed January 27, 2017.

164 **"But he did not like it":** Ibid.

165 **"absolutely no problem in it":** Ibid.

165 **"I do not know why the people don't like it":** Ibid.

165 **"complementary description":** Ibid.

165 **positivism-inspired arguments:** Whether Bohr himself was a positivist was and is a subject of much debate. Cushing and many others argue that he was; Howard and others argue that he was not. But the particulars of Bohr's views are far less significant, historically, than the fact that his views were obscure—which hardly anyone would dispute—that positivist reasoning was ubiquitously deployed in defense of the Copenhagen interpretation, and such defenses were often presented as the views of Bohr himself.

166 **visiting Stanford for the past term:** *Stanford Daily* 1928, "Dr. Moritz Schlick to Be Visiting Professor Next Summer Quarter," July 31, p. 1, http://stanforddailyarchive.com/cgi-bin/stanford?a=d&d=stanford19280731-01.2.6.

166 **"token of gratitude and joy":** Hans Hahn, Rudolf Carnap, and Otto Neurath 1973, "The Scientific Conception of the World: The Vienna Circle," in Otto Neurath 1973, *Empiricism and Sociology* (Reidel), p. 299.

166 **most senior members:** See Ayer 1982, *Philosophy in the Twentieth Century* (Vintage), p. 127, for a little more on the authorship of the manifesto.

167 **"this *spirit of a scientific conception of the world* is alive":** Hahn, Carnap, and Neurath 1973, p. 301. Emphasis in original.

167 **"sets the material in motion":** Peter Godfrey-Smith 2003, *Theory and Reality: An Introduction to the Philosophy of Science* (University of Chicago Press), p. 23.

167 **"unfathomable depths rejected":** Ibid., p. 306.

168 **"this view is rejected":** Ibid., p. 309.

168 **"no realm of ideas that stands over or beyond experience":** Ibid., p. 316.

168 **"*logical analysis*":** Ibid., p. 309. Emphasis in original.

170 **"*the one empirical science*":** Ibid., p. 316. Emphasis in original.

170 **"according to rational principles":** Ibid., pp. 317–318.

170 **Bauhaus school of architecture and design:** See Peter Galison 1990, "Aufbau/Bauhaus: Logical Positivism and Architectural Modernism," *Critical Inquiry* 16:709–752.

170 **"inner link with the scientific world-conception":** Hahn, Carnap, and Neurath 1973, pp. 304–305.

171 **"removing the metaphysical and theological debris of millennia":** Ibid., p. 317.

171 **"turn away from metaphysics and theology":** Ibid., p. 305.

171 **"drawing of an elephant":** Ayer 1982, p. 123.

172 **"quantities [that] cannot, in principle, be observed experimentally":** Pauli 1921, *Theory of Relativity*, trans. G. Field (Dover), p. 4.

172 **"fictitious and without physical meaning":** Ibid., p. 206.
172 **"how many angels are able to sit on the point of a needle":** Born 2005, p. 218.
172 **someone else both groups held in similar regard:** See Cushing 1994, pp. 110–111, 114.
173 **not later in life:** There's some controversy about whether Einstein was an adherent of Mach's ideas when he was younger and later changed his mind, or whether he never liked Mach's philosophy much, with a great deal of (fascinating!) literature weighing in on either side. But nearly all are agreed that by the 1920s, Einstein was solidly out of Mach's camp.
173 **"only exterminate harmful vermin":** Isaacson 2007, p. 334.
173 **"A good joke should not be repeated too often":** Cushing 1994, pp. 110–111.
173 **"has the sole purpose of determining what *is*":** Kumar 2008, p. 262. Emphasis in original.
173 **"supposedly exists irrespective of any act of observation or substantiation":** Einstein 1949b, p. 667.
174 **"such a bloodless ghost":** Ibid.
174 **source of positivist inspiration:** Cushing 1994, pp. 110, 114.
174 **"through these theories physics is permanently changed":** Bridgman 1927, *The Logic of Modern Physics* (Macmillan), p. 1.
174 **"what the concepts useful in physics are and should be":** Ibid., pp. 2–4.
174 **"*the concept is synonymous with the corresponding set of operations*":** Ibid., p. 5. Emphasis in original.
175 **"basic attitudes which agree with mine":** Jan Faye 2007, "Niels Bohr and the Vienna Circle," preprint, http://philsci-archive.pitt.edu/3737/, accessed December 23, 2016.
175 **their views were not too far apart:** Ibid.
176 **"Quantum Theory and the Knowability of Nature":** The original title, in German, was "Quantentheorie und Erkennbarkeit der Natural." Following William H. Werkmeister 1936, "The Second International Congress for the Unity of Science," *Philosophical Review* 45 (6): 593–600, in translating "Erkennbarkeit," in this context, as "knowability."
176 **"neither true nor false, but *meaningless*":** Ibid. Emphasis in original.
176 **distasteful, such as vitalism:** Abraham Pais 1991, *Niels Bohr's Times in Physics, Philosophy, and Polity* (Oxford University Press), p. 443.
176 **"express[es] himself somewhat unclearly":** Faye 2007.
176 **"the sense of my efforts":** Ibid.
176 **"only when the experiment includes a position determination":** Schiff 1955, *Quantum Mechanics*, 2nd ed. (McGraw-Hill), p. 6.
177 **"there is no orbit in the ordinary sense":** Heisenberg 1958, p. 48.
177 **"misuse of language which . . . cannot be justified":** Ibid.

177 **couldn't actually justify most formulations of the Copenhagen interpretation:** At least one prominent positivist, Hans Reichenbach, recognized that the verification theory of meaning couldn't be used to straightforwardly argue for the Copenhagen interpretation. "It would be wrong to argue that statements about the value of an entity before a measurement are meaningless because they are not verifiable. Statements about the value after the measurement are not verifiable either. If, in the [Bohr-Heisenberg interpretation], the one sort of statement are [*sic*] forbidden and the other admitted, this must be considered as a rule which, logically speaking, is arbitrary, and which can be judged only from the standpoint of expediency" (Reichenbach 1944, *Philosophic Foundations of Quantum Mechanics* [Dover], p. 142). Reichenbach dismissed the Copenhagen interpretation as problematic, because it promotes an ad hoc principle about what statements are meaningless into a law of physics. He argued instead for an interpretation based on a three-valued logic system, but this was later discovered to have its own problems related to the boundary between the microscopic and the macroscopic.

178 **"suffered for National Socialism":** Friedrich Stadler 2001, "Documentation: The Murder of Moritz Schlick," in *The Vienna Circle: Studies in the Origins, Development, and Influence of Logical Empiricism*, edited by Friedrich Stadler (Springer), p. 906.

179 **"the icy slopes of logic":** For more on this, see George Reisch 2005, *How the Cold War Transformed Philosophy of Science: To the Icy Slopes of Logic* (Cambridge).

179 **"intellectually fired by a living teacher":** Willard Van Orman Quine 1976, *The Ways of Paradox* (Harvard University Press), p. 42.

179 **"an ardent disciple of Carnap":** Willard Van Orman Quine 2008, *Quine in Dialogue*, edited by Dagfinn Føllesdal and Douglas B. Quine (Harvard University Press), p. 25.

180 **verification theory of meaning:** The other "dogma of empiricism" Quine attacked was the analytic-synthetic distinction, but in the course of the paper Quine argued that the two dogmas were actually two sides of the same coin and deployed effective arguments against both. Quine's paper is usually remembered more for attacking the analytic-synthetic distinction, but, for our story, his argument against the verification theory of meaning is much more significant.

181 **"only as a corporate body":** Willard Van Orman Quine 1953, *From a Logical Point of View*, Harper Torchbooks ed. (Harper and Row), p. 41.

181 **doubts about logical positivism:** The impact of Quine's paper on the philosophical community is undeniable. But it's also odd, because Quine was not the first to point out that it was impossible to verify individual statements, nor that the distinction between the analytic and synthetic (as mentioned earlier) was problematic. In fact, several leading positivists

had pointed both of these things out before, including Carnap himself. See Godfrey-Smith 2003, pp. 32–33, for more on this. Why, then, did Quine's paper have the impact it did? This is a puzzle with many conjectured solutions in the literature. It seems likely that the positivists did not fully appreciate the scope of these two problems for their program. And Quine's lucid, lively writing made these problems explicit and memorable—and thereby unavoidable.

181 **"wrestling already with the problem of meaning":** Thomas S. Kuhn 2000, *The Road Since Structure*, edited by James Conant and John Haugeland (University of Chicago Press), p. 279.

181 **"made Aristotle's philosophy make sense":** Skúli Sigurdsson 1990, "The Nature of Scientific Knowledge: An Interview with Thomas S. Kuhn," *Harvard Science Review*, Winter, pp. 18–25, http://www.edition-open-access.de/proceedings/8/3/index.html.

182 **over twenty years earlier:** Kuhn 2000, pp. 291–292.

182 **his working title for it was *The Structure of Scientific Revolutions*:** James A. Marcum 2015, *Thomas Kuhn's Revolutions* (Bloomsbury), p. 13.

182 **"what atoms and molecules, compounds and mixtures, were":** Thomas S. Kuhn 1996, *The Structure of Scientific Revolutions*, 3rd ed. (University of Chicago Press), p. 40.

183 **day-to-day practice of science itself:** That being said, Kuhn did not see any particular problem with the Copenhagen interpretation; indeed, he took much of his inspiration for *Structure* from the work of Norwood Hanson, who was both intensely antipositivist and pro-Copenhagen.

183 **didn't catch on with professional philosophers of science:** This idea did catch on with many sociologists and historians of science (and it certainly lit up the public imagination). And there are some philosophers who are sympathetic to Kuhn's ideas: largely, these are philosophers in the intellectual tradition of Hegel. Modern philosophy is (broadly) split into two camps: those in the tradition of Hegel, known as the *Continental philosophers*, and those in the tradition of Russell and the positivists, known as the *analytic philosophers*. This is not to say that the Continental philosophers all agree with Hegel, any more than the analytic philosophers all agree with the positivists—this chapter is largely about a revolution within analytic philosophy in which most analytic philosophers rejected positivism. But analytic and Continental philosophers tend to follow their intellectual forebears in the *problems* they tend to go after, and especially in the *style* in which they approach them. Analytic philosophers are more concerned with questions about the philosophy of science; Continental philosophers tend to write about questions of politics and personal experience. They do have some overlapping interests: philosophy of language, ethics, and ancient philosophy, to name a few. And there are analytic political philosophers and Continental philosophers of science. The place

where the analytic-Continental divide is most stark is in methodology. Analytic philosophers generally value clear writing and logical analysis and have a healthy appreciation of science. Among Continental philosophers, argumentation is often based on introspection, political considerations, and aesthetics; and they tend to be much more skeptical of the validity of any scientific (or mathematical, or logical) results than their analytic colleagues.

A. J. Ayer summed up the continuing influence of positivism among analytic philosophers well. In 1982, long after logical positivism of the sort the Vienna Circle advocated had been discarded by the philosophical community, Ayer wrote, "Few of the principal theses of the Vienna Circle survive intact. . . . [But] I think it can be said that the spirit of Viennese positivism survives. In its re-accommodation of philosophy with science, its logical techniques, its insistence on clarity, its banishment of what I can best describe as a strain of wooly uplift from philosophy, it gave a new direction to the subject which is not now likely to be reversed" (1982, pp. 140–141).

As Continental philosophers are generally in the minority among philosophers of physics, and as they have not contributed nearly as much to scientific discussions about the interpretation of quantum physics, I will be almost entirely ignoring them in this book. Elsewhere, where I refer to "philosophers," you should mentally annotate that as "analytic philosophers."

183 ***scientific realism:*** Kuhn, Feyerabend, and Hanson were not realists, but Smart, Putnam, Popper, Maxwell, and much of the rest of the philosophical community—including some former members of the Vienna Circle, like Herbert Feigl—were convinced by the arguments in favor of realism. Today, philosophers of physics are, by a wide majority, realists of some stripe.

184 **"what we see through an ordinary windowpane":** Grover Maxwell 1962, "The Ontological Status of Theoretical Entities." *Minnesota Studies in the Philosophy of Science* 3:7.

184 **"separating the observable from the unobservable":** Ibid., p. 11.

186 **"no longer seem surprising":** J. J. C. Smart 1963, *Philosophy and Scientific Realism* (Routledge and Kegan Paul), p. 39.

187 **"if there really were a criminal":** Ibid., p. 47.

187 **"doesn't make the success of science a miracle":** Hilary Putnam 1979, *Mathematics, Matter, and Method*, 2nd ed. (Cambridge University Press), p. 73.

187 **"alternatives to the prevailing Copenhagen interpretation":** Smart 1963, p. 40.

187 **"too much of a coincidence to be believed":** Ibid., p. 47.

187 **"obeyed by *all* physical interactions":** Hilary Putnam 1965, "A Philos-

opher Looks at Quantum Mechanics," in Putnam 1979, p. 132. Emphasis in original.

188 **"it must apply to macro-systems":** Ibid., p. 148.

188 **"becomes once again observable from the earth":** Ibid., p. 149.

188 **"outcome of the two-slit experiment":** Smart 1963, p. 48.

188 **"foreshadowed by such writers as D. Bohm and J.-P. Vigier":** Ibid., pp. 43–44.

189 **"something is wrong with the [quantum] theory":** Putnam 1979, p. 81.

189 **languishing in an editor's desk:** It is particularly unfortunate that Bell's proof languished so long that it is unlikely Hanson ever saw it. Hanson died prematurely in a plane crash in 1967, the year after Bell's paper appeared. Hanson, not a positivist and not a realist—he was more aligned with Kuhn—was a rabid defender of the Copenhagen interpretation, but much of his defense rested on the validity of von Neumann's proof.

189 **"*no* satisfactory interpretation of quantum mechanics exists today":** Putnam 1965, p. 157. Emphasis in original.

189 **"nature and magnitude of the difficulties":** Ibid., pp. 157–158.

189 **"found unsatisfactory by physicists":** Smart 1963, p. 41.

Chapter 9

194 **"if you come to the same conclusions":** John Clauser, interview by Joan Bromberg, May 20, 2002, Walnut Creek, CA, USA, courtesy of the Niels Bohr Library & Archives, American Institute of Physics, College Park, MD, USA, http://www.aip.org/history-programs/niels-bohr-library/oral-histories/25096, accessed March 6, 2017.

194 **"[C]ommon wisdom is frequently a poor interpretation":** Wick 1995, p. 116.

194 **"solve the problem that he couldn't solve":** Clauser interview, May 20, 2002.

194 **"undoubtedly as a quack by others":** John F. Clauser 2002, "Early History of Bell's Theorem," in *Quantum [Un]speakables: From Bell to Quantum Information*, edited by R. A. Bertlmann and A. Zeilinger (Springer, 2002), pp. 77–78.

194 **"Jesus Christ, this is a very important result":** Clauser, interview with the author, Walnut Creek, CA, USA, August 12, 2015.

195 **couldn't easily be adapted to perform such a test:** Wu did try to do a Bell experiment like this with her students Kasday and Ullman a few years later, but it didn't work out well—it involved a lot of extra assumptions. See Whitaker 2012, p. 179.

195 **"'exactly what I'm looking for'":** Clauser 2015, interview.

195 **"hadn't really been aware of what Bell's theorem said":** Ibid.

195 **"'This is a waste of time'":** Ibid.

196 **first correspondence of any kind:** Bell 1964 was actually published in 1965, despite the publication date. See Freire 2015, p. 237.

196 **"which would shake the world!":** Letter from John Bell to John Clauser, March 5, 1969. Courtesy of John Clauser.

196 **"a young student living in this era of revolutionary thinking":** Clauser 2002, p. 80.

196 **"these people are all nuts":** H. Dieter Zeh, interview with the author, Neckargemünd, Germany, October 23, 2015.

197 **"a very important step":** Ibid.

197 **"that solves the measurement problem":** Ibid.

198 **"even to think about that":** Ibid.

198 **"directing your attention to this misfortune":** Olival Freire Jr. 2009, "Quantum Dissidents: Research on the Foundations of Quantum Theory Circa 1970," *Studies in History and Philosophy of Modern Physics* 40:282, doi:10.1016/j.shpsb.2009.09.002.

198 **"some very negative comments":** Zeh 2015, interview.

198 **extinguish his academic career:** Freire 2009.

198 **"our relationship deteriorated":** Ibid., p. 282.

199 **"not fully understood the problem":** Ibid., p. 281.

199 **"does not apply to macroscopic objects":** Kristian Camilleri 2009, "A History of Entanglement: Decoherence and the Interpretation Problem," *Studies in History and Philosophy of Modern Physics* 40:292n5.

199 **first to use the term "measurement problem":** E. P. Wigner 1963, "Problem of Measurement," *American Journal of Physics* 31 (1): 6–15.

200 **"I have never been *invited* to Copenhagen":** Zeh interview, 2015.

200 **"whose future careers such statements may hurt":** Freire 2015, p. 157.

200 **certain details in the quantum theory of measurement:** Ibid., p. 161.

201 **"encouraged me to get this published":** Zeh 2015, interview.

202 **"'That was your finest hour'":** Abner Shimony, interview by Joan Bromberg, September 9 and 10, 2002, Wellesley, MA, USA, courtesy of the Niels Bohr Library & Archives, American Institute of Physics, College Park, MD, USA, http://www.aip.org/history-programs/niels-bohr-library/oral-histories/25643, accessed March 6, 2017.

202 **"indulgence and understanding":** Letter from Abner Shimony to W. David Wick, June 27, 1993. Courtesy of W. David Wick.

202 **"I wanted to do a thesis with Wightman":** Shimony 2002, interview.

202 **"find the flaw in the argument":** Letter from Shimony to Wick, 1993.

202 **"I never saw anything wrong with it":** Shimony 2002, interview.

202 **"the measurement problem had not been solved":** Letter from Shimony to Wick, 1993.

202 **"long adhered to a version of realism":** Ibid.

202 **"independently observe physical systems":** Abner Shimony 1963, "Role

of the Observer in Quantum Theory," *American Journal of Physics* 31:772, doi:10.1119/1.1969073.

203 **spirited defense of evolutionary theory:** Shimony 2002, interview.

203 **"'importance of research in foundations of quantum mechanics'":** Letter from Shimony to Wick, 1993.

203 **"'This is something very grea'":** Shimony 2002, interview.

203 **"relevant piece of literature":** Ibid.

203 **"the less convinced I was":** Ibid.

204 **"test Bell's Inequality":** Letter from Shimony to Wick, 1993.

204 **"I was wrong":** Ibid.

204 **describing precisely the experiment that he and Horne were preparing to do:** John Clauser 1969, "Proposed Experiment to Test Local Hidden-Variable Theories." *Bulletin of the American Physical Society* 14:578.

204 **"phone call from Abner Shimony":** Clauser 2015, interview.

204 **"civilized way to handle the matter of independent discovery":** Letter from Abner Shimony to Eugene Wigner, August 8, 1969. Courtesy of W. David Wick.

205 **"we'd keep swapping drafts":** Clauser 2002, interview.

205 **determine whether Bell's inequality was violated:** John F. Clauser, Michael A. Horne, Abner Shimony, and Richard A. Holt 1969, "Proposed Experiment to Test Local Hidden-Variable Theories," *Physical Review Letters* 23:880, doi:10.1103/PhysRevLett.23.880.

206 **"outcome in favor of [local] hidden variables":** Letter from Shimony to Wigner, 1969.

206 **"Commins thought it was a total crock":** Clauser 2002, interview.

206 **"I would have been dead":** Clauser 2015, interview.

206 **"'It looks like a very interesting experiment to me'":** Clauser 2002, interview.

207 **"pretty good at dumpster diving":** Kaiser 2011, p. 47.

207 **control the motions of the polarizers:** Whitaker 2012, p. 174.

207 **something awfully strange was going on in nature:** Stuart J. Freedman and John F. Clauser 1972, "Experimental Test of Local Hidden-Variable Theories," *Physical Review Letters* 28:9389–41, doi:10.1103/PhysRevLett.28.938.

207 **foundations of quantum mechanics:** See Freire 2015, Chapter 6, for more on the origins of the Varenna summer school.

208 **"Woodstock of quantum dissidents":** Freire 2015, p. 197.

208 **"I had never heard of them":** H. Dieter Zeh 2006, "Roots and Fruits of Decoherence," arXiv:quant-ph/0512078v2.

208 **"which is of course quite wrong":** Zeh 2015, interview.

208 **"'Now I can just do what I like'":** Freire 2009.

208 **"my students never had a chance":** Zeh 2015, interview.

209 **"I will never be ready to forgive":** Zeh 2015, interview.
209 **"dark ages of decoherence":** Zeh 2006.
209 **"I was just having fun":** Clauser 2002, interview.
209 **"junk science":** Ibid.
209 **"strong letter answering the question in your favor":** Freire 2015, p. 271.
209 **turning on, tuning in, and dropping out:** Kaiser 2011.
210 **grand total of fifty-three jobs:** Kaiser 2002, pp. 150–152; Kaiser 2011, pp. 22–23.
210 **Quantum nonlocality was real:** Freire 2015.
211 **wasn't sure where he'd be next year:** Letter from John Bell to John Clauser, May 30, 1975; Letter from John Clauser to John Bell, July 1, 1975. Courtesy of John Clauser.
211 **"MAY WE PUT YOUR NAME ON THE POSTER?":** Telex from John Bell to John Clauser, June 30, 1975. Courtesy of John Clauser.
212 **sources of funding that kept the lights on:** Bohm, of course, is the obvious example. For another example, see Hans Freistadt's discussion group on quantum foundations in New York City in the 1950s, discussed in Kaiser 2011, pp. 20–21.
212 **"ruin his career by doing so":** Clauser 2002, p. 72.
213 **"I want to work on the measurement problem":** David Albert, interview with the author, New York, NY, USA, February 4, 2015.
213 **"interested in these things in philosophy":** Ibid.
213 **"by snail mail in those days":** Ibid.
213 **"or you can leave the program":** Ibid.
214 **"no more talk about the measurement problem":** Ibid.
214 **not so lucky:** For example, see Freire 2015 on the short-lived physics career of Klaus Tausk.
214 **"unless it can be related to experimental data":** Samuel Goudsmit 1973, "Important Announcement Regarding Papers About Fundamental Theories," *Physical Review D*, 8:357.
214 **informal collection of editors, including Shimony:** Kaiser 2011, p. 122.
214 **"confrontation and ripening of ideas":** Freire 2015, p. 268.
215 **"list of recipients":** Ibid., p. 269.
215 **"changing the orientation of the polarizers":** Alain Aspect, interview with the author, Palaiseau, France, November 4, 2015.
216 **"possible discrepancies with quantum predictions":** Clauser 2015, interview.
216 **"possibility to do it in my lab":** Aspect 2015, interview.
217 **"real experiment to do":** Ibid.
217 **"I did not realize that so much":** Ibid.
218 **"the way I was understanding it":** Ibid.

Chapter 10

219 **"or they laughed about it and thought I am crazy":** Reinhold Bertlmann, interview with the author, Vienna, Austria, November 2, 2015.

220 **"Now you are famous!":** Ibid.

220 **read and reread the title of the paper:** Bell 1981.

220 **"Bertlmann's Socks and the Nature of Reality":** J. S. Bell 1980, "Bertlmann's Socks and the Nature of Reality," CERN Preprint CERN-TH-2926, https://cds.cern.ch/record/142461?ln=en.

221 **"is not the EPR business just the same?":** Ibid., p. 139.

221 **"do not *have* any definite properties in advance of observation":** Ibid., p. 142.

221 **"How does the second sock know what the first has done?":** Ibid., p. 143.

221 **"correlations cry out for explanation":** Ibid., pp. 151–152.

222 **"for all practical purposes":** Ibid., p. 214.

222 **"but on Sundays I have principles":** Nicholas Gisin 2002, "Sundays in a Quantum Engineer's Life," in Bertlmann and Zeilinger 2002, p. 199.

222 **"the speaker could just dissolve and liquify":** Nicholas Gisin, interview with the author, Vienna, Austria, October 24, 2015.

223 **"The Great John Bell":** Ibid.

223 **"I had to dig into this field":** Bertlmann 2015, interview.

224 **"'Get out of here, I'm not interested'":** Clauser 2015, interview.

224 **"He made interesting comments":** Aspect 2015, interview.

224 **"your talk was excellent":** Freire 2015, p. 278.

224 ***The Tao of Physics*:** This book came out in 1975, but it didn't mention Bell in the first edition—that came in the afterword to the second edition, published in 1983.

224 **standard way of teaching the subject:** Mermin's papers are also the basis for the explanation of Bell's theorem in Chapter 7.

224 **"when your ideally pristine presentation appeared":** Feynman to Mermin, March 30, 1984, in Richard P. Feynman 2005, *Perfectly Reasonable Deviations from the Beaten Path*, edited by Michelle Feynman (Basic Books), p. 367. Short parenthetical note in the final sentence elided without an ellipsis.

225 **"I leave that open":** Richard P. Feynman 1982, "Simulating Physics with Computers," *International Journal of Theoretical Physics* 21 (6/7): 467–488.

226 **$20 million initiative in quantum information:** Kaiser 2011, p. 232; Jennifer Ouellette 2005, "Quantum Key Distribution," *Industrial Physicist*, January/February, pp. 22–25, https://people.cs.vt.edu/~kafura/cs6204/Readings/QuantumX/QuantumKeyDistribution.pdf, accessed July 14, 2017.

226 **funding quantum information technology:** Interagency Working Group on Quantum Information Science of the Subcommittee on Physical Sciences 2016, *Advancing Quantum Information Science: National Challenges*

and Opportunities, joint report of the Committee on Science and Committee on Homeland and National Security of the National Science and Technology Council, July, https://www.whitehouse.gov/sites/whitehouse.gov/files/images/Quantum_Info_Sci_Report_2016_07_22%20final.pdf, accessed July 14, 2017.

226 **€1 billion of research and development:** http://www.nature.com/news/europe-plans-giant-billion-euro-quantum-technologies-project-1.19796, accessed July 14, 2017.

226 **quantum communication satellite:** http://www.nature.com/news/chinese-satellite-is-one-giant-step-for-the-quantum-internet-1.20329, accessed July 14, 2017.

227 **"started talking about this in earnest":** Interview of Basil Hiley by Olival Freire on January 11, 2008, Birkbeck College, London, England, courtesy of the Niels Bohr Library & Archives, American Institute of Physics, College Park, MD, USA, https://www.aip.org/history-programs/niels-bohr-library/oral-histories/33822, accessed July 14, 2017.

228 **back to his former interests:** Freire 2015, pp. 165, 319–320.

229 **"important and largely open":** Schlosshauer 2011, *Elegance and Enigma: The Quantum Interviews* (Springer), pp. 35–36.

229 **"the whole set of problems":** Camilleri 2009, p. 294.

229 **forerunner of his own:** Ibid., p. 295.

229 **"without any interpretational baggage attached":** Ibid., p. 295.

230 **not normally have been accessible to such a junior researcher:** Ibid., p. 294.

230 **"times were a-changin'.":** Schlosshauer 2011, p. 37.

230 **"without ever mentioning that":** Interview with Zeh by the author, 2015.

230 **"safeguard Joos's career":** Camilleri 2009, p. 296.

231 **"'Decoherence? What is that?'":** Interview with Zeh by the author, 2015.

231 **"decoherence destroys superpositions":** W. H. Zurek 1991, "Decoherence and the Transition from Quantum to Classical," *Physics Today* 44 (October): 36–44.

231 **"environment-induced decoherence by itself":** Zeh 2002, "Decoherence: Basic Concepts and Their Interpretation," https://arxiv.org/abs/quant-ph/9506020.

231 **"system-apparatus-environment combined wave function?":** W. H. Zurek 1981, "Pointer Basis of Quantum Apparatus: Into What Mixture Does the Wave Packet Collapse?," *Physical Review D* 24 (6): 1517, http://dieumsnh.qfb.umich.mx/archivoshistoricosmq/ModernaHist/Zurek%20b.pdf.

231 **"ill-fated trip to Copenhagen":** Camilleri 2009, p. 298.

232 **"beautiful atomic beam techniques quantifying the whole process":** P. W. Anderson 2001, "Science: A 'Dappled World' or a 'Seamless Web'?," *Studies in History and Philosophy of Modern Physics* 32:487–494.

232 **"'gentle pillow for the true believer'":** Jeffrey Bub 1999, *Interpreting the Quantum World*, rev. ed. (Cambridge University Press), p. 6.

233 **"'this is of course what Bohr always meant'":** Freire 2015, p. 307.

234 **led inevitably to solipsism:** Whitaker 2016, p. 41.

234 **ruling out determinism, rather than locality:** Wheeler and Zurek 1983, p. 188.

235 **inspiration for "it from bit":** Charles W. Misner, Kip S. Thorne, and Wojciech H. Zurek 2009, "John Wheeler, Relativity, and Quantum Information," *Physics Today*, April 2009, pp. 40–46.

235 **"deliberate theoretical choice?":** Bell 2004, p. 160.

236 **contradict existing experiments:** William Feldmann and Roderich Tumulka 2012, "Parameter Diagrams of the GRW and CSL Theories of Wavefunction Collapse," *Journal of Physics A: Mathematical and Theoretical* 45 065304 (13pp.), doi:10.1088/1751-8113/45/6/065304; Angelo Bassi et al. 2013, "Models of Wave-Function Collapse, Underlying Theories, and Experimental Tests," *Reviews of Modern Physics* 85 (2), doi:10.1103/RevModPhys.85.471.

The limit on how frequently collapse happens is also dependent on how tightly collapse is localized in space (i.e., the two parameters are degenerate). The figure of "tens of thousands of years" assumes that collapse localizes wave functions to within about one hundred nanometers—small on the scale of everyday objects but still a thousand times larger than a hydrogen atom.

236 **share a single wave function:** This is not *quite* accurate—in the spontaneous-collapse model I'm describing (GRW), there are no particles per se. So, technically, the number of "slot machines" involved is determined by the number of dimensions of the configuration space the wave function lives in. But that number of dimensions is in turn tied to the number of particles that "inhabit" the wave function, so this description is not quite wrong, either—I'm just glossing a few details. (Different spontaneous-collapse theories have different ontologies anyhow, though none of them take particles to be fundamental.)

236 **"is not both dead and alive for more than a split second":** Bell 2004, p. 204.

237 **GRW model:** G. C. Ghirardi, A. Rimini, and T. Weber 1986, "Unified Dynamics for Microscopic and Macroscopic Systems," *Physical Review D* 34:470.

237 **"only a change which is very small":** Ibid., p. 209.

237 **"social deviance" among physicists:** Philip Pearle 2009, "How Stands Collapse II," in *Quantum Reality, Relativistic Causality, and Closing the Epistemic Circle*, edited by W. C. Myrvold and J. Christian (Springer, 2009), p. 257.

238 **"saints of ancient religions . . . by introspection":** Bell 2004, p. 170.

238 **"as if [they] were not made of atoms and not ruled by quantum mechanics":** Ibid., p. 213.

238 **"any conception of locality which works with quantum mechanics":** John S. Bell 1990, "Indeterminism and Non Locality" (talk given in Geneva, January 22, 1990), https://cds.cern.ch/record/1049544?ln=en, accessed, July 21, 2017; transcript: http://www.quantumphil.org/Bell-indeterminism-and-nonlocality.pdf.

238 **"tenacity to push through his questions":** Shimony 2002, interview.

239 **"still so brimful of vitality":** Kurt Gottfried and N. David Mermin 1991, "John Bell and the Moral Aspect of Quantum Mechanics," *Europhysics News* 22 (4): 67–69.

239 **"He would have been destroyed":** Gisin 2015, interview.

240 **"his work in quantum physics was not appreciated":** Bertlmann 2015, interview.

240 **he was shortlisted for the Nobel Prize the year before he died:** Whitaker 2016, p. 374.

240 **"he could not see the fruits of his work":** Bertlmann 2015, interview.

240 **"the most spell-binding lecture I have ever heard":** Bertlmann and Zeilinger 2002, p. 271.

240 **"better qualified system . . . with a PhD?":** Bell 2004, p. 216.

240 **"redeveloped in a [way consistent with special relativity]":** Ibid., p. 230.

241 **"a highly improbable one":** Ibid., p. 194.

Chapter 11

243 **"whether they had ever read it or not":** Bryce S. DeWitt 1970, "Quantum Mechanics and Reality," *Physics Today* 23 (9): 30–35, doi:10.1063/1.3022331.

243 **"general review of different interpretations of quantum mechanics":** Freire 2015, pp. 226–227.

243 **"objective world that obviously exists all around us?":** DeWitt 1970.

244 **"The answer is that we can":** Ibid.

244 **"myriads of copies of itself":** Ibid.

244 **"schizophrenia with a vengeance":** Ibid.

244 **"begun by Heisenberg in 1925":** Ibid.

245 **"resolve the logical difficulties":** All these quotes are from the *Physics Today* article from 1971 with replies to DeWitt and his reply to the replies. Leslie E. Ballentine et al. 1971, "Quantum-Mechanics Debate," *Physics Today* 24 (4), doi:10.1063/1.3022676.

245 **"one of the best kept secrets of this century":** See Jammer 1974, p. 509.

245 **"play a role at the very foundations of cosmology":** DeWitt (the reply to replies) in Ballentine et al. 1971.

246 **"look elsewhere for interesting physics challenges":** Kip Thorne, *Black Holes and Time Warps: Einstein's Outrageous Legacy* (W. W. Norton), p. 268.

246 **"gravely detrimental to the beauty of the theory":** John D. Norton 2015, "Relativistic Cosmology," http://www.pitt.edu/~jdnorton/teaching/HPS_0410/chapters_2017_Jan_1/relativistic_cosmology/index.html, accessed July 24, 2017.

247 **"gravitational waves do not exist":** Born 2005, p. 122. This letter is undated, but it is in reply to a letter of Born's from August 1936 and references a paper written in late 1936, so it is highly likely that the letter was from 1936.

247 **a confusion that persisted for decades:** Daniel Kennefick 2005, "Einstein Versus the Physical Review," *Physics Today* 58 (9): 43, doi:10.1063/1.2117822.

248 **"black holes"—must be real:** Wheeler was not the first to call them black holes, but he pioneered the usage of the term. Misner, Thorne, and Zurek 2009.

249 **"the only way of doing this":** DeWitt and DeWitt-Morette interview, 1995.

249 **"Everett had been given a raw deal":** Freire 2015, p. 130.

249 **"was deliberately written in a sensational style":** Cécile DeWitt-Morette 2011, *The Pursuit of Quantum Gravity: Memoirs of Bryce DeWitt from 1946 to 2004* (Springer), p. 95.

249 ***Analog* even ran an article on the many-worlds interpretation:** Byrne 2010, p. 319.

250 **security clearance that far outstripped their own:** Ibid., pp. 3–4. This information came from Everett's FBI file.

250 **on an endless loop, drink in hand:** Byrne 2010, p. 196.

251 **"without his efforts it would never have been presented at all":** Letter from Everett to William Harvey, June 20, 1977, Everett Papers, http://hdl.handle.net/10575/1150, accessed July 23, 2017. NB: Harvey is the same sociologist who interviewed Philip Pearle for his thesis on "social deviance."

251 **"concerning the nature of physical theory":** Everett to Frank, May 31, 1957, Everett Papers, http://hdl.handle.net/10575/1153, accessed July 23, 2017.

251 **"essentially different from all other physical facts":** Frank to Everett, August 3, 1957, Everett Papers, http://hdl.handle.net/10575/1173, accessed July 23, 2017.

251 **"punking the measurement problem made him laugh":** Peter Byrne, personal communication, October 13, 2016.

251 **never spoke of quantum physics again:** Byrne 2010, p. 339.

252 **"the price of seriousness":** Evelyn Fox Keller 1979, "Cognitive Repression in Contemporary Physics," *American Journal of Physics* 47 (8): 720.

252 **"not the Everett Wheeler interpretation":** Byrne 2010, p. 323.

252 **"heavy load of metaphysical baggage":** Ibid., p. 332.

252 **"was always implacably opposed to the theory":** Ibid., p. 322.

253 **"did not speak in terms of 'relative states' or any other euphemism":** Ibid., pp. 321–322.

253 **"*Where was it computed?*":** D. Deutsch 1985, "Quantum Theory, the Church-Turing Principle, and the Universal Quantum Computer," *Proceedings of the Royal Society of London* A 400:114. Emphasis in original. (Note that Penrose was the sponsor of the paper.)

253 **"the early universe when neither existed":** In Freire 2015, p. 322.

253 **"Bohr brainwashed a whole generation of theorists":** Douglas Huff and Omer Prewett, eds., 1979, *The Nature of the Physical Universe: 1976 Nobel Conference* (Wiley), p. 29.

254 **left his ashes out with the trash:** Byrne 2010, p. 347. Everett's family kept the ashes for a year after the cremation, but ultimately did leave them out with the trash, as Everett had specified.

255 **"quantum mechanics writ large across the sky":** National Aeronautics and Space Administration 2013, "Wilkinson Microwave Anisotropy Probe," https://map.gsfc.nasa.gov/, accessed July 24, 2017.

255 **no need to develop a theory to address such unobservable phenomena:** L. Rosenfeld 1963, "On Quantization of Fields," *Nuclear Physics* 40:353.

260 **"It is hard to imagine a more radical violation of Occam's razor":** Martin Gardner 2001, "Multiverses and Blackberries," *Skeptical Inquirer*, September/October, http://www.csicop.org/si/show/multiverses_and_blackberries, accessed July 24, 2017.

260 **"theory that we think is just rock-solid":** David Wallace, interview with the author, Santa Cruz, CA, USA, June 27, 2013.

261 **"deeper dialogue between scientists and philosophers":** George Ellis and Joe Silk 2014, "Defend the Integrity of Physics," *Nature* 516 (December 18): 321–323, doi:10.1038/516321a.

263 **it took a new theory to displace it, rather than an alleged "falsification":** See Thomas Levenson 2015, *The Hunt for Vulcan* (Random House), for a detailed and entertaining account of the story of Le Verrier, Einstein, and Vulcan.

264 **"neat, plausible, and wrong":** "The Divine Afflatus," *New York Evening Mail*, November 16, 1917. Also https://en.wikiquote.org/wiki/H._L._Mencken.

265 **"psychotic denial of this deep logical problem right at the center of this whole project!":** Albert 2013 (lecture at the UCSC Institute for the Philosophy of Cosmology), http://youtu.be/gjvNkPmaILA?t=1h28m40s.

Chapter 12

267 **discuss his idea of an *International Encyclopedia of Unified Science*:** Hans-Joachim Dahms 1996, "Vienna Circle and French Enlightenment: A Comparison of Diderot's *Encyclopédie* with Neurath's *International*

Encyclopedia of Unified Science," in *Encyclopedia and Utopia: The Life and Work of Otto Neurath (1882–1945)*, edited by E. Nemeth and Friedrich Stadler (Springer), p. 53.

268 **between La Palma and Tenerife in the Canary Islands:** Xiao-Song Ma et al. 2012, "Quantum Teleportation over 143 Kilometres Using Active Feed-Forward," *Nature* 489 (September 13): 269–273, doi:10.1038/nature11472.

268 **"The rest is mathematics":** Anton Zeilinger, interview with the author, Vienna, Austria, November 2, 2015.

268 **coax a buckyball . . . to interfere with itself:** Markus Arndt et al. 1999, "Wave-Particle Duality of C_{60} molecules," *Nature* 401 (October 14): 680–682, doi:10.1038/44348.

269 **"I don't think that you can even define it precisely":** Zeilinger 2015, interview.

269 **"no human reads the result?":** Steven Weinberg 2014, "Quantum Mechanics Without State Vectors," arXiv:1405.3483; Steven Weinberg 2013, *Lectures on Quantum Mechanics* (Cambridge University Press), p. 82.

270 **"asking questions is useful":** Gerard 't Hooft, interview with the author, Vienna, Austria, October 24, 2015.

270 **"not only may but *must* break down":** Jorrit de Boer, Erik Dal, and Ole Ulfbeck, eds., 1986, *The Lesson of Quantum Theory* (Elsevier), p. 53. Emphasis in original.

270 **preferred interpretation of quantum physics:** Max Tegmark 1997, "The Interpretation of Quantum Mechanics: Many Worlds or Many Words?," arXiv:quant-ph/9709032; Maximillian Schlosshauer et al. 2013, "A Snapshot of Foundational Attitudes Toward Quantum Mechanics," arXiv:1301.1069; Christoph Sommer 2013, "Another Survey of Foundational Attitudes Towards Quantum Mechanics," arXiv:1303.2719; Travis Norsen and Sarah Nelson 2013, "Yet Another Snapshot of Foundational Attitudes Toward Quantum Mechanics," arXiv:1306.4646; Sujeevan Sivasundaram and Kristian Hvidtfelt Nielsen 2016, "Surveying the Attitudes of Physicists Concerning Foundational Issues of Quantum Mechanics," arXiv:1612.00676.

270 **a massive sample bias in the results:** As Norsen and Nelson (2013) say, "The [surveys] reveal much more about the processes by which it was decided who should be invited to a given conference, than they reveal about trends in the thinking of the community as a whole." This kind of sample bias also explains the two surveys where Copenhagen did not win a plurality, as they were conducted at unusual conferences: one (Norsen and Nelson 2013) was a conference organized by Bohmians, and unsurprisingly pilot-wave theory came out on top; the other one (Sommer 2013) was a very small conference composed primarily of students, where there was no clear preference other than "undecided." The

one recent survey conducted on this subject outside the confines of a conference (Sivasundaram and Nielsen 2016, which is also the largest survey to date) showed the clearest results in favor of Copenhagen, with nearly 40 percent of respondents preferring Copenhagen, and no other interpretation garnering the support of more than 6 percent of respondents. But this survey has serious methodological problems as well—despite the fact that it was conducted by mail rather than at a conference, the sample of physicists who received the survey was still unrepresentative of the field as a whole. And the response rate was only 10 percent, with no correction for response bias by the survey designers. In short: if any sociologists of science are reading this, please conduct a proper survey of physicists on this subject! It's a slam dunk—you'd get a guaranteed journal publication, and you'd likely get good media coverage too.

270 **"should write a clear paper about quantum mechanics":** Zeilinger 2015, interview.

271 **"That's a different story":** Schweber 2016, interview.

271 **"they go back to talking about [measurements causing collapse]":** Emery, personal communication, January 10, 2017; phone interview with the author, May 5, 2017.

271 **"distinction between reality and our knowledge of reality":** Anton Zeilinger 2005, "The Message of the Quantum," *Nature* 438 (December 8): 743.

271 **"untestable and scientifically meaningless":** Emanuel Derman 2012, "2012: What Is Your Favorite Deep, Elegant, or Beautiful Explanation?," *Edge*, https://www.edge.org/responses/what-is-your-favorite-deep-elegant-or-beautiful-explanation, accessed July 28, 2017.

272 **standard defenses of Copenhagen doesn't work:** Much of the resurgence in empiricism since 1980 is due to the work of the philosopher Bas van Fraassen, who champions a position he calls "constructive empiricism." Unsurprisingly, van Fraassen is more sympathetic to the Copenhagen interpretation than most philosophers of physics. But, he admitted, "by today's standards [Copenhagen] is not an interpretation." He prefers the "transactional interpretation" of physicist Carlo Rovelli, which is an attempt to update the Copenhagen interpretation with an antirealist spirit (van Fraassen, interview with the author).

272 **philosophers usually take physics very seriously indeed:** As in Chapter 8, I'm talking about analytic philosophers here; Continental philosophers are a different breed altogether. But most philosophers of physics are analytic, anyhow: many Continental philosophers deal with philosophy of science, but there are few who specialize in philosophy of *physics*.

272 **"Philosophy is dead":** Matt Warman 2011, "Stephen Hawking Tells

Google 'Philosophy Is Dead,'" *Telegraph*, May 17, http://www.telegraph.co.uk/technology/google/8520033/Stephen-Hawking-tells-Google-philosophy-is-dead.html, accessed July 28, 2017.

272 **"there is a lot of brainpower there":** Massimo Pigliucci 2014, "Neil deGrasse Tyson and the Value of Philosophy," *Scientia Salon*, May 12, https://scientiasalon.wordpress.com/2014/05/12/neil-degrasse-tyson-and-the-value-of-philosophy/, accessed July 28, 2017.

272 **"hard to understand what justifies it":** Ross Andersen 2012, "Has Physics Made Philosophy and Religion Obsolete?," *Atlantic*, April 23, https://www.theatlantic.com/technology/archive/2012/04/has-physics-made-philosophy-and-religion-obsolete/256203/, accessed July 28, 2017.

273 **"no understanding of logical and philosophical arguments":** Isaacson 2007, p. 514.

275 **"'Shut up and calculate!'":** Mermin 1990, p. 199.

275 **he was the source of the phrase:** N. David Mermin 2004b, "Could Feynman Have Said This?," *Physics Today* 57 (5): 10–11, doi:http://dx.doi.org/10.1063/1.1768652.

275 **"calculate measurable quantities with unprecedented precision":** Mermin 1990, p. 200.

277 **"a real trauma, in the psychological sense of the word, for physics":** Albert, interview with the author, February 4, 2015.

278 **"but that's kind of uncomfortable":** Fraser, interview with the author, May 24, 2017.

282 **"an idea that has gone wrong":** Deepak Chopra 1995, "Interviews with People Who Make a Difference: Quantum Healing," by Daniel Redwood, Healthy.net, http://www.healthy.net/scr/interview.aspx?Id=167, accessed September 20, 2017.

283 **"not *weird*, it's gibberish, it's unintelligible":** Albert, interview with the author, February 4, 2015. Emphasis in original.

284 **affects the new ideas and new theories we develop:** "Feynman: Knowing Versus Understanding," YouTube, posted by TehPhysicalist, May 17, 2012, https://www.youtube.com/watch?v=NM-zWTU7X-k. This is Feynman at Cornell in 1964, giving the Messenger Lectures. These lectures were later turned into a book, *The Character of Physical Law*.

285 **bridging far-distant points with wormholes:** For example, Chunjun Cao, Sean M. Carroll, and Spyridon Michalakis 2016, "Space from Hilbert Space: Recovering Geometry from Bulk Entanglement," https://arxiv.org/abs/1606.08444, and many other papers by van Raamsdonk, Susskind, and Maldacena.

285 **same fundamental truth about the cosmos:** For example, Laura Mersini-Houghton 2008, "Thoughts on Defining the Multiverse," https://arxiv.org/abs/0804.4280.

285 **doesn't break relativity:** For example, Elizabeth S. Gould and Niyaesh Afshordi 2014, "A Non-local Reality: Is There a Phase Uncertainty in Quantum Mechanics?," https://arxiv.org/abs/1407.4083.

286 **new and strange foundational problems of its own:** Note for physicists and other specialists: The argument that the commutation relations in QFT ensure locality holds no water at all, because those relations don't apply to measurement processes. When a measurement occurs in QFT, collapse occurs, and just as in standard nonrelativistic quantum mechanics, that collapse must occur instantaneously across all space in order to account for the violation of Bell's inequality. So "measurement" is still a problem, and nonlocality is still there (unless you take advantage of a loophole, as the many-worlds interpretation arguably does).

As for the special interpretational problems that QFT has, see Laura Ruetsche 2011, *Interpreting Quantum Theories* (Oxford University Press); and Paul Teller 1995, *An Interpretive Introduction to Quantum Field Theory* (Princeton University Press). In particular, Haag's theorem seems to pose a problem for the theory.

286 **ongoing challenge:** These theories are difficult to extend to QFT in part because of the troubling foundational problems that QFT has regarding its own consistency (see previous note).

288 **"a mere artisan or specialist and a real seeker after truth":** Albert Einstein, letter to Robert Thornton, December 7, 1944, EA 61-574, https://plato.stanford.edu/entries/einstein-philscience/.

Appendix

289 **"very machinery of the universe":** Wheeler and Ford 1998, p. 334.

289 **don't behave the same way in both directions:** The beam splitter has at its heart a half-silvered mirror, which lets through half the light that hits it and bounces the other half off. And when half of the beam coming in from below bounced off to the right, the beam splitter gave it a twist (a 180° phase shift), sending it out of sync with the half-beam that passed through from the left.

291 **"we make the whole idea of following a single path meaningless":** Wheeler and Ford 1998, p. 336.

292 **"assigned by the measurements we make—now!":** Ibid., p. 337. Emphasis in original.

292 **"uncertainty collapses to certainty":** Ibid., p. 338.

292 **the "essence" of quantum physics, "as the delayed-choice experiment shows, is *measurement*":** Ibid., p. 339. Emphasis in original.

293 **the world does not branch:** In most versions of many-worlds, there are no particles as such; the same is true for most versions of spontaneous-collapse theories. So my use of the word "photon" is a bit of a fudge, but you can just read it as "wave packet" if you're really persnickety.

References

Interviews Conducted by the Author

Aharonov, Yakir. Vienna, Austria, October 24, 2015.
Albert, David. New York, NY, USA, February 4, 2015.
Albert, David. Telephone interview, May 17, 2017.
Aspect, Alain. Palaiseau, France, November 4, 2015.
Bell, Mary. Geneva, Switzerland, October 19 and 20, 2015.
Bertlmann, Reinhold. Vienna, Austria, November 2, 2015.
Bub, Jeffrey. Telephone interview, February 2 and 7, 2017.
Carroll, Sean. Malibu, CA, USA, November 14, 2015.
Clauser, John. Walnut Creek, CA, USA, August 12, 2015.
Emery, Nina. Telephone interview, May 5, 2017.
Esfeld, Michael. Geneva, Switzerland, October 21, 2015.
Fraser, Doreen. Waterloo, ON, Canada, May 24, 2017.
Gisin, Nicholas. Vienna, Austria, October 24, 2015.
Goldstein, Sheldon, and Nino Zanghì. New Brunswick, NJ, USA, February 3, 2015.
Grangier, Phillip. Palaiseau, France, November 4, 2015.
Hardy, Lucien. Waterloo, ON, Canada, May 23, 2017.
Hiley, Basil. London, UK, October 29, 2015.
Kaiser, David. Cambridge, MA, USA, January 19, 2016.
Leggett, Anthony. Telephone interview, May 4, 2017.
Leifer, Matthew. Vienna, Austria, October 24, 2015.
't Hooft, Gerard. Vienna, Austria, October 24, 2015.
Maudlin, Tim. New York, NY, USA, January 28, 2015.
Mermin, N. David. Ithaca, NY, USA, January 11 and 12, 2016.
Myrvold, Wayne. London, ON, Canada, May 24, 2017.
Nauenberg, Michael. Santa Cruz, CA, USA, August 6, 2015.
Ney, Alyssa. Davis, CA, USA, May 8, 2017.
Penrose, Roger. London, UK, October 27, 2015.

Rudolph, Terence. London, UK, October 29, 2015.
Saunders, Simon. Oxford, UK, October 26, 2015.
Schweber, Silvan Samuel. Telephone interview, September 7, 2016.
Sebens, Charles. Telephone interview, May 3, 2017.
Smolin, Lee. Toronto, ON, Canada, May 22, 2017.
Spekkens, Robert. Waterloo, ON, Canada, May 23, 2017.
Steinberg, Aephraim. Toronto, ON, Canada, May 25, 2017.
Vaidman, Lev. Vienna, Austria, October 24, 2015.
van Fraassen, Bas. Pinole, CA, USA, May 20, 2017.
Wallace, David. Santa Cruz, CA, USA, June 27, 2013.
Wallace, David. Oxford, UK, October 26, 2015.
Wiseman, Howard. Vienna, Austria, October 24, 2015.
Wüthrich. Christian. Saig, Germany, July 20, 2015.
Zeh, H. Dieter. Neckargemünd, Germany, October 23, 2015.
Zeilinger, Anton. Vienna, Austria, November 2, 2015.

Other Interviews

Bohm, David. Interview by Lillian Hoddeson, May 8, 1981. Edgware, London, England. Courtesy of the Niels Bohr Library & Archives, American Institute of Physics, College Park, MD, USA. https://www.aip.org/history-programs/niels-bohr-library/oral-histories/4513.

Bohm, David. Interview by Martin J. Sherwin, June 15, 1979, New York, NY, USA. Atomic Heritage Foundation, "Voices of the Manhattan Project." http://manhattanprojectvoices.org/oral-histories/david-bohms-interview. Accessed August 28, 2016.

Bohm, David. Interview by Maurice Wilkins, July 7, 1986. Courtesy of the Niels Bohr Library & Archives, American Institute of Physics, College Park, MD, USA. http://www.aip.org/history-programs/niels-bohr-library/oral-histories/32977-3. Accessed August 28, 2016.

Bohr, Niels. Interview by Thomas S. Kuhn, Aage Petersen, and Erik Rudinger, November 17, 1962, Copenhagen, Denmark. Courtesy of the Niels Bohr Library & Archives, American Institute of Physics, College Park, MD, USA. http://www.aip.org/history-programs/niels-bohr-library/oral-histories/4517-5. Accessed January 27, 2017.

Clauser, John. Interview by Joan Bromberg, May 20, 21, and 23, 2002, Walnut Creek, CA, USA. Courtesy of the Niels Bohr Library & Archives, American Institute of Physics, College Park, MD, USA. http://www.aip.org/history-programs/niels-bohr-library/oral-histories/25096. Accessed March 6, 2017.

DeWitt, Bryce, and Cecile DeWitt-Morette. Interview by Kenneth W. Ford, February 28, 1995, Austin, TX, USA. Courtesy of the Niels Bohr Library & Archives, American Institute of Physics, College Park, MD, USA. http://www.aip.org/history-programs/niels-bohr-library/oral-histories/23199. Accessed October 26, 2016.

Dirac, Paul. Interview by Thomas S. Kuhn, May 14, 1963. Cambridge, England. Courtesy of the Niels Bohr Library & Archives, American Institute of Physics, College Park, MD, USA. https://www.aip.org/history-programs/niels-bohr-library/oral-histories/4575-5. Part 5.

Hiley, Basil. Interview by Olival Freire, January 11, 2008, Birkbeck College, London, England. Courtesy of the Niels Bohr Library & Archives, American Institute of Physics, College Park, MD, USA. https://www.aip.org/history-programs/niels-bohr-library/oral-histories/33822. Accessed, July 14, 2017.

Shimony, Abner. Interview by Joan Bromberg, September 9 and 10, 2002, Wellesley, MA, USA. Courtesy of the Niels Bohr Library & Archives, American Institute of Physics, College Park, MD USA. http://www.aip.org/history-programs/niels-bohr-library/oral-histories/25643. Accessed March 6, 2017.

Stern, Otto. Interview by Thomas S. Kuhn, May 29 and 30, 1962, Berkeley, CA, USA. Courtesy of the Niels Bohr Library & Archives, American Institute of Physics, College Park, MD, USA. http://www.aip.org/history-programs/niels-bohr-library/oral-histories/4904. Accessed October 26, 2016.

Whitman, Marina [von Neumann's daughter]. Interview by Gray Watson. January 30, 2011. https://web.archive.org/web/20110428125353/http://256.com/gray/docs/misc/conversation_with_marina_whitman.shtml.

Wigner, Eugene. Interview by Charles Weiner and Jagdish Mehra, November 30, 1966, Princeton, NJ, USA. Courtesy of the Niels Bohr Library & Archives, American Institute of Physics, College Park, MD, USA. http://www.aip.org/history-programs/niels-bohr-library/oral-histories/4964. Accessed April 6, 2016.

Bibliography

Abers, Ernest S. 2004. *Quantum Mechanics*. Pearson.

Albert, David. 2013. Lecture at the UCSC Institute for the Philosophy of Cosmology. http://youtu.be/gjvNkPmaILA? t=1h28m40s.

Anderson, P. W. 2001. "Science: A 'Dappled World' or a 'Seamless Web'?" *Studies in History and Philosophy of Modern Physics* 32:487–494.

Andersen, Ross. 2012. "Has Physics Made Philosophy and Religion Obsolete?" *Atlantic*, April 23. https://www.theatlantic.com/technology/archive/2012/04/has-physics-made-philosophy-and-religion-obsolete/256203/. Accessed July 28, 2017.

Arndt, Markus, et al. 1999. "Wave-Particle Duality of C_{60} molecules." *Nature* 401 (October 14): 680–682. doi:10.1038/44348.

Ayer, A. J. 1982. *Philosophy in the Twentieth Century*. Vintage.

Bacciagaluppi, Guido, and Antony Valentini. 2009. *Quantum Theory at the Crossroads: Reconsidering the 1927 Solvay Conference*. arXiv:quant-ph/0609184v2.

Ball, Philip. 2013. *Serving the Reich: The Struggle for the Soul of Physics Under Hitler.* Vintage.

Ballentine, Leslie E., et al. 1971. "Quantum-Mechanics Debate." *Physics Today* 24 (4). doi:10.1063/1.3022676.

Barnett, Lincoln. 1949. *The Universe and Dr. Einstein.* Victor Gollancz.

Barrett, Jeffrey Alan, and Peter Byrne, eds. 2012. *The Everett Interpretation of Quantum Mechanics: Collected Works 1955–1980 with Commentary.* Princeton University Press.

Bassi, Angelo, et al. 2013. "Models of Wave-Function Collapse, Underlying Theories, and Experimental Tests." *Reviews of Modern Physics* 85 (2). doi:10.1103/RevModPhys.85.471.

Bell, John S. 1964. "On the Einstein-Podolsky-Rosen Paradox." *Physics* 1:195–200. Reprinted in Bell 2004.

———. 1966. "On the Problem of Hidden Variables in Quantum Mechanics." *Reviews of Modern Physics* 38:447–452. Reprinted in Bell 2004.

———. 1980. "Bertlmann's Socks and the Nature of Reality." CERN Preprint CERN-TH-2926. https://cds.cern.ch/record/142461?ln=en.

———. 1981. "Bertlmann's Socks and the Nature of Reality." *Journal de Physique*, Seminar C2, suppl., 42 (3): C2 41–61. Reprinted in Bell 2004.

———. 1990. "Indeterminism and Non Locality." Talk given in Geneva, January 22, 1990. https://cds.cern.ch/record/1049544?ln=en, accessed July 21, 2017. Transcript: http://www.quantumphil.org./Bell-indeterminism-and-nonlocality.pdf.

———. 2004. *Speakable and Unspeakable in Quantum Mechanics.* 2nd ed. Cambridge University Press.

Bell, John, Antoine Suarez, Herwig Schopper, J. M. Belloc, G. Cantale, John Layter, P. Veija, and P. Ypes. 1990. "Indeterminism and Non Locality." Talk given at Center of Quantum Philosophy of Geneva, January 22. http://cds.cern.ch/record/1049544?ln=en. Transcript, http://www.quantumphil.org./Bell-indeterminism-and-nonlocality.pdf.

Beller, Mara. 1999a. "Jocular Commemorations: The Copenhagen Spirit." *Osiris* 14:252–273.

———. 1999b. *Quantum Dialogue: The Making of a Revolution.* University of Chicago Press.

Bernstein, Jeremy. 1991. *Quantum Profiles.* Princeton University Press.

———. 2001. *Hitler's Uranium Club: The Secret Recordings at Farm Hall.* 2nd ed. Copernicus.

Bertlmann, R. A., and A. Zeilinger, eds. 2002. *Quantum [Un]speakables: From Bell to Quantum Information.* Springer.

Bird, Kai, and Martin J. Sherwin. 2005. *American Prometheus: The Triumph and Tragedy of J. Robert Oppenheimer.* Vintage.

Blackmore, John T. 1972. *Ernst Mach; His Work, Life, and Influence.* University of California Press.

Bohm, David. 1957. *Causality and Chance in Modern Physics*. Harper Torchbooks ed. Harper and Row.

Bohr, Niels. 1934. *Atomic Theory and the Description of Nature*. Cambridge University Press.

———. 1949. "Discussion with Einstein on Epistemological Problems in Atomic Physics." In Schilpp 1949, 201–241.

———. 2013. *Collected Works*. Vol. 7, *Foundations of Quantum Physics II (1933–1958)*. Edited by J. Kalckar. Elsevier.

Born, Max. 1978. *My Life: Recollections of a Nobel Laureate*. Scribner's Sons.

———. 2005. *The Born-Einstein Letters: Friendship, Politics and Physics in Uncertain Times*. Macmillan.

Bricmont, Jean. 2016. *Making Sense of Quantum Mechanics*. Springer International.

Bridgman, Percy W. 1927. *The Logic of Modern Physics*. Macmillan.

Bub, Jeffrey. 1999. *Interpreting the Quantum World*. Rev. ed. Cambridge University Press.

Byrne, Peter. 2010. *The Many Worlds of Hugh Everett III: Multiple Universes, Mutual Assured Destruction, and the Meltdown of a Nuclear Family*. Oxford University Press.

Camilleri, Kristian. 2009. "A History of Entanglement: Decoherence and the Interpretation Problem." *Studies in History and Philosophy of Modern Physics* 40:290–302.

Cao, Chunjun, Sean M. Carroll, and Spyridon Michalakis. 2016. "Space from Hilbert Space: Recovering Geometry from Bulk Entanglement." https://arxiv.org/abs/1606.08444.

Cassidy, David. 1991. *Uncertainty: The Life and Science of Werner Heisenberg*. W. H. Freeman.

———. 2009. *Beyond Uncertainty: Heisenberg, Quantum Physics, and the Bomb*. Bellevue Literary Press.

Chopra, Deepak. 1995. "Interviews with People Who Make a Difference: Quantum Healing," by Daniel Redwood. Healthy.net. http://www.healthy.net/scr/interview.aspx?Id=167. Accessed September 20, 2017.

Clauser, John F. 1969. "Proposed Experiment to Test Local Hidden-Variable Theories." *Bulletin of the American Physical Society* 14:578.

———. 2002. "Early History of Bell's Theorem." In Bertlmann and Zeilinger 2002, 61–98.

Clauser, John F., Michael A. Horne, Abner Shimony, and Richard A. Holt. 1969. "Proposed Experiment to Test Local Hidden-Variable Theories." *Physical Review Letters* 23:880–884. doi:10.1103/PhysRevLett.23.880.

Cushing, James. 1994. *Quantum Mechanics: Historical Contingency and the Copenhagen Hegemony*. University of Chicago Press.

Dahms, Hans-Joachim. 1996. "Vienna Circle and French Enlightenment: A Comparison of Diderot's *Encyclopédie* with Neurath's *International*

Encyclopedia of Unified Science." In *Encyclopedia and Utopia: The Life and Work of Otto Neurath (1882–1945)*, edited by E. Nemeth and Friedrich Stadler, 53–61. Springer.

de Boer, Jorrit, Erik Dal, and Ole Ulfbeck, eds. 1986. *The Lesson of Quantum Theory*. Elsevier.

Derman, Emanuel. 2012. "2012: What Is Your Favorite Deep, Elegant, or Beautiful Explanation?" *Edge*. https://www.edge.org/responses/what-is-your-favorite-deep-elegant-or-beautiful-explanation. Accessed July 28, 2017.

Deutsch, D. 1985. "Quantum Theory, the Church-Turing Principle, and the Universal Quantum Computer." *Proceedings of the Royal Society of London* A 400:97–117.

DeWitt, Bryce S. 1970. "Quantum Mechanics and Reality." *Physics Today* 23 (9): 30–35. doi:10.1063/1.3022331.

DeWitt-Morette, Cécile. 2011. *The Pursuit of Quantum Gravity: Memoirs of Bryce DeWitt from 1946 to 2004*. Springer.

Discussion Sections at Symposium on the Foundations of Modern Physics: The Copenhagen Interpretation 60 Years after the Como Lecture. 1987.

Dresden, Max. 1991. "Letters: Heisenberg, Goudsmit and the German 'A-Bomb.'" *Physics Today* 44 (5): 92–94. doi:10.1063/1.2810103.

Einstein, Albert. 1949a. "Autobiographical Notes." In Schilpp 1949, 2–94.

———. 1949b. "Reply to Criticisms." In Schilpp 1949, 665–688.

———. 1953. "Elementary Considerations on the Interpretation of the Foundations of Quantum Mechanics." Translated by Dileep Karanth. http://arxiv.org/abs/1107.3701.

Ellis, George, and Joe Silk. 2014. "Defend the Integrity of Physics." *Nature* 516 (December 18): 321–323. doi:10.1038/516321a.

Faye, Jan. 2007. "Niels Bohr and the Vienna Circle." Preprint. http://philsci-archive.pitt.edu/3737/. Accessed December 23, 2016.

Feldmann, William, and Roderich Tumulka. 2012. "Parameter Diagrams of the GRW and CSL Theories of Wavefunction Collapse." *Journal of Physics A: Mathematical and Theoretical* 45 (2012) 065304 (13pp.). doi:10.1088/1751-8113/45/6/065304.

Fermi, Laura. 1954. *Atoms in the Family: My Life with Enrico Fermi*. University of Chicago Press.

Feynman, Richard P. 1982. "Simulating Physics with Computers." *International Journal of Theoretical Physics* 21 (6/7): 467–488.

———. 2005. *Perfectly Reasonable Deviations from the Beaten Path*. Edited by Michelle Feynman. Basic Books.

"Feynman: Knowing Versus Understanding." YouTube. Posted by Teh Physicalist, May 17, 2012. https://www.youtube.com/atch?v=NM-zWTU7X-k.

———. 2015. *The Quantum Dissidents: Rebuilding the Foundations of Quantum Mechanics*. Springer-Verlag.

Feynman, Richard, Robert B. Leighton, and Matthew L. Sands. 1963. *The Feynman Lectures on Physics*. Vol. 1. Basic Books.

Fine, Arthur. 1996. *The Shaky Game*. 2nd ed. University of Chicago Press.

Forman, Paul. 1971. "Weimar Culture, Causality, and Quantum Theory: Adaptation by German Physicists and Mathematicians to a Hostile Environment." *Historical Studies in the Physical Sciences* 3:1–115.

———. 1987. "Behind Quantum Electronics: National Security as Basis for Physical Research in the United States, 1940–1960." *Historical Studies in the Physical and Biological Sciences* 18 (1): 149–229.

Freedman, Stuart J., and John F. Clauser. 1972. "Experimental Test of Local Hidden-Variable Theories." *Physical Review Letters* 28:938–941. doi:10.1103/PhysRevLett.28.938.

Freire, Olival, Jr. 2009. "Quantum Dissidents: Research on the Foundations of Quantum Theory Circa 1970." *Studies in History and Philosophy of Modern Physics* 40:280–289. doi:10.1016/j.shpsb.2009.09.002.

French, A. P., and P. J. Kennedy, eds. 1985. *Niels Bohr: A Centenary Volume*. Harvard University Press.

Galison, Peter. 1990. "Aufbau/Bauhaus: Logical Positivism and Architectural Modernism." *Critical Inquiry* 16:709–752.

Gamow, George. 1988. *The Great Physicists from Galileo to Einstein*. Dover.

Gardner, Martin. 2001. "Multiverses and Blackberries." *Skeptical Inquirer*, September/October 2001. http://www.csicop.org/si/show/multiverses_and_blackberries. Accessed July 24, 2017.

Ghirardi, G. C., A. Rimini, and T. Weber. 1986. "Unified Dynamics for Microscopic and Macroscopic Systems." *Physical Review D* 34:470.

Gisin, Nicholas. 2002. "Sundays in a Quantum Engineer's Life." In Bertlmann and Zeilinger 2002, 199–207.

Godfrey-Smith, Peter. 2003. *Theory and Reality: An Introduction to the Philosophy of Science*. University of Chicago Press.

Gottfried, Kurt, and N. David Mermin. 1991. "John Bell and the Moral Aspect of Quantum Mechanics." *Europhysics News* 22 (4): 67–69.

Goudsmit, Samuel. 1947. *Alsos*. AIP Press.

———. 1973. "Important Announcement Regarding Papers About Fundamental Theories." *Physical Review D* 8:357.

Gould, Elizabeth S., and Niyaesh Afshordi. 2014. "A Non-local Reality: Is There a Phase Uncertainty in Quantum Mechanics?" https://arxiv.org/abs/1407.4083.

Griffiths, David J. 2005. *Introduction to Quantum Mechanics*. 2nd ed. Pearson Education.

Hahn, Hans, Rudolf Carnap, and Otto Neurath. 1973. "The Scientific Conception of the World: The Vienna Circle." In Neurath 1973, 299–318.

Hawking, Stephen. 1988. *A Brief History of Time*. Bantam Dell.

———. 1999. "Does God Play Dice?" http://www.hawking.org.uk/does-god-play-dice.html. Accessed March 18, 2016.

Hearings Before the Committee on Un-American Activities, House of Representatives. 1949. Eighty-First Congress, First Session (March 31 and April 1). Statement of David Bohm.

Heidegger, Martin. 1996. *Being and Time: A Translation of "Sein und Zeit."* Translated by Joan Stambaugh. State University of New York Press.

———. 1999. *Contributions to Philosophy from Enowning*. Translated by Parvis Emad and Kenneth Maly. Indiana University Press.

Heilbron, John L. 1985. "The Earliest Missionaries of the Copenhagen Spirit." *Revue d'histoire des sciences* 38 (3–4): 195–230. doi:10.3406/rhs.1985.4005.

Heisenberg, Werner. 1958. *Physics and Philosophy*. Harper Torchbooks, ed. Harper and Row.

———. 1971. *Physics and Beyond*. HarperCollins.

Holton, Gerald. 1988. *Thematic Origins of Scientific Thought*. Rev. ed. Harvard University Press.

———. 1998. *The Advancement of Science, and Its Burdens*. Harvard University Press.

Howard, Don. 1985. "Einstein on Locality and Separability." *Studies in History and Philosophy of Science* 16:171–201.

———. 1990. "'Nicht sein kann was nicht sein darf,' or the Prehistory of EPR, 1909–1935: Einstein's Early Worries About the Quantum Mechanics of Composite Systems." In *Sixty-Two Years of Uncertainty: Historical, Philosophical, and Physical Inquiries into the Foundations of Quantum Mechanics*, edited by Arthur I. Miller, 61–111. Plenum Press.

———. 2004. "Who Invented the 'Copenhagen Interpretation'? A Study in Mythology." *Philosophy of Science* 71 (5): 669–682.

———. 2007. "Revisiting the Einstein-Bohr Dialogue." *Iyyun: The Jerusalem Philosophical Quarterly* 56:57–90.

———. 2015. "Einstein's Philosophy of Science." In *The Stanford Encyclopedia of Philosophy*, Winter 2015 ed., edited by Edward N. Zalta. http://plato.stanford.edu/archives/win2015/entries/einstein-philscience/.

Huff, Douglas, and Omer Prewett, eds. 1979. *The Nature of the Physical Universe: 1976 Nobel Conference*. Wiley.

Incandenza, James O. 1997. *Kinds of Light*. Meniscus Films.

Interagency Working Group on Quantum Information Science of the Subcommittee on Physical Sciences. 2016. *Advancing Quantum Information Science: National Challenges and Opportunities*. Joint report of the Committee on Science and Committee on Homeland and National Security of the National Science and Technology Council. July. https://www.whitehouse.gov/sites/whitehouse.gov/files/images/Quantum_Info_Sci_Report_2016_07_22%20final.pdf. Accessed July 14, 2017.

Isaacson, Walter. 2007. *Einstein: His Life and Universe*. Simon and Schuster.

Jaki, Stanley L. 1978. "Johann Georg von Soldner and the Gravitational Bending of Light, with an English Translation of His Essay on It Published in 1801." *Foundations of Physics* 8 (11/12): 927–950.

Jammer, Max. 1974. *The Philosophy of Quantum Mechanics*. John Wiley & Sons.

———. 1989. *The Conceptual Development of Quantum Mechanics*. 2nd ed. Tomash.

Kaiser, David. 2002. "Cold War Requisitions, Scientific Manpower, and the Production of American Physicists After World War II." *Historical Studies in the Physical and Biological Sciences* 33 (1): 131–159.

———. 2004. "The Postwar Suburbanization of American Physics." *American Quarterly* 56 (4): 851–888.

———. 2007. "Turning Physicists into Quantum Mechanics." *Physics World*, May, 28–33.

———. 2011. *How the Hippies Saved Physics: Science, Counterculture, and the Quantum Revival*. W. W. Norton.

———. 2012. "Booms, Busts, and the World of Ideas: Enrollment Pressures and the Challenge of Specialization." *Osiris* 27 (1): 276–302.

———. 2014. "History: Shut Up and Calculate!" *Nature* 505 (January 9): 153–155. doi:10.1038/505153a.

Keller, Evelyn Fox. 1979. "Cognitive Repression in Contemporary Physics." *American Journal of Physics* 47 (8): 718–721.

Kennefick, Daniel. 2005. "Einstein Versus the Physical Review." *Physics Today* 58 (9): 43–48. doi:10.1063/1.2117822.

Kuhn, Thomas S. 1996. *The Structure of Scientific Revolutions*. 3rd ed. University of Chicago Press.

———. 2000. *The Road Since Structure*. Edited by James Conant and John Haugeland. University of Chicago Press.

Kumar, Manjit. 2008. *Quantum: Einstein, Bohr, and the Great Debate About the Nature of Reality*. Icon Books.

Lang, Daniel. 1953. "A Farewell to String and Sealing Wax." Reprinted in *From Hiroshima to the Moon: Chronicles of Life in the Atomic Age*, by Daniel Lang, 215–246. Simon and Schuster, 1959.

———. 1959. *From Hiroshima to the Moon: Chronicles of Life in the Atomic Age*. Simon and Schuster.

Levenson, Thomas. 2015. *The Hunt for Vulcan*. Random House.

Lindley, David 2001. *Boltzmann's Atom*. Free Press.

———. 2007. *Uncertainty: Einstein, Heisenberg, Bohr, and the Struggle for the Soul of Science*. Anchor.

Ma, Xiao-Song, et al. 2012. "Quantum Teleportation over 143 Kilometres Using Active Feed-Forward." *Nature* 489 (September 13): 269–273. doi:10.1038/nature11472.

Mann, Charles, and Robert Crease. 1988. "Interview: John Bell." *OMNI*, May, 85–92, 121.

Marcum, James A. 2015. *Thomas Kuhn's Revolutions*. Bloomsbury.

Margenau, Henry. 1950. *The Nature of Physical Reality: A Philosophy of Modern Physics*. McGraw-Hill.

———. 1954. "Advantages and Disadvantages of Various Interpretations of the Quantum Theory." *Physics Today* 7 (10): 6–13. doi:10.1063/1.3061432.

———. 1958. "Philosophical Problems Concerning the Meaning of Measurement in Physics." *Philosophy of Science* 25 (1): 23–33. doi:10.1086/287574.

Maudlin, Tim. 2002. *Quantum Non-locality and Relativity*. 2nd ed. Blackwell.

Maxwell, Grover. 1962. "The Ontological Status of Theoretical Entities." *Minnesota Studies in the Philosophy of Science* 3:3–27.

Mencken, H. L. 1917. "The Divine Afflatus." *New York Evening Mail*, November 16.

Mermin, N. David. 1985. "Is the Moon There When Nobody Looks? Reality and the Quantum Theory." *Physics Today* 38 (4): 38–47.

———. 1990. *Boojums All the Way Through: Communicating Science in a Prosaic Age*. Cambridge University Press.

———. 1993. "Hidden Variables and the Two Theorems of John Bell." *Reviews of Modern Physics* 65 (3): 803–815.

———. 2004a. "What's Wrong with This Quantum World?" *Physics Today*, February, 10–11.

———. 2004b. "Could Feynman Have Said This?" *Physics Today* 57 (5): 10–11. doi:http://dx.doi.org/10.1063/1.1768652.

Mersini-Houghton, Laura. 2008. "Thoughts on Defining the Multiverse." https://arxiv.org/abs/0804.4280.

Miller, Arthur I. 2012. *Insights of Genius: Imagery and Creativity in Science and Art*. Springer.

Misner, Charles W. 2015. "A One-World Formulation of Quantum Mechanics." *Physica Scripta* 90 (088014), 6pp.

Misner, Charles W., Kip S. Thorne, and Wojciech H. Zurek. 2009. "John Wheeler, Relativity, and Quantum Information." *Physics Today*, April, 40–46.

National Aeronautics and Space Administration. 2013. "Wilkinson Microwave Anisotropy Probe." https://map.gsfc.nasa.gov/. Accessed July 24, 2017.

Neurath, Otto. 1973. *Empiricism and Sociology*. Reidel.

New York Times. 1935. "Einstein Attacks Quantum Theory." Science Service, May 4, p. 11.

New York Times. 1935. "Statement by Einstein," May 7, p. 21.

Nielsen, Michael A., and Isaac L. Chuang. 2000. *Quantum Computation and Quantum Information*. Cambridge University Press.

Norsen, Travis. 2007. "Against 'Realism.'" *Foundations of Physics* 37 (3): 311–340. doi:10.1007/s10701-007-9104-1.

Norsen, Travis, and Sarah Nelson. 2013. "Yet Another Snapshot of Foundational Attitudes Toward Quantum Mechanics." arXiv:1306.4646.

Norton, John D. 2015. "Relativistic Cosmology." http://www.pitt.edu/~jdnorton/teaching/HPS_0410/chapters_2017_Jan_1/relativistic_cosmology/index.html. Accessed July 24, 2017.

O'Connor, J. J., and E. F. Robertson. 2003. "Erwin Rudolf Josef Alexander Schrödinger." http://www-groups.dcs.st-and.ac.uk/~history/Biographies/Schrodinger.html. Accessed September 25, 2017.

Olwell, Russell. 1999. "Physical Isolation and Marginalization in Physics: David Bohm's Cold War Exile." *Isis* 90 (4): 738–756.

Ouellette, Jennifer. 2005. "Quantum Key Distribution." *Industrial Physicist*, January/February, 22–25. https://people.cs.vt.edu/~kafura/cs6204/Readings/QuantumX/QuantumKeyDistribution.pdf. Accessed July 14, 2017.

Pais, Abraham. 1991. *Niels Bohr's Times in Physics, Philosophy, and Polity*. Oxford University Press.

Pauli, Wolfgang. 1921. *Theory of Relativity*. Translated by G. Field. Dover.

———. 1994. *Writings on Physics and Philosophy*. Edited by Charles P. Enz and Karl von Meyenn. Translated by Robert Schlapp. Springer-Verlag.

Pearle, Philip. 2009. "How Stands Collapse II." In *Quantum Reality, Relativistic Causality, and Closing the Epistemic Circle*, edited by W. C. Myrvold and J. Christian, 257–292. Springer.

Peat, F. David. 1997. *Infinite Potential: The Life and Times of David Bohm*. Addison Wesley Longman.

Pigliucci, Massimo. 2014. "Neil deGrasse Tyson and the Value of Philosophy." *Scientia Salon*, May 12. https://scientiasalon.wordpress.com/2014/05/12/neil-degrasse-tyson-and-the-value-of-philosophy/. Accessed July 28, 2017.

Powers, Thomas. 2001. "Heisenberg in Copenhagen: An Exchange." *New York Review of Books*, February 8, 2001.

Putnam, Hilary. 1965. "A Philosopher Looks at Quantum Mechanics." In Putnam 1979, 130–158.

———. 1979. *Mathematics, Matter, and Method*. 2nd ed. Cambridge University Press.

Quine, Willard Van Orman. 1953. *From a Logical Point of View*. Harper Torchbooks ed. Harper and Row.

———. 1976. *The Ways of Paradox*. Harvard University Press.

———. 2008. *Quine in Dialogue*. Edited by Dagfinn Føllesdal and Douglas B. Quine. Harvard University Press.

Reichenbach, Hans. 1944. *Philosophic Foundations of Quantum Mechanics*. Dover.

Reisch, George. 2005. *How the Cold War Transformed Philosophy of Science: To the Icy Slopes of Logic*. Cambridge University Press.

Rhodes, Richard. 1986. *The Making of the Atomic Bomb*. Simon and Schuster.

Rosenfeld, L. 1963. "On Quantization of Fields." *Nuclear Physics* 40:353.

Ruetsche, Laura. 2011. *Interpreting Quantum Theories*. Oxford University Press.

Sarkar, Sahotra, ed. 1996a. *Science and Philosophy in the Twentieth Century*. Vol. 1, *The Emergence of Logical Positivism*. Garland.

———, ed. 1996b. *Science and Philosophy in the Twentieth Century*. Vol. 5, *Decline and Obsolescence of Logical Positivism*. Garland.

Schiff, Leonard I. 1955. *Quantum Mechanics*. 2nd ed. McGraw-Hill.

Schilpp, Paul Arthur, ed. 1949. *Albert Einstein: Philosopher-Scientist*. MJF Books.

Schlosshauer, Maximilian, ed. 2011. *Elegance and Enigma: The Quantum Interviews*. Springer.

Schlosshauer, Maximillian, et al. 2013. "A Snapshot of Foundational Attitudes Toward Quantum Mechanics." arXiv:1301.1069.

Seevinck, M. P. 2012. "Challenging the Gospel: Grete Hermann on von Neumann's No-Hidden-Variables Proof." Radboud University, Nijmegen, the Netherlands. http://mpseevinck.ruhosting.nl/seevinck/Aberdeen_Grete_Hermann2.pdf. Accessed September 20, 2017.

Shimony, Abner. 1963. "Role of the Observer in Quantum Theory." *American Journal of Physics* 31:755–773. doi:10.1119/1.1969073.

Sigurdsson, Skúli. 1990. "The Nature of Scientific Knowledge: An Interview with Thomas S. Kuhn." *Harvard Science Review*, Winter, 18–25. http://www.edition-open-access.de/proceedings/8/3/index.html.

Sivasundaram, Sujeevan, and Kristian Hvidtfelt Nielsen. 2016. "Surveying the Attitudes of Physicists Concerning Foundational Issues of Quantum Mechanics." arXiv:1612.00676.

Smart, J. J. C. 1963. *Philosophy and Scientific Realism*. Routledge and Kegan Paul.

Smyth, Henry D. 1951. "The Stockpiling and Rationing of Scientific Manpower." *Physics Today* 4 (2): 18. doi:10.1063/1.3067145.

Sommer, Christoph. 2013. "Another Survey of Foundational Attitudes Towards Quantum Mechanics." arXiv:1303.2719.

Stadler, Friedrich. 2001. "Documentation: The Murder of Moritz Schlick." In *The Vienna Circle: Studies in the Origins, Development, and Influence of Logical Empiricism*, edited by Friedrich Stadler, 866–909. Springer.

Stanford Daily. 1928. "Dr. Moritz Schlick to Be Visiting Professor Next Summer Quarter," July 31, p. 1. http://stanforddailyarchive.com/cgi-bin/stanford?a=d&d=stanford19280731-01.2.6.

Talbot, Chris, ed. 2017. *David Bohm: Causality and Chance, Letters to Three Women*. Springer.

Tegmark, Max. 1997. "The Interpretation of Quantum Mechanics: Many Worlds or Many Words?" arXiv:quant-ph/9709032.

Teller, Paul. 1995. *An Interpretive Introduction to Quantum Field Theory*. Princeton University Press.

Thorne, Kip. 1994. *Black Holes and Time Warps: Einstein's Outrageous Legacy*. W. W. Norton.

Von Neumann, John. 1955. *Mathematical Foundations of Quantum Mechanics*. Translated by Robert T. Beyer. Princeton University Press.

Warman, Matt. 2011. "Stephen Hawking Tells Google 'Philosophy Is Dead.'" *Telegraph*, May 17. http://www.telegraph.co.uk/technology/google/8520033/Stephen-Hawking-tells-Google-philosophy-is-dead.html. Accessed July 28, 2017.

Weinberg, Steven. 2003. *The Discovery of Subatomic Particles*. 2nd ed. Cambridge University Press.

———. 2012. "Collapse of the State Vector." *Physical Review* A 85, 062116.

———. 2013. *Lectures on Quantum Mechanics*. Cambridge University Press.

———. 2014. "Quantum Mechanics Without State Vectors." arXiv:1405.3483.

Werkmeister, William H. 1936. "The Second International Congress for the Unity of Science." *Philosophical Review* 45 (6): 593–600.

Wheeler, John A. 1957. "Assessment of Everett's 'Relative State' Formulation of Quantum Theory." In Barrett and Byrne 2012, 197–202.

———. 1985. "Physics in Copenhagen in 1934 and 1935." In French and Kennedy 1985, 221–226.

Wheeler, John A., and Kenneth Ford. 1998. *Geons, Black Holes, and Quantum Foam: A Life in Physics*. W. W. Norton.

Wheeler, John A., and Wojciech H. Zurek, eds. 1983. *Quantum Theory and Measurement*. Princeton University Press.

Whitaker, Andrew. 2012. *The New Quantum Age: From Bell's Theorem to Quantum Computation and Teleportation*. Oxford University Press.

———. 2016. *John Stewart Bell and Twentieth-Century Physics*. Oxford University Press.

Wick, W. David. 1995. *The Infamous Boundary*. Copernicus.

Wigner, E. P. 1963. "Problem of Measurement." *American Journal of Physics* 31 (1): 6–15.

Wigner, Eugene, and Andrew Szanton. 1992. *The Recollections of Eugene P. Wigner: As Told to Andrew Szanton*. Plenum Press.

Wise, M. Norton. 1994. "Pascual Jordan: Quantum Mechanics, Psychology, National Socialism." In *Science, Technology, and National Socialism*, edited by Monika Renneberg and Mark Walker. Cambridge University Press.

Zeh, H. Dieter. 2002. "Decoherence: Basic Concepts and Their Interpretation." https://arxiv.org/abs/quant-ph/9506020.

———. 2006. "Roots and Fruits of Decoherence." arXiv:quant-ph/0512078v2.

Zeilinger, Anton. 2005. "The Message of the Quantum." *Nature* 438 (December 8): 743.

Zurek, W. H. 1981. "Pointer Basis of Quantum Apparatus: Into What Mixture Does the Wave Packet Collapse?" *Physical Review D* 24 (6): 1516–1525.

———. 1991. "Decoherence and the Transition from Quantum to Classical." *Physics Today* 44 (October): 36–44.

Index

JOHN CASTILLO

Adam Becker is a science writer with a PhD in astrophysics from the University of Michigan and a BA in philosophy and physics from Cornell. He has written for the BBC and *New Scientist*. He has also recorded a video series with the BBC and several podcasts with the *Story Collider*. Adam is a visiting scholar at UC Berkeley's Office for History of Science and Technology and lives in Oakland, CA.